# Religion, science, and worldview

Richard S. Westfall.

# Religion, science, and worldview

## ESSAYS IN HONOR OF RICHARD S. WESTFALL

Edited by

MARGARET J. OSLER
*The University of Calgary*

and

PAUL LAWRENCE FARBER
*Oregon State University*

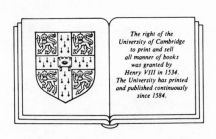

*The right of the
University of Cambridge
to print and sell
all manner of books
was granted by
Henry VIII in 1534.
The University has printed
and published continuously
since 1584.*

## CAMBRIDGE UNIVERSITY PRESS

CAMBRIDGE

LONDON   NEW YORK   NEW ROCHELLE

MELBOURNE   SYDNEY

*Burgess*

*Q*
*126.8*
*.R45*
*1985*

Published by the Press Syndicate of the University of Cambridge
The Pitt Building, Trumpington Street, Cambridge CB2 1RP
32 East 57th Street, New York, NY 10022, USA
10 Stamford Road, Oakleigh, Melbourne 3166, Australia

First published 1985

Printed in the United States of America

*Library of Congress Cataloging in Publication Data*
Main entry under title:
Religion, science, and worldview.
Bibliography: p.
Includes index.
1. Science – History – Addresses, essays, lectures.
2. Newton, Isaac, Sir, 1642–1727 – Addresses, essays,
lectures.   3. Religion and science – History of controversy
– Addresses, essays, lectures.   4. Science – Social
aspects – Addresses, essays, lectures.   5. Westfall,
Richard S.   I. Westfall, Richard S.   II. Osler,
Margaret J., 1942– .   III. Farber, Paul Lawrence,
1944– .
Q126.8.R45   1985      509      85-3783
ISBN 0 521 30452 0

# CONTENTS

# ILLUSTRATIONS

Frontispiece. Richard S. Westfall

# CONTRIBUTORS

J. BRUCE BRACKENRIDGE
Department of Physics, Lawrence University, Appleton, Wisconsin 54911

B. J. T. DOBBS
Department of History, Northwestern University, Evanston, Illinois 60201

PAUL LAWRENCE FARBER
Department of General Science, Oregon State University, Corvallis, Oregon 97331

EDWARD GRANT
Department of History and Philosophy of Science, 130 Goodbody Hall, Indiana University, Bloomington, Indiana 47405

PETER M. HARMAN
Department of History, Furness College, Bailrigg, Lancaster LA1 4YG, England

R. W. HOME
Department of History and Philosophy of Science, University of Melbourne, Parkville, Victoria 3052, Australia

ERNAN MCMULLIN
Program in the History and Philosophy of Science, University of Notre Dame, Notre Dame, Indiana 46556

RON MILLEN
Department of History and Philosophy of Science, 130 Goodbody Hall, Indiana University, Bloomington, Indiana 47405

MARGARET J. OSLER
Department of History, University of Calgary, Calgary, Alberta T2N 1N4, Canada

EDWARD G. RUESTOW
Department of History, University of Colorado, Boulder, Colorado 80302

JAMES A. RUFFNER
Science Library, Wayne State University, Detroit, Michigan 48202

STEPHEN M. STRAKER
Department of History, University of British Columbia, Vancouver, British Columbia v6t 1w5, Canada

VICTOR E. THOREN
Department of History and Philosophy of Science, 130 Goodbody Hall, Indiana University, Bloomington, Indiana 47405

# PREFACE: WESTFALL AS TEACHER AND SCHOLAR

The scientific revolution of the sixteenth and seventeenth centuries is central to understanding modern culture and the modern world. Pivotal in the development of the concepts of nature and of human nature that dominate contemporary life, this intellectual upheaval was deeply enmeshed in the religious, philosophical, and political currents of the time. Richard S. Westfall's contributions to history have focused on this crucial period in the development of modern science, and he has constantly addressed the fundamental issues and the most significant individuals. Always the intellectual historian, Westall has been concerned with examining in depth the wider context of ideas and individuals. His first book, *Science and Religion in Seventeenth-Century England* (1958), now a classic, discussed the theological and ideological contexts in which the new science was evolving. His two award-winning books on the life and work of Isaac Newton, *Force in Newton's Physics* (1971) and *Never at Rest: A Biography of Isaac Newton* (1980), place Newton's monumental work in the context of the intellectual and social currents of the time. Never one to avoid delicate issues, Westfall probes to the heart of the matter and goes beyond the history of ideas to a richer intellectual history, unclouded with ideology. His most recent work on the development of the European scientific community provides a case in point. Setting aside the twentieth-century categories that others have attempted to impose on this material, Westfall approaches this conflict-ridden field by examining the issues in terms of the day. His efforts have already produced insights that have eluded sociologists and historians for fifty years.

In each area to which Westfall has turned his attention, he has exerted a profound and on-going influence. The present collection of essays, which is divided into three parts, reflects that influence, both on his former graduate students and on the larger community of scholars studying matters related to his areas of interest.[1] The diversity and sophistication of studies reflected in the first part, Newtonian Studies, owe much to Westfall's seminal research. Indeed, the wealth of Newton's manuscripts that has become available in recent decades is accessible to younger scholars in part because of Westfall's careful study of them.

Westfall's interest in Newton and his insistence on the importance of Newton's theology for understanding his scientific work can be traced directly back to his concern with science and religion in seventeenth-century England. The chapters in the second part, Science and Religion, reaffirm the historical complexity of the issues Westfall raised in his early book. The problems had deep roots in medieval thought, they were of acute concern to seventeenth-century thinkers, and they continued to influence the development of scientific ideas throughout the seventeenth century and beyond.

The chapters in the third part, Historiography and the Social Context of Science, reflect not only other aspects of Westfall's interest, notably, methodological issues about the history of science, but also one of his most striking successes as a teacher: He has imparted to his students both the skills needed for pursuing detailed historical research and the breadth of vision to apply these skills to new areas of study.

Westfall's influence on his students is deeper and more personal than the influence of his scholarship on their intellectual development. No one who has listened to the introductory graduate course on the scientific revolution will ever forget those finely crafted lectures, full of wit and historical insight, delivered in a booming, sometimes thundering, voice and compared by at least one neophyte historian of science to fine preaching in some seventeenth-century church. Despite the formality of his lectures and the clearly articulated point of view they convey, Westfall has an openness to dissent and discussion that fosters the intellectual growth of each

---

[1] The following authors were doctoral students under Westfall's supervision in the Department of History and Philosophy of Science at Indiana University: Roderick W. Home, Ron Millen, Jim Ruffner, Victor E. Thoren, Stephen M. Straker, Margaret J. Osler, and Edward Ruestow. The others have either worked closely with him or have been deeply influenced by his work.

student. This openness is even more evident in Westfall's advanced seminars on various specialized topics in seventeenth-century science, seminars dealing with such issues as the mechanical philosophy, seventeenth-century mechanics, and, of course and inevitably, Newton. Meeting weekly with a small group of students who have struggled through thoughtfully chosen reading assignments, Westfall directs these seminars by asking a few, carefully prepared questions, which lead the discussion to significant and unexpected insights about the material at hand. Westfall exerts enough control to guide the students in the direction he considers important without stifling the students' own imagination and creative interaction with the material. As many of us have subsequently struggled to find stimulating ways of presenting difficult material to recalcitrant undergraduates, we can still look back to these lectures and seminars as models for what fine teaching can achieve.

Term papers and thesis chapters always come back full of comments ranging from matters of historical interpretation to strongly worded suggestions about literary style. Often the comments are supplemented by a personal conference. None of us who have experienced these criticisms can confront a split infinitive as an emotionally neutral matter of grammar: Some will go to extreme lengths of syntactical acrobatics to avoid one; others, with the perverse delight of adolescent rebellion, will do everything in their powers to, by all means and, of course, as frequently as possible, make sure that their infinitives are split! In an era when too many university graduates are functionally illiterate, Westfall's students remain grateful for his concern with style and form as well as the content of scholarly work. His own writings display a brilliant facility with language.

In addition to his work as a scholar and a teacher, Westfall the man has had a profound effect on colleagues and students who have had the good fortune to interact with him. He has always striven to be fair in his dealings with other people: whether as president of the History of Science Society, dealing with some contentious issue at one of those interminable council meetings; whether as department chairman in an era when universities across the continent were in a state of turmoil; whether as teacher and friend, helping his students establish themselves in the profession at a time when jobs and resources were rapidly disappearing. Emblematic is an incident that occurred some years ago when he found himself in the dubious position of recommending two of his students for the same job. Writing to one of the students, he said that he did not know whether he should rejoice or weep, for the success of the one or the failure of the other in securing the position. His attitude of fairness and his feeling for others has led practically everyone, even adversaries in controversies, to feel respect and affection for him. One characteristic that has endeared him to his students and other scholars is the fact that his quest for knowledge, although undertaken with high seriousness, is tempered with a subtly self-effacing humor.

As Westfall's students, friends, and colleagues, we are grateful for all that he has given us. We have produced this volume in his honor to express that gratitude and to wish him well for a continuing creative relationship with history and with life.

# ACKNOWLEDGMENTS

We are grateful to Gloria Westfall for pirating Sam's personal computer in order to provide us with the starting point for the bibliography of Westfall's writings and for purloining the photograph of him that we have used in this volume. We appreciate the many hours Susan Fitch has devoted to helping us with technical matters.

The Department of History at the University of Calgary and the Department of General Science at Oregon State University have provided considerable support for various aspects of this project. In particular, Olga Leskiw and Karla Russell have been extremely cooperative in providing secretarial help with the project.

Cambridge University Press has supported this project and has provided the means for the editors of this volume to work together on several occasions and to continue their systematic survey of the varieties of single malt Scotch.

M.J.O.
P.L.F.

# Part I

NEWTONIAN STUDIES

# 1

## Conceptual problems in Newton's early chemistry: a preliminary study

B. J. T. DOBBS

Isaac Newton's chemistry has been studied primarily in the material he published in his old age in the Queries of the *Opticks*.[1] Valuable though the analyses of that material are, they tell us only about Newton's chemical thought thirty to forty years after he took up the study of the field and almost nothing about the devel-

This material is based upon work supported by the National Science Foundation under Grant No. SES-8408624. Any opinion, findings, and conclusions or recommendations expressed in this publication are those of the author and do not necessarily reflect the views of the National Science Foundation.

[1] Marie Boas (Hall), "Newton and the Theory of Chemical Solution," *Isis, 43* (1952), 123; Marie Boas (Hall) and A. Rupert Hall, "Newton's Theory of Matter," *Isis, 51* (1960), 131–44; G. Daniel Goehring, "Isaac Newton's Theory of Matter: A program for chemistry," *Journal of Chemical Education, 53* (1976), 423–5; William J. Green, "Models and Metaphysics in the Chemical Theories of Boyle and Newton," *Journal of Chemical Education, 55* (1978), 434–6; Joshua C. Gregory, "The Newtonian Hierarchic System of Particles," *Archives internationales d'histoire des sciences, 7* (1954), 243–7; Thomas S. Kuhn, "Newton's '31st Query' and the Degradation of Gold," *Isis, 42* (1951), 296–8; idem, "Reply to Marie Boas (Hall)," *Isis, 43* (1952), 123–4; Douglas McKie, "Some Notes on Newton's Chemical Philosophy Written upon the Occasion of the Tercentenary of his Birth," *Philosophical Magazine, 33* (1952), 847–70; Hélène Metzger, *Newton, Stahl, Boerhaave et la doctrine chimique* (reprint of the 1930 ed.; Paris: Librairie scientifique et technique Albert Blanchard, 1974); Lyman C. Newall, "Newton's Work in Alchemy and Chemistry," in *Sir Isaac Newton, 1727–1927. A Bicentenary Evaluation of His Work. A Series of Papers Prepared under the Auspices of The History of Science Society in Collaboration with The American Astronomical Association of America and Various Other Organizations* (Baltimore: Williams & Wilkins, 1928), pp. 203–55; J. R. Partington, *A History of Chemistry,* 4 vols., (London: Macmillan Press; New York: St. Martin's Press, 1961–70), *II,* 468–77; Arnold Thackray, *Atoms and Powers. An Essay on Newtonian Matter-Theory and the Development of Chemistry,* Harvard Monographs in the History of Science (Cambridge, Mass.: Harvard University Press, 1970); idem, " 'Matter in a nut-shell': Newton's *Opticks* and Eighteenth-Century Chemistry," *Ambix, 15* (1968), 29–53; S. I. Vavilov, "Newton and the Atomic Theory," in *The Royal Society Tercentenary Celebrations 15–19 July 1946* (Cambridge: Cambridge University Press, 1947), pp. 43–55.

opment of his concepts. Since Newton left behind a huge legacy of unpublished manuscripts, his developing thought in mathematics and in many areas of science has been exhaustively explored by philosophers and historians, but it has heretofore been impossible to study his development as a chemist because his chemical papers had been misclassified as alchemical. The recent isolation from the mass of his alchemical manuscripts of some early Newton chemical papers, however, provides a unique opportunity to launch a study of his chemical development, and it is the purpose of this article to offer a preliminary excursion into the conceptual problems Newton faced in his early chemical work.

## NEWTON'S EARLY CHEMICAL PAPERS AND THE CHALLENGE THEY PRESENT

The manuscripts that record Newton's early chemistry were, for a considerable period, confused with his alchemical papers. It is sometimes admittedly difficult to distinguish between seventeenth-century chemistry and alchemy in the texts and manuscripts of the period, but here we shall follow a relatively simple distinction. Chemistry, even in the seventeenth century, was concerned with the practical operations of metallurgy, pharmacy, and food preservation; the manufacture of glass, porcelain, pottery, or mortar; the dyeing of cloth and the tanning of leather; distillation, and so forth. Alchemy, on the other hand, though it might employ ordinary chemical techniques, had as its overarching goal the preparation of an agent of perfection (the philosopher's stone) or the achievement of some form of perfection itself – in metals (gold), in medicine (a universal medicine), in soul (salvation), in cosmos (the redemption of matter). Newton himself made a differentiation between chemistry and alchemy that was similar to this, calling the first type common, vulgar, or mechanical chemistry. The second type, our alchemy, he called "vegetable" chemistry.[2] With both the vulgar and the vegetable chemistries treating extensively of transformations in matter, however, Newton's distinction has not always been apparent to those who have preserved and orga-

[2] Isaac Newton, "Of nature's obvious laws and processes in vegetation," Smithsonian Institution, Washington, D.C., Burndy MS 16, fol. 5$^{r.v}$; cf. B. J. T. Dobbs, "Newton Manuscripts at the Smithsonian Institution," *Isis, 68* (1977), 105–7; idem, "Newton's Alchemy and His Theory of Matter," *Isis, 73* (1982), 511–28.

nized his papers since his death in 1727, and the manuscripts with chemical content of any sort have been dispersed according to other criteria.

Newton died intestate, and virtually all of his papers were retained by the family until the nineteenth century when the earl of Portsmouth, to whom they had descended, undertook to present the scientific materials among them to the University of Cambridge.[3] A university-appointed syndicate examined the entire collection of books and manuscripts in the possession of the family, separating the collection into "scientific" and "nonscientific" portions. We can now see that the judgments of members of the syndicate were, to a certain extent, based on nineteenth-century preconceptions about what was "scientific," and even then were not entirely consistent. Thus, the syndicate placed the laboratory record of Newton's own experimentation on the transformations of matter in the "scientific" portion, where it remains with the Portsmouth Collection in University Library, Cambridge.[4] Newton's record of his experiments was almost totally empirical, with no stated rationales for most of the experimental procedures, and most of it is cryptic to the point of indecipherability in the present state of our knowledge. Newton even encoded the names of many of his chemicals in his own idiosyncratic symbolic system. Yet some of the experiments and some of the terminology have now been correlated with other papers of Newton's that derive from alchemical sources, and on balance it seems likely that much, perhaps all, of that experimental work was directed toward alchemical ends.[5] Presumably the syndicate made its decision to retain the laboratory record because of its obviously experimental character and never recognized the alchemical nature of the experiments, for it returned all of the other alchemical papers to the family as being "nonscientific" and thus of no interest to the university. Similarly,

[3] Isaac Newton, *The Mathematical Papers of Isaac Newton,* ed. Derek T. Whiteside with the assistance in publication of M. A. Hoskin, 8 vols. (Cambridge: Cambridge University Press, 1967–80), I, xv–xxxvi.

[4] The record consists of a notebook and several loose sheets, University Library, Cambridge, Portsmouth Collection, MSS Add. 3973 and 3975. Cf. Marie Boas (Hall) and A. Rupert Hall, "Newton's Chemical Experiments," *Archives internationales d'histoire des sciences,* 11 (1958), 113–52.

[5] B. J. T. Dobbs, *The Foundations of Newton's Alchemy, or "The Hunting of the Greene Lyon"* (Cambridge: Cambridge University Press, 1975), esp. pp. 16–17, 139–86, and 249–55; Richard S. Westfall, *Never at Rest: A Biography of Isaac Newton* (Cambridge: Cambridge University Press, 1980), esp. pp. 290–301 and 361–71.

drafts of the Queries for the *Opticks* went into the Portsmouth Collection, even though modern scholarship finds some profound affinities between alchemical doctrine and the "scientific" ideas expressed in the drafts, and indeed in the published Queries.[6]

Lumped together with the papers that were so clearly and distressingly alchemical[7] and were returned to the family as being "nonscientific," were a few chemical manuscripts composed by Newton in the 1660s before he became immersed in the alchemical enterprise. Along with personal papers and theological manuscripts, these chemical/alchemical materials remained with the family until they were dispersed at auction in 1936.[8] It was not until I systematically surveyed all of the scattered "alchemical" papers to which I could gain access in the 1970s that the orthodox chemical nature of a few of them was recognized. Of the 121 lots of "alchemical" materials auctioned by Sotheby's in 1936, at least three (lot nos. 16, 49, 79) and possibly four more (lot nos. 36, 88, 96, 115) should now be reclassified as chemical, for, according to Newton's own distinction, they are concerned with common or vulgar chemistry only.

These manuscripts demonstrate that Newton made himself into an accomplished chemist, probably not long after his *annus mirabilis* of 1666. The most extensive and systematic of the chemical papers is Sotheby lot no. 16, now in the Bodleian Library, Oxford.[9] Organized alphabetically as a sort of chemical dictionary of sixteen small quarto pages, it is written in Newton's tiny early handwriting and appears to be the production of a young man who has set out to master a new field of inquiry and is now busy reducing the information he has acquired into an orderly and useful form. In it are brief but graphic explanations of commercial as

---

[6] Dobbs, "Newton's Alchemy and His Theory of Matter" [see note 2].

[7] For the classic statement of nineteenth-century horror at Newton's interest in alchemy, see David Brewster, *Memoirs of the Life, Writings, and Discoveries of Sir Isaac Newton*, 2 vols. (Edinburgh: Thomas Constable; Boston: Little, Brown, 1855), II, 374–5: ". . . we cannot understand how a mind of such power, and so nobly occupied with the abstractions of geometry, and the study of the material world, could stoop to be even the copyist of the most contemptible alchemical poetry, and the annotator of a work, the obvious production of a fool and a knave."

[8] *Catalogue of the Newton Papers sold by order of The Viscount Lymington to whom they have descended from Catherine Conduitt, Viscountess Lymington, Greatniece of Sir Isaac Newton* (London: Sotheby, 1936); the chemical/alchemical materials from the sale are also listed in Dobbs, *Foundations* [see note 5], pp. 235–48.

[9] Bodleian MS Don. b. 15, quotations by permission of the Bodleian Library, Oxford.

well as laboratory preparations, what the best ores are and how to refine them, the practical uses of various substances, and much more. Citations to Robert Boyle's *Origine of Formes and Qualities* suggest that Newton was working on the dictionary about 1666–8,[10] and it will be our primary point of reference for Newton's early chemical knowledge and understanding.

Chemical postulates about matter were in some ways sharply at odds with the assumptions about matter employed by the mechanical philosophers, of which Newton was one. The mechanical philosophers had come to think of matter in terms of minute *particles,* whereas the chemists thought of matter in terms of specific *substances* with distinct chemical properties. The two strands of thought were never adequately fused before the early nineteenth century, when John Dalton finally identified each variety of chemical substance with its own specific type of particle. What conceptual blockage prevented that fruitful identification of substance and particle in the seventeenth and eighteenth centuries? The puzzle is present in an especially acute form in the Newton papers. For although Newton had become a mechanical philosopher before the later 1660s, as his student notebook shows,[11] yet in his earliest chemical papers one finds hardly a hint of a particulate theory of matter.

Basic to this dichotomy of thought is the general problem of the appearance of "forms and qualities" in matter. According to the chemists, forms and qualities inhere in chemical substances, and in that sense the chemical thinking of the seventeenth century remained somewhat Aristotelian. However, the mechanical philosophers preferred to discuss matter as if odor, taste, color, crystalline form, and chemical reactivity did not inhere in the basic corpuscles. For most of the mechanical philosophers, the primitive particles were made of one stuff, "one catholic and universal

---

[10] Robert Boyle, *The Origine of Formes and Qualities, (According to the Corpuscular Philosophy,) Illustrated by Considerations and Experiments, (Written formerly by way of Notes upon an Essay about Nitre)* (Oxford: Printed by H. Hall Printer to the University, for Ric: Davis, 1666). Newton's references are to this first edition. A later version of the book may be found in Robert Boyle, *The Works of the Honourable Robert Boyle. To which is prefixed The Life of the Author,* 6 vols. (London: Printed for J. and F. Rivington, L. Davis, W. Johnston, S. Crowder, T. Payne, G. Kearsley, J. Robson, B. White, T. Becket and P. A. De Hondt, T. Davies, T. Cadell, Robinson and Roberts, Richardson and Richardson, J. Knox, W. Woodfall, J. Johnson, and T. Evans, 1772), *III,* 1–137.

[11] J. E. McGuire and Martin Tamny, *Certain Philosophical Questions: Newton's Trinity Notebook* (Cambridge: Cambridge University Press, 1983).

matter," as they so frequently argued, the corpuscles of which had only "primary" attributes such as extension, shape, impenetrability, motion, and perhaps weight. The philosophical problem of moving from the "primary" qualities of mechanical thought to the "secondary" qualities to which our senses respond and by which the chemist classifies his substances and distinguishes them from each other is a significant one, and one which Newton's contemporaries addressed at length.

Newton's later solution to the problem, as we know from the *Opticks,* was to postulate a hierarchical internal structure of parts and pores for chemical substances, the ultimate parts being the universal corpuscles of mechanical philosophy. In his solution, secondary qualities emerged from the internal structure and larger size of the complex hierarchies he envisioned, but not without the addition of certain "active principles," which to the orthodox mechanical philosopher were unacceptable. Even so, Newton's solution was not really adequate. It will eventually be instructive to examine his speculative structures anew in the context of the complex issues that engendered them, but that is beyond the scope of the present article, for we must first see just what those issues were.

### NEWTON'S FIRST PHYSICALIST VIEWS

Newton's student notebook records his first encounters with the mechanical philosophy in sections entitled "Of the first matter," "Of atoms," "Of a vacuum and atoms," and "Of quantity."[12] The "Certain Philosophical Questions" he raised were centered in the seventeenth-century revival of atomism but were reflective of ancient controversies, and, although Newton examined his contemporaries' views critically, it is clear that he was engaged in these passages with the speculative and logical tradition going back to Leucippus and Democritus and not with natural manifestations of matter in the phenomenal world. The problems that concerned him were, for example, whether the first matter "be mathematical points, or mathematical points and parts, or a simple entity before division indistinct, or individuals, i.e., atoms." Concluding for

---

[12] Ibid., pp. 336–47.

atoms, he argued against the infinite divisibility of matter and shied away from the profound difficulties Aristotle had had with the idea of indivisible extension implied by atomism.[13]

Nowhere in his initial inquiries did Newton attempt to relate his indivisible particles to the sensory world, but we may safely assume that the distinction between primary and secondary qualities was already fixed in his mind. Walter Charleton's *Physiologia Epicuro-Gassendo-Charltoniana* restated the traditional Democritean/Epicurean doctrine, and Charleton was one of Newton's first sources. Atoms have *"Consimilarity of Substance,"* Charleton said; they also have magnitude or quantity, determinate figure, and gravity or weight. To those four properties, Epicurus had added resistance, but Charleton preferred to conflate resistance with gravity since both depended on the atom's solidity. These attributes, and only these, are primary and inseparable from the atoms; the qualities of compound bodies, i.e., secondary qualities, emerge only from the *"Concurse, Connexion, Position, Order, Number,* etc.*"* of the atoms and are not essential characteristics of the atoms themselves.[14]

### THE CHEMICAL CONCEPT OF MATTER

Contemporary chemists were not unaware of the revived corpuscular philosophy, but at least one of them had scant patience with it, dismissing the speculative tradition on epistemological grounds and opting for a radical empiricism that defined the parts of bodies by the direct "testimony of the senses." The chemical physician

[13] Aristotle, *Physica*, 231$^a$21–231$^b$18; cf. Lillian U. Pancheri, "Greek Atomism and the One and the Many," *Journal of the History of Philosophy, 13* (1975), 139–44; Friedrich Solmsen, *Aristotle's System of the Physical World. A Comparison with his Predecessors*, Cornell Studies in Classical Philology, ed. by Harry Caplan, James Hutton, G. M. Kirkwood, and Friedrich Solmsen, vol. *XXXIII* (Ithaca, New York: Cornell University Press, 1960), 199–204; McGuire and Tamny, "Commentary," in *Certain Philosophical Questions* [see note 11], pp. 49–60.

[14] Walter Charleton, *Physiologia Epicuro-Gassendo-Charltoniana: or a Fabrick of Science Natural, Upon the Hypothesis of Atoms, Founded by Epicurus, Repaired by Petrus Gassendus, Augmented by Walter Charleton*, with indexes and introduction by Robert Hugh Kargon (reprint of the London ed. of 1654; The Sources of Science, No. 31; New York: Johnson Reprint, 1966), pp. 111–12; Andrew G. Van Melsen, *From Atomos to Atom: The History of the Concept Atom* (reprint of the 1952 ed.; Harper Torchbooks/The Science Library; New York: Harper & Brothers, 1960); Robert Hugh Kargon, *Atomism in England from Hariot to Newton* (Oxford: Oxford University Press [Clarendon Press], 1966).

or "Naturalist," said Nicolas le Fèvre, will not rely "upon bare and naked contemplation" but will

> endeavour to bring his demonstrations under your sight, and satisfie also your other senses, by making you to touch, smell, and taste the very parts which enter'd in the composition of the body in question, knowing very well that what remains after the resolution of the mixt, according to the rules of Art, was that very substance that constituted it.[15]

The "resolution of the mixt" (a mixt being what we should call a chemical compound) was generally effected by some sort of fire analysis. Though fire analysis raised other problems, which we will consider below, one could hardly deny the efficacy of the procedure in producing substances that differed from the original mixt, and most of the chemists maintained the position that, since the new substances came "out of" the mixt, they must have been present in it before analysis and so must be the principles of which the body was composed. To le Fèvre, this chemical argument was vastly more meaningful than the speculative one, and he belabored some of the philosophical questions that had interested Newton at considerable length in order to make his point more explicit.

> But if you ask from the School Philosopher, What doth make the compound of a body? He will answer you, that it is not yet well determined in the Schools: That, to be a body, it ought to have quantity, and consequently to be divisible; that a body ought to be composed of things divisible and indivisible, that is to say, of points and parts: but it cannot be composed of points, for a point is indivisible, and without quantity, and consequently cannot communicate any quantity to the body, since it hath none in its self; so that the answer should have concluded the body to be composed of divisible parts. But against this also will be objected, If it be so, let us know, whether the minutest part of the body is divisible or no, if it be answered, Divisible, then it is instanced again, that it is not the minutest, since there is yet a place left for division: but if this minu-

---

[15] Nicolas le Fèvre, *A Compleat Body of Chymistry: Wherein is contained whatsoever is necessary for the attaining to the Curious Knowledge of this Art; Comprehending in General the whole Practice thereof: and Teaching the most exact Preparations of Animals, Vegetables and Minerals, so as to preserve their Essential Vertues. Laid open in two Books, and Dedicated to the Use of all Apothecaries, &c. By Nicasius le Febure, Royal Professor in Chymistry to his Majesty of England, and Apothecary in Ordinary to His Honourable Houshold. Fellow of the Royal Society. Rendred into English, by P. D. C. Esq; one of the Gentlemen of His Majesties Privy-Chamber. Part I. Corrected and amended; with the Additions of the late French copy* (London: Printed for O. *Pulleyn* Junior, and are to be sold by *John Wright* at the Sign of the *Globe* in *Little-Brittain,* 1670), p. 9.

test part be affirmed to be indivisible, then the answer falleth again into the former difficulty, since it returns to affirm it a point, and consequently without quantity; of which being deprived, it is impossible it should communicate the same to the body, since divisibility is an essential property to quantity.[16]

Le Fèvre was demonstrator in chemistry at the Jardin du Roi in Paris for a number of years in midcentury, then after 1660, chemist and apothecary to Charles II in London, becoming Fellow of the Royal Society in 1663. His chemical treatise was part of the French text tradition of the seventeenth century initiated by Beguin's *Tyrocinium chymicum* and culminating in Lémery's *Cours de chymie*. In its several editions, le Fèvre's treatise was perhaps one of the most significant chemical publications of the 1660s.[17] English editions of 1662 and 1664 appeared in time for Newton to have used them in compiling his dictionary of 1666–8, and in that case Newton would have been made vividly aware of le Fèvre's intensely chemical approach to matter theory, with its emphasis on the epistemological importance of the secondary qualities, for le Fèvre concluded his diatribe against the philosophers thus:

You see then, that Chymistry doth reject such airy and notional Arguments, to stick close to visible and palpable things, as it will appear by the practice of this Art: For if we affirm, that such a body is compounded of an acid spirit, a bitter or pontick salt, and a sweet earth; we can make manifest by the touch, smell, taste, those parts which we extract, with all those conditions we do attribute unto them.[18]

Based as it is on a naively realistic approach to matter, le Fèvre's position must have seemed reactionary to most mechanical philosophers, whose program emphasized the quantitative at the expense of the qualitative. Only the primary characteristics of matter were supposed to have objective existence; the subjective sensory qualities did not "really" exist in nature and were to be reduced to quantitative determinations of particulate magnitude, figure, configuration, and motion. The naive assumption that le Fèvre made – that color, taste, odor, etc., were the essential properties of sub-

---

[16] Ibid., pp. 9–10.
[17] Partington, *History of Chemistry* [see note 1], III, 1–48, esp. pp. 17–24; Hélène Metzger, *Les doctrines chimiques en France du début du XVII^e à la fin du XVIII^e siècle* (reprint of the 1923 ed.; Paris: Librairie scientifique et technique Albert Blanchard, 1969), esp. pp. 62–82.
[18] le Fèvre, *Compleat Body of Chymistry* [see note 15], p. 10.

stances – could have seemed to seventeenth-century mechanists only as a throwback to a discredited Aristotelianism. But in fact, the seventeenth-century mechanical philosophers did not have so clear a vision of the future development of science as they thought, and on their limitations Paneth has made a number of relevant observations from the chemist's point of view.[19]

Chemistry was, and to a certain extent still is, a subject in which interest focuses on the secondary qualities of substances. Though the early mechanists expected to reduce chemistry to physics in short order, that proved not to be possible. Even yet, an emphasis upon the qualitative characteristics of matter in its many forms pervades the discipline of chemistry in spite of the fact that most of the so-called secondary qualities have in the twentieth century been given quantitative particulate explication. Using their naive–realistic approach, the chemists slowly unraveled complex manifestations of matter to isolate the simple substances of which they were composed. Procedures for finding the "principles of bodies," that is, the simple elementary substances comprising them, were refined of course, but later characterizations of substances did not differ in any important way from that of le Fèvre. When sodium was first isolated in the nineteenth century, it was classed as a metal precisely because it had the same secondary qualities that had defined the metals since antiquity. Sodium chloride continued to taste like salt and hydrogen sulfide continued to smell bad, even as new nomenclature was devised to reflect their composition more accurately. Even at the beginning of the twentieth century, chemistry itself could still be defined as "bangs and stinks," yet by then it had discovered and organized into families almost all of the naturally occurring chemical elements – work all done, as Paneth has observed, on the basis of that naive, primitive conception of substance that insists on the primacy of secondary qualities.[20]

Given chemistry's subsequent successes, one can now hardly agree with the seventeenth-century mechanists who devalued the chemist's approach to matter, and, as far as one can tell from his chemical dictionary, Newton may at first have accepted the chemical concept at face value, without bringing the mechanists' cri-

---

[19] F. A. Paneth, "The Epistemological Status of the Chemical Concept of Element (I)," *The British Journal for the Philosophy of Science, 13* (1962), 1–14, and "The Epistemological Status of the Chemical Concept of Element (II)," ibid., 144–60.

[20] Paneth, "Chemical Concept of Element (I)," [see note 19], esp. pp. 1–9.

tique to bear upon it, for in his discussion of the fire analysis of several materials, he simply gives the substances into which each is resolved – substances defined by their secondary qualities.

Harts horne in sand yeilds Urinous spirit, volatile salt stinking oyle & flegme. Woods in a retort in naked fire yeild an acid spirit & fixt salt, & most of them an oyle espetially the heavy ones as box &c: Tartar yeilds a very little spirit more flegme, a great quantity of foetid oyle & a fixed Salt. Wood-soot yeilds an urinous spirit, a yellow oyle, & a white & very volatile salt.[21]

### PARTICLES AND SUBSTANCES

Another objection to the position of the mechanical philosophers turns on the distinction they wished to make between primary and secondary qualities, for their arguments masked an arbitrary stopping point. As was realized in the eighteenth century, if the secondary qualities were not present in nature, then neither were the primary ones the seventeenth century had defined.[22] That never became an issue for the early mechanists, however, to whom the primitive particles with their primary qualities seemed surely to have objective existence. Their problem was to relate their postulated particles to sensory phenomena.

Since the particles were supposed to be devoid of secondary qualities and also too small to be perceived by the human sensory apparatus, the problem became one of transdiction – or transduction, as it is sometimes called: relating the perceived qualities of the sensory world to the qualities of the corpuscles that were, in principle, imperceptible. It was the problem first adequately solved by Dalton when he demonstrated that the relative weights of substances measured in the macrorealm could be ascribed to the microrealm as the relative weights of the individual particles of the substances.[23] But in the seventeenth century, two fundamentally

[21] Bodleian MS Don. b. 15, fol. 6ᵛ.
[22] Paneth, "Chemical Concept of Element (I)" [see note 19]; *The Concept of Matter in Modern Philosophy*, ed. Ernan McMullin (revised ed.; Notre Dame, Ind.: University of Notre Dame Press, 1978).
[23] Henry Guerlac, "The Background to Dalton's Atomic Theory," in *John Dalton & the progress of science. Papers presented to a conference of historians of science held in Manchester September 19–24 1966 to mark the bicentenary of Dalton's birth,* ed. D. S. L. Cardwell (Manchester: Manchester University Press; New York: Barnes & Noble, 1968), pp. 57–91.

different approaches were put forward, and neither was capable of solving the problem.

The *matter* comprising all the particles was assumed to be the same, but traditionally the corpuscular systems allowed variety in size and shape. Especially in those systems with infinitely divisible particles, such as that of Descartes, it was inevitably a temptation for the mechanists to ascribe to the particles speculative shapes and sizes that would translate directly into the perceived qualities of bodies in the macrorealm. Good examples of this tendency may be found in Descartes's *Meteora,* where water has long, flexible particles and hot spirits have small, spherical or oval ones.[24] Descartes's matter was all the same in the beginning, but motion had reduced it to three varieties. It was to particles of the third of these – terrestrial matter – that Descartes assigned his ad hoc shapes and sizes.

Newton encountered Descartes's mode of explaining macro-properties by the sizes and shapes of the particles in his early explorations of mechanical thought and was skeptical from the first of some of Descartes's specific attributions. Raising the question "Whether fresh water consists of long bending parts and salt [water] of stiff and long ones," as Descartes had said in *Meteora,* Newton found six reasons for the falsity of the first suggestion and did not even bother to consider the second. But certain "branchy" types of particles seemed acceptable, as did the notion that "burning waters" must have "many such globuli as fire is made of . . . because they are most easily agitated and so heat and enliven men. . . ."[25]

Nor were other mechanical philosophers immune to that sort of thinking when the correlation between macro- and micro-spheres seemed "obvious" to them, the most notorious case being the attribution of the sharp taste and corrosive chemical action of acids to knifelike or needlelike particles. Even Boyle fell into the trap with acids, though he usually made the generalized "texture" of groups of corpuscles responsible for secondary qualities and did not specify the exact shape of individual particles. But in speaking of the change of sweet grape juice into vinegar he had this to say:

[24] René Descartes, *Oeuvres de Descartes publiées par Charles Adam & Paul Tannery,* 11 vols. (Paris: Librairie Philosophique J. Vrin, 1964–74), *VI,* 651–720; J. F. Scott, *The Scientific Work of René Descartes (1596–1650),* with foreword by H. W. Turnbull (reprint of the 1952 ed.; London: Taylor & Francis, 1976), pp. 65–9.

[25] McGuire and Tamny, *Certain Philosophical Questions* [see note 11], pp. 372–5.

It is obvious, that, though the recently expressed juice of grapes be sweet, whilst it retained the texture, that belongs to it, as it is new, . . . yet, after fermentation, it will . . . degenerate into vinegar. In which liquor, to a multitude of the more solid corpuscles of the must, their frequent and mutual attritions may be supposed to have given edges like those of the blades of swords or knives. . . . Now this liquor . . . does so abound with corpuscles, which, on account of their edges, or other-wise sharp and penetrative shape, are acid and corrosive, that the better sort of it will . . . dissolve coral, crab's-eyes, and even some stones. . . .[26]

Similarly, the Cartesian chemist Nicolas Lémery attributed the action of acids to their pointed particles. According to Lémery, when an acid combines with an alkali, the acid points enter the pores of the alkali and break off, stopping up the pores, and the result is a neutral substance.[27]

Easy to visualize as such speculative shapes were, they never-theless raised significant problems. It is clear that their appeal was much greater for those corpuscularians who were not also ato-mists. The particles of Descartes's third matter could theoretically acquire any shape from natural processes of attrition, so for him particulate shape could literally account for all types of secondary characteristics. But, though he recognized that some types of matter seemed to be more common than others, it was not clear why nature had been so selective with the infinite gradations of theo-retical possibility. Nor had he any way to account for the stability of recognized forms of matter, given the fact that his particles were so easily ground down. As Newton later observed in the *Opticks*, earth and water made from old, worn particles would not be the same as the earth and water made in the first creation, understanding that as a reason to opt for indivisible particles.[28]

[26] Robert Boyle, *Experiments, Notes, &c. about the mechanical Origin or Production of divers particular Qualities: Among which is inserted, a Discourse of the Imperfection of the Chemist's Doctrine of Qualities; together with some Reflections upon the Hypothesis of Alkali and Acidum*, in *Works* (note 10), *IV*, 230–354, quotation from p. 315.

[27] Nicolas Lémery, *A Course of Chymistry, containing An easie Method of Preparing those Chymical Medicins which are used in Physick. With Curious Remarks and Useful Discourses upon each Preparation, for the benefit of such as desire to be instructed in the Knowledge of this Art. By Nicholas Lemery, M.D. The Third Edition, Translated from the Eighth Edition in the French, which is very much enlarged beyond any of the former* (London: Printed by R. N. for Walter Kettilby, at the Bishop's-Head in S. Paul's Church-yard, 1698), pp. 22–8. Newton owned a copy of this edition: John Harrison, *The Library of Isaac Newton* (Cambridge: Cambridge University Press, 1978), p. 177.

[28] Isaac Newton, *Optics, or a Treatise of the Reflections, Refractions, Inflections & Colours of Light*, foreword by Albert Einstein, intro. by Sir Edmund Whittaker, preface by I. Ber-

If one's particles are infinitely divisible and if shape accounts for their properties, there is also a problem in explaining the recovery of the original substances from chemical compounds. It might be plausible to explain the neutralization of an acid by the breaking off of its points, as Lémery did, but it is much less plausible to argue that each point gets stuck back on precisely where it should be when the original acid is regenerated from its neutral salt. Lémery was actually forced to take the position that the acid could not be recovered, though he must surely have known better.

For example, if you mix an Acid Spirit with the *Salt* of Tartar or some other Alkali, the edges of the Acid will so insinuate into the pores of the *Salt,* that if by distillation you would separate the Acid Spirit again from the *Salt,* you'll never be able to effect it, the *Acid* will have lost almost all its strength, because the edges of these *Spirits* are so far destroyed or changed, that they no longer preserve their former Figure.[29]

Charleton thought it better to insist that whatever shape the particle had, that shape – with all its peculiar points or hooks or corkscrews – must be as permanent as the particle itself.[30] Here the atomists differed most fundamentally from those mechanical philosophers who allowed the infinite divisibility of matter, since the atomists held the primitive particles to comprise a permanent substratum, with change accounted for by the coalitions and separations of changeless individual atoms. Therefore, Charleton said, one cannot attribute any secondary quality whatsoever to the atom.

Because, were any Colour, Odour, &c. essentially inhaerent in Atoms; that Colour, or Odour must be no less intransmutable than the subject of its inhaesion: and that Principles are Intransmutable, is implied in the notion of their being Principles; for it is of the formal reason of Principles, constantly to persever the same in all the transmutations of Concretions. Otherwise, all things would inevitably, by a long succession of Mutations, be reduced to clear Adnihilation.[31]

Since the effort to account for secondary qualities by the size and shape of the particles was a kind of sub rosa imputation of

nard Cohen, analytical table of contents by Duane H. D. Roller (based on the 4th London ed. of 1730; New York: Dover, 1952), p. 400.

[29] Lémery, *Course of Chymistry* [see note 27], p. 9.
[30] Charleton, *Physiologia* [see note 14], p. 104.
[31] Ibid., p. 130.

secondary qualities to the particles, the atomist was not so drawn to that form of speculation. For him, it did no good to correlate size and shape with macroproperties, for then he could not readily account for the changing secondary qualities of bodies described by the chemist. One could either be a divisibilist and account for change by the changing size and shape of the basic particles, or one could be an atomist and account for change by the coalition and separation of permanent basic particles. One could not reasonably combine the two approaches, and we may here gain some insight into the lack of particulate explication in Newton's chemical dictionary. Since he was already an indivisibilist, we may perhaps assume that he recognized the irrelevancy of speculation about the size and shape of the atoms. Though the fundamental particles might indeed differ in size and shape, we could never know exactly how, for only in association did the particles generate secondary qualities. Therefore, there was no direct correlation between what our senses revealed to us and the physical forms of the atoms: Acids did not necessarily have sharp particles at all.

The divisibilist and the atomist thus had quite distinctive ways of approaching the issue of transdiction. Those who held the infinite divisibility of matter did not strictly adhere to the classical dichotomy between primary and secondary qualities, though they presumably thought they did. Their preference was to translate secondary qualities directly into "obvious" particulate sizes and shapes and to account for change in the secondary qualities by changes in the particles themselves. The atomist might well agree that particles differ in size and shape, but he could not emphasize such differences nor correlate them with the sensory world. Since his first tenet was the permanence of the primary particles, he could not allow change in them at all. And since change is recognized by changing secondary qualities, the secondary qualities cannot be placed in the unchanging atoms but only in their coalition and separation. The seventeenth-century atomist was thus left with no method of transdiction at all, as far as the ultimate particles were concerned.

## PARTICLES AND PROPERTIES

In his eagerness to demonstrate that atoms are *"Naked"* and *"Unqualified"* principles, Charleton argued that sensible bodies lose their capacity to occasion sensory perception as they are more and more finely divided.

> Besides, all things become so much the more Decoloured, by how much the smaller the parts are into which they are divided; as may be most promptly experimented in the pulverization of painted Glass, and pretious stones: which is demonstration enough, that their Component Particles, in their Elementary and discrete capacity, are perfectly destitute of Colour. Nor is the force of this Argument restrained only to Colour, as the most eminent of Qualities sensible: but extensible also to all others, if examined by an obvious insistence upon particulars.[32]

He has already estimated that a grain of frankinsense must contain at least $7.776 \times 10^{17}$ elemental atoms,[33] and since his pulverized glass loses its color long before it is ground down to particles of atomic dimensions, he supposes it to be obvious that the even smaller atoms are themselves colorless. Stated baldly in this way, his argument seems to reduce to a matter of sensory threshold: We perceive no color because the effects generated by the tiny particles are subliminal with respect to our sensory apparatus. But that was not really what he meant, for in that case, could a single atom be somehow magnified to a sensible dimension, it would have color, and that is actually what he wants to deny.

Color and all the other secondary qualities arise from the "con-texture" of the particles in combination, Charleton said. His exemplars are chemical reactions in which colors are produced by the combination of substances themselves devoid of color, which can be, he said,

> only by a determinate Commixture, and position of their insensible particles: no otherwise than as the same Feathers in the neck of a Dove, or train of a Peacock, upon a various position of their parts both among themselves, and toward the incident Light, praesent various Colours to the eye. . . .[34]

In some ways, Charleton's approach was quite similar to that taken by Boyle a few years later, in associating color exclusively with the textures formed by insensible particles and deemphasizing color as a characteristic property of substance. Some examples used by Boyle in his *Experiments and Considerations Touching Col-*

---

[32] Ibid.    [33] Ibid., p. 114.    [34] Ibid., pp. 132–3.

*ours* of 1664 will clarify the point. One notable instance is offered to demonstrate the power a liquid has to alter colors by putting the corpuscles in motion.

And though you rubb Blew *Vitriol,* how Venereal and Unsophisticated soever it be, upon the Whetted Blade of a Knife, it will not impart to the Iron its Latent Colour, but if you moisten the *Vitriol* with your Spittle, or common Water, the Particles of the Liquor disjoyning those of the *Vitriol,* and thereby giving them the Various Agitation requisite to Fluid Bodies, the Metalline Corpuscles of the thus Dissolv'd *Vitriol* will Lodge themselves in Throngs in the Small and Congruous Pores of the Iron they are Rubb'd on, and so give the Surface of it the Genuine Colour of the Copper.[35]

Conversely, if aqua fortis be dropped on the surface of a copper plate, the corpuscles of the menstruum joining with those of the metal "will produce a very sensible Asperity upon the Surface of the Plate," and will "Concoagulate" into tiny grains of pale blue vitriol.[36] Although, having been trained as a chemist, he names the new material as a specific substance, it seems that Boyle is thinking of the appearance of the blue color as a change in the texture of the surface of the copper plate, just as he had treated the appearance of the copper color on the knife as a surface phenomenon of the iron. He concentrated on mechanical rearrangements of corpuscles instead of describing the reactions as the combination or separation of acid and metallic principles because he thought "a Chymical Explication" not to be as useful as "one that is truly Philosophical or Mechanical."[37]

Yet in fact, Boyle had associated particles with properties. He talked about the metalline corpuscles of the vitriol, the saline corpuscles of menstrua, the "Alcalizate particles of the Salt of Tartar that swim up and down in the Oyl."[38] In speaking of salts, he actually attributed to them a "Saline principle" and noted that they are mixed bodies having "besides what is Saline, both Sulphu-

[35] Robert Boyle, *Experiments and Considerations Touching Colours First occasionally Written, among some other Essays, to a Friend; and now suffer'd to come abroad as The Beginning of an Experimental History of Colours,* introd. by Marie Boas Hall (facsimile reprint of the 1664 ed.; The Sources of Science; New York: Johnson Reprint, 1964), pp. 62–3. A later edition is in *Works* (note 10), *I,* 662–799.

[36] Ibid., p. 65.    [37] Ibid., p. 307.    [38] Ibid., p. 306.

reous, Aqueous, and Gross or Earthly parts."[39] Also this particular book of Boyle's is most famous for its systematic description of color tests by which the chemist may decide whether a given substance be acid, alkaline, or neutral, and if alkaline, whether "urinous" or "alcalizate."[40]

Despite his protestations that he knows the difference between a chemical and a mechanical explication, in some ways in these passages he has combined the two ways of thought, for he has made certain particles to be the bearers of secondary qualities. Albeit he does not conceive them to be the ultimate units of mechanical philosophy, for he thinks they themselves are complex concretions of the ultimate particles, still these chemical corpuscles of his can be recognized and classified by the ordinary human senses. Acid corpuscles redden syrup of violets, alkaline ones turn it green. Metallic particles in solution, though there invisible, have latent in them all the characteristics of metals, and all those qualities are made manifest when the particles are freed again. The philosopher is here given a partial entry into the corpuscular world. By chemical inference, he reaches beyond the direct evidence of sense to classify and organize a portion of the microrealm. Though he still knows nothing of the shape, size, weight, or motion of the ultimate particles, he knows something about the intermediate-level concretions of them that take part in chemical reactions.

Newton's student notebook bears extensive notes from Boyle's *Experiments and Considerations Touching Colours,* and much of that material is carried forward in a condensed form into the chemical dictionary.[41] Though we have seen that Newton early toyed with ideas of shaped corpuscles, we know that he did not later rely on a geometric explication of secondary qualities. But did he in the mid- to late-1660s absorb Boyle's notion of particles of intermediate complexity that could be identified by their chemical reactivity and were at least in some respects amenable to sensory recognition? Newton did not make explicit comments on Boyle's chemical corpuscles in either the student notebook or the dictio-

---

[39] Ibid., p. 261.

[40] Ibid., esp. pp. 245–54; Marie Boas (Hall), *Robert Boyle and Seventeenth-Century Chemistry* (Cambridge: Cambridge University Press, 1958), pp. 133–4.

[41] McGuire and Tamny, *Certain Philosophical Questions* [see note 11], pp. 386–7, 440–3, 452–63; Bodleian MS Don. b. 15, fds. 4$^v$–5$^r$, for example, summarizes the color tests for salts.

nary, and his analyses in the dictionary are given in terms of chemical principles.

When Newton says, for example, that the "Spirit of Sal Armoniack" may be drawn from the salt mixed with quicklime or salt of tartar or lapis calaminaris, for any of those additives "lay hold on $y^e$ salt & soe let loose the Urinous spirit," he does *not* say that the salt is comprised of saline *particles* that somehow hold the volatile *particles* of the urinous spirit in bondage.[42] Even his contrast between "grosser" and "subtiler" avoids the common mechanical interpretation of particle size: "grosser substances are called bodys in relation to $y^e$ subtiler $w^{ch}$ they call spirits."[43] The terminology is of "substances" and "spirits." Newton nowhere in the dictionary actually uses the terminology of particles, corpuscles, or atoms. Even though we know of his background in mechanical philosophy and his recent study of some of Boyle's works, he seems in the dictionary to be thinking wholly within the discourse of chemistry. His avoidance of mechanical terms in the dictionary must surely mean that he kept an open mind in the matter of chemical corpuscles for a period of time, but in the early 1670s, Newton does explicitly call "vulgar chemistry" nothing but "mechanical coalitions or separations of particles," and, sounding very much like Boyle, describes the differing particles as watery, earthy, saline, airy, oily, spiritous, etc.[44] Newton's particles are not always specifically identical with Boyle's, but the approach to the problem of correlating particles with properties is the same. Newton in the 1670s gave chemical principles a particulate embodiment just as Boyle had done, and, since secondary qualities were always conceived actually to inhere in the chemical principles, they were likewise inherent in Newton's chemical particles – as the very names of the particles imply.

Such a potentially fruitful union of chemical knowledge and corpuscularianism may have seemed a fairly obvious one in the seventeenth century. Hooykaas has given early examples of it, and indeed argues that some aspects of the chemist's experience tend to promote corpuscular theorizing. The facts of chemical analysis, in which metals, for example, are recovered from their com-

[42] Bodleian MS Don. b. 15, fol. $5^r$.   [43] Ibid., fol. $2^r$.

[44] Burndy MS 16 [see note 2], fols. $5^r$–$6^r$; cf. Dobbs, "Newton's Alchemy and his Theory of Matter" [see note 2], esp. pp. 518–19.

pounds unchanged, notwithstanding their apparent destruction in the compound, may have promoted ideas of permanent chemical particles.[45] Then there are other contemporary examples, such as John Mayow's "igneo-aërial particles" that support combustion both in nitre and in air,[46] as well as the chemical particles of Thomas Willis that are so similar to Boyle's and Newton's.[47] But in reality the problems in associating particles and properties were somewhat too complex to be resolved in this manner just then. For convenience of discussion, we may divide our remaining problems into three sections, though in the seventeenth century they were all interrelated: problems of size, subliminality, and texture; problems of the latency of properties; and problems of the chemical principles themselves.

### PROBLEMS OF SIZE, SUBLIMINALITY, AND TEXTURE

Boyle wanted to show how very small propertied particles might be divided and still be detectable by the senses, and to that end dissolved a grain of cochineal, which he found to give a noticeable color to water even when diluted by 125,000 times its own weight.[48] Presumably each particle of cochineal so dispersed still retains its color, but does one know even so how large the particle is? Could the human eye detect color in a single particle of that size, whatever the size is, or are many particles required? Would smaller particles really be devoid of color, or have the limits of human vision simply been reached? If the latter, can there be particles of cochineal so small that an objective limit is reached, after which the particles no longer have color and indeed are no longer particles of cochineal but rather belong to another substance that has particles of a smaller size and different color or no color at all? The mechanical philosopher cannot answer such questions, and New-

[45] R. Hooykaas, "The Experimental Origin of Chemical Atomic and Molecular Theory before Boyle," *Chymia*, 2 (1949), 65–80.

[46] John Mayow, *Medico-Physical Works. Being a Translation of Tractatus Quinque Medico-Physici (1674)* (Alembic Club Reprints, no. 17; re-issue ed.; Edinburgh/London: Published for The Alembic Club by E. & S. Livingstone, 1957), pp. 10–15.

[47] Thomas Willis, *De fermentatione*, as translated and summarized in Hansruedi Isler, *Thomas Willis 1621–1675. Doctor and Scientist. With 11 illustrations* (New York: Hafner, 1968), pp. 46–52.

[48] Boyle, *Experiments Touching Colours* [see note 35], pp. 255–7.

ton, when he recorded Boyle's experiment, merely noted the facts and made no comment on their meaning.[49]

But, as we have seen, Charleton and Boyle gave the answer that the "texture" or "contexture" of many particles was the ground of sensory perception. In some ways, their answer was eminently practical. Even today, though our language casually assumes that we can assign properties to single atoms or single molecules, in fact we cannot detect their properties directly, and many properties so assigned really have no meaning when applied to the single unit. We say that an atom of silver is the smallest particle that has all the properties of silver. Then we say that silver has a silvery color, is malleable, and is a good conductor of electricity. Malleability and conductivity are properties that arise only from the "contexture" of large numbers of silver atoms together, even though the root of those properties may be traced to atomic structure. We feel confidence in the calculations that tell us the dimensions of a single atom, but we cannot isolate one to ascertain its color even with our most sophisticated equipment.

Yet even if modern language is not strictly accurate, we can assign properties to large groups of atoms and molecules in a meaningful way, for they (and their associated ions and radicals) are the smallest units of substances that interact at the chemical level, whereas the mechanical philosopher tended to jump conceptually from the small particles of sensible bodies to the "naked" ultimate particles of "one catholic and universal matter." He had no way to decide whether there was an objective limit to the size of a particle of cochineal. As cochineal was divided and dispersed so that its secondary qualities passed below his threshold of perception, the mechanical philosopher seemed to think it had simply reverted to common, propertyless matter. Although I have no desire to pass retrospective judgment on the mechanical philosopher, a comparison with modern theory may help to clarify this point. If his common matter is taken as comparable to our protons, neutrons, and electrons, we might say that he failed to see that the common matter was differentiated below the sensory threshold into stable units of *the quite different types of matter* that constitute the chemical elements, and that such differentiated matter was neither common to all things nor without properties.

[49] McGuire and Tamny, *Certain Philosophical Questions* (note 11), pp. 456–7; Bodleian MS Don. b. 15, fol. 2ᵛ.

The mechanical philosopher could perhaps never have achieved that insight by reasoning "up" the ladder of complexity from his a priori postulate of a common matter. We have seen how Charleton leaped directly from pulverized glass that had lost its color to the state of ultimate nakedness. We have seen how Descartes failed to lend stability to the forms of common matter at any level. But the chemist, reasoning downward from the complex substances of the phenomenal world, did not move "down" the ladder of complexity so quickly. Though he might well believe in a prime matter underlying all things that was itself without properties, all his training led him also to believe that there were intermediate-level elements and/or principles that were not only differentiated from each other but were in fact the bearers of the properties that appeared in more complex substances. These elements and principles were themselves substances and had qualities inhering in them; they were the products of chemical analysis and could be identified by taste, color, smell, and so forth, just as le Fèvre had said.

That the approaches of the mechanist and the chemist might be fruitfully united was apparent to Boyle, Newton, Mayow, and Willis. And the assignment of particulate structure to chemical principles did eventually yield the ripest sort of fruit, but it did not really do so in the seventeenth century. Not only was there a flaw in the mechanists' a priori assumption of a common matter without properties at the atomic level, but there were decided flaws in the ways contemporary chemists related properties to the presence of their elements and principles, and also in contemporary lists of the elements and principles themselves.

### PROBLEMS OF THE LATENCY OF PROPERTIES

The chemical elements and/or principles were conventionally considered not only to have properties but also to convey properties to the substances compounded of them. Consequently, if a mixt or compounded substance evinced a property such as fluidity or inflammability, the original assumption had been that the correlate element or principle must dominate in the compound; in these instances, water and fire for the Aristotelians, mercury and sulfur for the Paracelsians.

Robert Boyle never tired of criticizing this assumption, and we may easily appreciate its defectiveness from his amusing "Aristotelian" analysis of "dephlegmed spirit of wine." Ask them, he said,

whence it has its great fluidness, they will tell you, from water, which yet is far less fluid than it. . . . But if you ask, whence it becomes totally inflammable, they must tell you, from the fire; and yet the whole body, at least, as far as sense can discover, is fluid, and the whole body becomes flame, (and then is most fluid of all;) so that fire and water, as contrary as they make them, must both be, by vast odds, predominant in the same body. This spirit of wine also, being a liquor, whose least parts, that are sensible, are actually heavy, and compose a liquor, which is seven or eight hundred times as heavy as air of the same bulk, which yet experience shews not to be devoid of weight, must be supposed to abound with earthy particles; and yet this spiritous liquor may, in a trice, become flame, which they would have to be the lightest body in the world.[50]

So much for the earth, air, fire, and water of the Aristotelians. Similar critiques could be made for the mercury, salt, and sulfur of the Paracelsians, and Boyle indeed had often attacked both camps, beginning with his *Sceptical Chymist* of 1661 (see note 52).

An additional assumption of ancient origin is also relevant to our discussion, though its influence was fading in the seventeenth century. All of the Aristotelian elements had been thought to be present in every substance. One can still see the impact of this idea in the matter theory of Sir Kenelm Digby, a first-generation mechanical philosopher whose *Two Treatises* of 1644 was probably read by Newton.[51] Even though one element was expected to dominate, the others were always there also in any mixt, and similarly for the Paracelsian *tria prima*. One of Boyle's critiques involved showing that this assumption was false because it was not possible to obtain a complete set of elements or principles from every mixt body.[52] But in actual chemical practice, there had been a shift away from the assumption that the complete set was necessarily present in everything, toward a more pragmatic definition

---

[50] Boyle, *Mechanical Origin of Qualities*, in *Works* (note 10), *IV*, 283.

[51] B. J. T. Dobbs, "Studies in the Natural Philosophy of Sir Kenelm Digby: Part I," *Ambix*, 18 (1971), 1–25; McGuire and Tamny, *Certain Philosophical Questions* [see note 11], pp. 392–3.

[52] Robert Boyle, *The Sceptical Chymist: or Chymico-Physical Doubts and Paradoxes, Touching the Experiments, Whereby vulgar Spagyrists are wont to endeavour to evince their Salt, Sulphur and Mercury, to be the true Principles of Things. To which, in this Edition, are subjoined divers Experiments and Notes about the Producibleness of Chymical Principles*, in *Works* [see note 10], *I*, 458–661, for example, p. 480.

of elements and principles as those substances that could actually be obtained through the analysis of a body. Thus the chemist had come to speak more and more of "an earth," defined by its color, taste, etc., rather than of "Earth." And although he might still think that his "earth" was a variant of "Earth," he made the point that he had gotten it out of the mixt by analytical techniques and so it must have been one of the constituents of the mixt. So even after Boyle's devastating criticisms, the chemists continued to of- fer lists of elements or principles, obtainable (more or less) by analysis. During the seventeenth century, the lists seemed usually to represent compromises between the Aristotelian and Paracel- sian systems and often contained five, six, or seven items. Not only was there much confusion and argument over the contents of the lists, but also over the meaning to be attached to them.

Part of the difficulty stemmed from the fact that properties change when substances combine and disengage, so that the ancient as- sumption that the elements or principles conveyed their own properties to the mixt was a slippery and difficult one to handle. Let us first illustrate this problem with a modern example. So- dium and chlorine are two substances now recognized as chemical elements. Sodium in its elementary form is a silvery metal that is so reactive it will spontaneously burst into flame if exposed to the air. Chlorine by itself exists as a deadly pale-green gas. According to the old theories, a combination of two such principles might yield an inflammable liquid with a silvery-greenish hue and poi- sonous qualities. Of course, neither sodium nor chlorine had been isolated nor named as a chemical element in Newton's time, so for him this particular example was not a problem, but in fact the combination of sodium and chlorine yields table salt, the very symbol of solidity and salubrious earthiness to the seventeenth century. The point is that *all* of the properties carried by the con- stituents of the salt have disappeared. And Newton was already aware of analogous cases. The *caput mortuum* remaining after the preparation of aqua regia, its volatile parts being evaporated, he said, is an alkali even though its ingredients were acid and vola- tile.[53] He also noted that oil of anniseeds, "though a hot & strong liquor," freezes sooner than water,[54] which presumably surprised

---

[53] Bodleian MS Don. b. 15, fol. 1ʳ.    [54] Ibid., fol. 4ʳ.

him because a substance with so much of the fiery principle in it should be more difficult to freeze.

In Newton's period, the arguments over the elements and principles turned to a great extent on issues of "fire analysis." It was often argued that fire "dissolves" bodies, but of course there are different ways of applying fire, and differences of technique result in different products. The products of burning are different from those of distillation; distillation products may differ when air is or is not accessible. A few well-known substances, such as gold and glass, stubbornly refuse analysis by fire. In any case, not everyone agreed that fire effects a dissolution of the mixt into its elements or principles, arguing that the substances obtained are new ones, *not* present in the mixt but actually generated by the action of the fire.[55]

Some aspects of these arguments were resolved in the eighteenth century, when the role of oxygen was elucidated and the criterion of weight applied, as is well known. The question of whether the acid vapor that comes "out of" sulfur when it is burned is a principle of sulfur or is something newly generated was central to Lavoisier's revolution in chemical thought. But there are other aspects to these old arguments in which both sets of protagonists had something valid to say, for, even when one substance does come "out of" another one through analytic processes, it can still be claimed in some sense that it is a new substance "generated" by the process, because it has been changed and its properties are no longer the same as those it had in the mixt. Boyle was clearly aware of this issue when he said his blue vitriol, no matter how pure and "venereal" it was, would not impart its "latent colour" to the iron. The latent color was that of the copper that he knew was "in" the vitriol, but only when the metalline corpuscles came "out of" the vitriol could the "genuine colour of the copper" become patent.

We now attribute such changes to changes in electron configuration as atoms enter into combination or revert to the elementary state, but for the seventeenth-century corpuscularian the most obvious explanation of latency may have lain in the notion that one type of particle had become completely surrounded by an-

---

[55] Allen G. Debus, "Fire Analysis and the Elements in the Sixteenth and the Seventeenth Centuries," *Annals of Science 23* (1967), 127–47.

other type. Thus in 1679, in a letter to Boyle, Newton suggested that when metals are dissolved in acids, the "saline particles" of the acid encompass the metallic particles "as a coat or shell does a kernell."[56] That does perhaps give a satisfactory mechanical explication for the latency of metallic properties but cannot at the same time account for the concurrent latency of saline properties since the saline particles remain exposed on the surface of the new combination particle, a difficulty Newton does not seem to have noticed.

At the conceptual level, a part of the difficulty with latency came from the conflation of two very different meanings in the chemical concept of element or principle. In F. A. Paneth's terminology, the two meanings are those of "basic substance" and "simple substance."[57] Paneth finds confusion between the two meanings even in post-Lavoisierian chemistry, and his considerations apply with even more force to the seventeenth century. The idea of "basic substances" is closely akin to the a priori elements and principles – the fundamental substances that constitute all compound bodies and provide them with their properties. Paneth observes that this concept carries a transcendental meaning that was not eliminated by the chemical revolution. Lavoisier considered oxygen to be the principle of acidity and so named it "acid-former." Even today, the halogens are called "salt-formers" by their very name. On the other hand, "simple substances" are much closer to the elements and principles defined as the products of analysis in the seventeenth century. They are the "basic substances" as they exist in uncompounded form, and they really do differ from the "basic substances" as they exist in compounds and so are, in a sense, "generated" by the analytic process.

The problems associated with latency were hardly new in the seventeenth century but probably had been intensified by the reintroduction of particulate theories of matter. Aristotle had recognized the problems, and the Aristotelians had always had a sophisticated answer in their doctrine of potency and act.[58] A substance

---

[56] Newton to Boyle, 28 February 1678/79, in Isaac Newton, *The Correspondence of Isaac Newton*, ed. H. W. Turnbull, J. P. Scott, A. R. Hall, and Laura Tilling, 7 vols. (Cambridge: Cambridge University Press, 1959–77), II, 288–96, quotation from p. 292.

[57] Paneth, "Chemical Concept of Element (II)" [see note 19].

[58] Aristotle, *De generatione et corruptione*, II, esp. 334$^a$16–334$^b$30; Hooykaas, "Experimental Origin" [see note 45], esp. pp. 67–8; Paneth, "Chemical Concept of Element (II)" [see note 19], esp. pp. 145–6.

loses its form when it enters into a mixt and is present in the mixt only potentially, not actually. But the Aristotelian answer would not do for chemistry in the long run, for there is in it no certainty that a substance would regain the same form when it emerged from the mixt into actuality. If copper becomes pure potency in blue vitriol, why should it not reemerge as silver? Why not indeed? The chemist cannot for a long time say why not, but he becomes painfully sure that it will not. If copper goes in, copper comes out. The Aristotelian could not account for that continuity. To put the same thing another way, there is a genuine and stable relationship between the isolated "simple substance" and the "basic substance" in the compound that the chemist was able eventually to establish. Nevertheless, the stability of that relationship had not yet been clearly established in the middle of the seventeenth century, and the flavor of Aristotelian variability in the emergence of new forms remains in Newton's question to Francis Aston in 1669. Newton wanted Aston to get information for him about transmutation in general, and especially wanted to know whether, in the Hungarian mining region of Schemnitz,

they change Iron into Copper by dissolving it in a Vitriolate water wch they find in cavitys of rocks in the mines & then melting the slymy solution in a strong fire wch in ye cooling proves copper.[59]

### CHEMICAL ELEMENTS AND PRINCIPLES

In his chemical dictionary, Newton gave signs of occasionally viewing the chemical elements and principles as "basic substances" that convey certain properties to mixt bodies. In connection with problems of latency, we have already noted some instances in which he seemed a little surprised not to find certain expected properties in the mixt. A more dramatic instance occurs in his entry on spirit of wine, which he calls "a vegitable Sulphur." Sulfur was the Paracelsian principle of inflammability, of course, and Newton has presumably applied the term to ethanol because when it is well rectified, "it burns all away w^{th}out leaving any moisture behind."[60] Such examples are, however, relatively rare in the dictionary, and there is much more evidence that Newton was firmly

[59] Newton to Aston, 18 May 1669, in Newton, *Correspondence* [see note 56], *I*, 9–13, quotation from p. 11.
[60] Bodleian MS Don. b. 15, fol. 5^r.

rooted in the pragmatic operational tradition of chemical principles as "simple substances."

Newton accepted the proposition that the chemical principles of bodies could be obtained by fire analysis. In his definition of "magistery," we see a mirror image of his theoretical understanding of analysis: "Magistery is a preparation wherein the bodys principles are not separated [sic] (as in distillation, Incineration &c) but only changed by having the p$^{ts}$ of another body united per minima to its p$^{ts}$."[61] In making a "magistery," a body's principles are combined; in fire analysis – distillation, burning, etc. – a body's principles are separated.

In his lengthy section on distillation, we encounter specific principles: spirits, which may be acid, corrosive, or urinous; oils; phlegms; salts both volatile and fixed.[62] Alum, for example, yields "after much flegm a white fume condensing into an acid spirit."[63] The *caput mortuum,* an earthy solid, "is that w$^{ch}$ stays behind in distillation."[64] Brimstone "inflamed by a hot iron" will give the spirit of sulfur; a pound of common salt yields nine or ten ounces of the spirit of salt; from vitriol one gets a "transparent spirit," a black oil, and (in the *caput mortuum*) a fixed salt of copper.[65]

Whether Newton had compiled his analyses from earlier sources or performed them himself, there is little to distinguish his principles from those of other chemists. One of the most popular compromise systems was that attributed to de Clave or Basso and carried forward by le Fèvre, Glaser, and Lémery: water or phlegm, earth, mercury or spirit, sulfur or oil, salt.[66] All of these principles were understood to be obtainable by the fire analysis of mixt bodies, and all were presumed to be simpler than the bodies from which they came. All bodies did not necessarily yield the entire set of principles. It is apparent that Newton, in the passages quoted above (and in many more as well), is working within the concept of elements and principles as "simple substances," a concept not different in principle from that enunciated by Lavoisier over a century later to form the cornerstone of modern chemistry. Although in the seventeenth century there was undoubtedly more conflation of the two aspects of the chemical concept that Paneth has defined,

---

[61] Ibid., fol. 7$^r$.    [62] Ibid., fols. 3$^r$ and 6$^v$.

[63] Ibid., fol. 1$^r$.    [64] Ibid., fol. 2$^r$.    [65] Ibid., fol. 8$^r$.

[66] Partington, *History of Chemistry* (note 1), *III*, 1–48; Metzger, *Doctrines chimiques* [see note 17].

so that "simple substances" were often also conceived as property-conveying "basic substances," still there is more conceptual continuity between the elements and principles of the seventeenth century and the "simple substances" of the post-Lavoisierian period than has been generally realized.[67]

The difficulties arose in the seventeenth century not so much from the concept of elements and principles as products of analysis, but from practicalities. All was not as it seemed to the seventeenth-century analyst. Of the few specific examples we have cited from Newton's dictionary, only the substances produced in the cases of alum and vitriol were really simpler than the bodies from which they came, and even they were not as simple as might be. The spirit of salt weighed less than its parent, which might be taken to mean that it was simpler, but it was exactly as complex as the salt itself, whereas the spirit of sulfur was actually more complex than brimstone. The rectification of the myriad of small mistakes being made in interpretation could come only after equipment for handling and weighing gases was devised and after the invention of that grand new analytic tool, the electric current. Then it was possible to carry analysis to a new level of simplicity, and radical changes resulted in the lists. In the matter of elements and principles, Newton worked perforce within the technical limitations of his time, and we may here have a partial explanation of the failure of efforts at transdiction in the seventeenth century. Perhaps it was not possible to find an effective solution until the chemists had solved the problems connected with their elements and principles; certainly, speaking historically, Dalton's solution came only after the list of "simple substances" had been thoroughly rectified.

CONCLUSION

We have attempted to set the conceptual problems Newton encountered in his early chemistry in contemporary context, drawing primarily upon his own manuscripts and the works of mechanical philosophers and of chemists that he probably or certainly read. The problems were many, diverse, and subtle. As Newton worked first within a mechanical discourse, then within a chemi-

[67] Robert Siegfried and Betty Jo Dobbs, "Composition, a Neglected Aspect of the Chemical Revolution," *Annals of Science 24* (1968), 275–93.

cal one, he found radically different views on the nature of matter, each associated with a genuinely different epistemological stance. Could the two be reconciled by giving chemical principles particulate embodiment? Apparently, he soon came to accept the idea that they could, but each side of the equation presented him with its own peculiar difficulties. With the benefit of hindsight, we may hazard the guess that the principal stumbling blocks were two in number, one from each side: the mechanists' a priori assumption of a common matter and the chemists' inability to provide an accurate list of the differentiated types of matter with their elements and/or principles.

The recently isolated chemical papers of Newton provide a starting point for new studies of his development as a chemist, but here we have worked principally within a very brief time period in the 1660s, when he had begun to study chemistry but before he launched his exhaustive exploration of alchemy. Any final assessment of his development must take into account his encounters with both the "vulgar" and the "vegetable" chemistries: a program for the future.

# 2

## The significance of Newton's *Principia* for empiricism

ERNAN MCMULLIN

Even before Newton's *Principia* appeared in print in July 1687, those aware of its contents were quite certain of the fundamental significance of the work for the new science of mechanics. And the passage of time showed them to be right. The *Principia* did mark a turning point in the history of mechanics, indeed in the history of science itself. I want to argue here that it also had far-reaching consequences for epistemology, and specifically for the empiricist theories of science then current. In Newton's own estimation, his epistemology was of course a straightforwardly empiricist one. But the long-range effect of the *Principia,* we shall see, was to challenge classical empiricism in some fundamental ways.

When I speak of "classical empiricism," I have two theses principally in mind. The first of these has to do with *meaning,* specifically with how terms derive their meaning from experience. The second concerns the *warrant* appropriate to a scientific hypothesis. The first is associated especially with Locke, the second with Bacon. The challenge in both cases comes from the unitary character of the theory of motion proposed in the *Principia.* The terms in this theory derive their meaning not from separate experiences of the qualities the terms denote (as Locke would suppose) but from the theory taken as a whole. And the theory itself derives its warrant from its explanatory power when applied as a whole to our observations of moving bodies, not from inductive generalizations supporting each of its constituent "laws" separately, as Bacon's theory of science might have led one to expect.

In order that my claim not be misunderstood, its scope must be further clarified. I do not claim that the *Principia* was the *only* work of that time in which the shortcomings of these two empiricist theses could be said to have been presaged. The development, for example, of the "method of hypothesis" in which the weight of an hypothesis is assumed to be given by the number and variety of its verified consequences had already to some extent called into question the simpler notion of inductive generalization prescribed by Bacon. Even in the *Novum Organon* itself, it was difficult to see how the imperceptibly small corpuscles and their "latent processes" could be arrived at by a gradual ascent from experience in which each "level" is established inductively; i.e., by the perception of similarities among particulars. But the *Principia* posed a far sharper challenge to empiricism than any other scientific theory up to that time had done.

To put it this way might suggest that this challenge was perceived by Newton or at least by his contemporaries. This was not, in fact, the case. As we look back on the *Principia* today, *we* can see how seriously empiricism was compromised by the shift from Aristotle to Newton. But it took a long time for this to be grasped. Indeed, it has been only in our own century that the significance in this regard of the *Principia* and of the kind of science it announced, has become clear. Why it took so long and what finally brought about the realization would be material for another essay.

My theme is thus of a complex sort. It is neither straightforwardly historical nor straightforwardly philosophical. I am not arguing that the *Principia* was a turning point in epistemology in the way it was in science. It had little direct effect on epistemology at the time or, indeed, for long afterwards. Its importance in this respect was of a less direct kind. Latent within it were more complex notions of meaning and of theory-validation than classical empiricism could accommodate. The fact that it took so long for these notions to be fully discerned in no way diminishes their philosophical significance, nor the historical significance of the fact that it is in the *Principia* that we, from *our* vantage point, can first clearly espy their appearance.

## AN EMPIRICIST STARTING-POINT

In an unfinished manuscript, *De gravitatione,*[1] Newton proposed to analyze the notion of body (or matter). He reminded the reader

---

[1] The manuscript is usually designated by its opening words: "De gravitatione et aequipondio fluidorum." It is published with a translation in *Unpublished Scientific Papers of*

that since God could perhaps bring about the appearance of matter in several ways, the analysis he would give should be taken as provisional, a description of a "kind of being similar in every way to bodies."[2] Suppose, he asked, God were to make a space impervious to bodies, "it seems impossible that we should not consider this space to be truly body from the evidence of our senses (which constitute our sole judges in this matter)." The only other additional property required is mobility. Such an entity would presumably also "operate upon our minds." These three properties, impenetrability, mobility, and the ability to act upon the human senses, are together sufficient to constitute body.[3] From them, other "universal qualities" of body (as he called them in Rule III of Reasoning in the *Principia*) can then be derived, such properties as shape, hardness, inertia, and even perhaps the quality of mutually gravitating towards other bodies.[4] He took it as obvious that such other qualities as color, taste, and smell do not inhere in the perceived object itself. A distinction between "primary" and "secondary" qualities had been an almost universal presupposition of seventeenth-century science. Newton did not develop it further, nor comment on it. From his early papers on light and colors to the last Queries he added to the *Opticks,* he simply took it for granted.

What constitutes a quality as primary? Newton would probably have agreed with his philosopher-friend, Locke, who devoted some attention to this issue. Primary qualities are "such as are utterly inseparable from the body, in what estate soever it be."[5] No matter how far matter be divided up, "each part still has solidity, extension, figure and mobility," even though the parts be so minute as to be no longer separately perceptible. But how do we *know* that color is not primary in this sense? Locke does not say. Indeed, he speculated about the possibility of a "microscopical eye," which would allow one to see the minutest parts of bodies.[6] But this

*Isaac Newton,* ed. A. R. and M. B. Hall (Cambridge: Cambridge University Press, 1962). The manuscript has no indication of the date of composition; the editors conclude that it is an early work from the mid-1660s.

[2] Ibid., p. 138.

[3] Ibid., pp. 139–40.

[4] He hesitated to call it an "essential" quality (because that would make matter essentially active), and settled, unsatisfactorily, for its being "universal" only. See chapter 3 of my *Newton on Matter and Activity* (Notre Dame, Ind.: University of Notre Dame Press, 1978).

[5] John Locke, *Essay Concerning Human Understanding,* ed. Peter H. Nidditch (Oxford University Press, 1975), book II, chap. 8, sec. 9.

[6] Ibid., chap. 23, sec. 12.

would, of course, entail that they have a quality analogous to color.

It would seem that the real basis of the distinction, both in Locke and in Newton, is not that primary qualities can be shown to be universally possessed whereas secondary qualities can be shown not to be. Rather, an implicit reductionist claim is being made that the latter can, in principle, be explained in terms of the former.[7] The primary properties are precisely those which a science of mechanics requires in the objects to which it can properly apply. And this science could then, in principle (or so it is assumed), explain all other properties of bodies, whether secondary (pertaining to the effects on human sense organs) or "tertiary" (pertaining to effects producible on other entities).[8] Newton believed that a knowledge of the configuration of the minute parts of bodies and of the forces acting between them would prove sufficient to account for the colors of bodies, their chemical interactions, and their coherence.[9]

If such "universal" qualities as extension, mobility, impenetrability, and inertia are so foundational to the science of nature, it is important to know how our ideas of these qualities are first formed. Locke's answer is well known:

Since the extension, figure, number and motion of bodies of an observable bigness may be perceived at a distance by the sight, it is evident some singly imperceptible bodies must come from them to the eyes, and thereby convey to the brain some motion which produces these ideas which we have of them in us.[10]

It may seem surprising that an empiricist like Locke would have deemed it "evident" that the origin of ideas was to be explained

[7] A further refinement could be added here: A property is *epistemically* primary if it plays an essential part in the reducing science; it is *ontologically* primary, if it exists in its own right. Recent discussions of reduction have made much of this distinction. I shall not need it here. See "Matter, Perception and Reduction," sec. 6 of the Introduction to my *Concept of Matter in Modern Philosophy* (Notre Dame, Ind.: University of Notre Dame Press, 1978), pp. 32–41.

[8] The hardness and solidity of perceptible bodies could presumably also be explained in this way. In draft-notes from his later years, Newton speculated about the extent to which apparently "solid" bodies might reduce to largely empty space, populated by corpuscles exerting strong forces of repulsion and attraction. Impenetrability might then cease to be primary though still universal. See A. Thackray, "Matter in a Nutshell," *Ambix, 15* (1968), 29–63.

[9] Newton had, of course, no real evidence for this belief, and history would later show his confidence to have been premature. The inability of his mechanics to explain the colors of bodies was, indeed, one of the factors that would motivate the replacement of that mechanics with the conceptually much richer quantum mechanics.

[10] Locke, *Human Understanding,* book II, chap. 8, sec. 12.

in this way, since there could be no direct empirical warrant for the "singly imperceptible bodies," nor could any inductive support be provided for the claim that bodily motions are the cause of ideas. Yet these suppositions were evident within the context of the mechanical philosophy, to which Locke on the whole subscribed. His next step would not, however, be so easily justified:

It is easy to draw this observation that the ideas of primary qualities of bodies are resemblances of them, and their patterns do really exist in the bodies themselves; but the ideas produced in us by these secondary qualities have no resemblance of them at all. . . . They are, in the bodies we denominate from them, only a power to produce those sensations in us; and what is sweet, blue, or warm in idea, is but the certain bulk, figure, and motion of the insensible parts in the bodies themselves, which we call so.[11]

There has been much discussion as to what Locke might have meant here by "resemblance."[12] It seems fair to say that he was postulating a likeness of pattern between, say, the particular extension of a perceived body and the "idea" of that extension as it exists in the perceiver's mind. But how about the idea of extension itself? Locke discussed how we make such particular ideas general. We abstract, he said, from the contingent circumstances of the particular idea, and form a general idea (to which the corresponding name is attached), which will serve as "general representative" for all the particular perceived extensions:

Such precise, naked appearances in the mind, without considering how, whence, or with what others they came there, the understanding lays up (with names commonly annexed to them) as the standards to rank real existences into sorts. . . . Thus, the same color being observed today in chalk or snow, which the mind yesterday received from milk, it considers that appearance alone, makes it a representative of all of that kind, and having given it the name "whiteness," it by that sound signifies the same quality wheresoever to be imagined or met with; and thus universals, whether ideas or terms, are made.[13]

There are two operations of the mind here: the first, that by which the particular "ideas" (perceptions) are formed and, in the case of the primary qualities, actually mirror in some way the spe-

---

[11] Ibid., sec. 15.
[12] See, for example, J. W. Yolton, "The science of Nature," in *Locke and the Compass of Human Understanding* (Cambridge: Cambridge University Press, 1970), chap. 2.
[13] Locke, *Human Understanding*, book II, chap. 11, sec. 9.

cific patterns in the world they convey to the mind; and the second, that of abstraction, by which general ideas (like extension, solidity) are formed by abstracting from the specific features of the particular perceptions. There are strong analogies between this analysis and the classical Aristotelian doctrine of abstraction, but there are some important differences too. And Berkeley was soon to point out some of the more obvious difficulties in Locke's account.

Our concern here, however, is not with the antecedents nor with the specifics of Locke's epistemology. The above is intended only to suggest in broad outline what the standard empiricist view was in Newton's day regarding the *origin* of such ideas as extension or body. Newton himself had little to say on this topic. His references to sense perception (in *De gravitatione,* the papers on light and colors, and the *Opticks*) are consonant with the mechanical philosophy.[14] In his response to Hooke of 1672, he acknowledged his use of the "hypothesis" that colors are "modes of sensation, excited in the mind by various motions, figures, or sizes of the corpuscles of light, making various mechanical impressions on the organ of sense."[15] More than forty years later, he drafted a fifth Rule as a possible addition to the four Rules of Reasoning for a new edition of the *Principia:*

Whatever things are not derived from objects themselves, whether by the external senses or by the sensation of internal thoughts, are to be taken for hypotheses. . . . And those things which follow from the phenomena neither by demonstration nor by the argument of induction, I hold as hypotheses.[16]

It seems likely, then, that the definitions of such terms as "body" and "hardness," over which he labored so much in the incessant drafts he made for revisions of the *Principia,* were for him quasi-inductive in nature. Individual bodies interact mechanically with the retina and produce "pictures, [which] propagated by motion along the fibers of the optic nerves in the brain, are the cause of

---

[14] See J. E. McGuire, "Body and Void and Newton's *De mundi systemate:* Some New Sources," *Archives History of the Exact Sciences, 3* (1967), 206–48, especially sec. 5.

[15] *Isaac Newton's Papers and Letters on Natural Philosophy,* ed. I. B. Cohen (Cambridge, Mass.: Harvard University Press, 1958), p. 119.

[16] A. Koyré, "Les *Regulae philosophandi,*" *Archives internationales d'histoire de sciences, 13* (1960), 3–14.

vision."[17] From this a perception of body is built up, and by something akin to generalization, the notion of body itself is formed.

## THE "DEFINITIONS" OF THE PRINCIPIA

The importance of this view derives, of course, from the crucial role that definitions play in Newtonian science. He treated them as nonproblematic, as part of the axiomatic basis of the science. Even though they had proved so difficult to formulate satisfactorily, he nowhere gave the impression that they could introduce a "hypothetical" (in the sense of problematic) element into his mechanics. In the *De gravitatione*, he began: "The foundations from which this science may be demonstrated are either definitions of certain words, or axioms and postulates denied by none."[18] He then went on to assert: "The terms, 'quantity,' 'duration,' and 'space' are too well-known to be susceptible of definitions by other words." The implication is that there is no *need* to define them since everyone understands them in the same way. When he went on to discuss the notion of space (or extension), he noted that it can be grasped only by the understanding and not by the imagination alone; it is our understanding that tells us that "there exists a greater extension than any we can imagine."[19] But the imagination plays a role too, for instance, "as when we may imagine spaces outside the world, or places empty of body," when showing that space is not an accident of body.[20] Still, though imagination and understanding help us to grasp a notion like space, it is our perception of particular extended bodies that furnishes the material for the notion which can then be elaborated in whatever detail we choose.

The process of definition seems, then, to be a matter of finding words for ideas antecedently grasped. Newton sought the authority of common usage, especially where there had been some dispute among philosophers (as in the case of "body" and "void," where he was at odds with both the Cartesians and Leibniz):

---

[17] Isaac Newton, *Opticks* (New York: Dover, 1952), p. 15. This edition is taken from the fourth edition of 1730.

[18] Ibid., p. 122. See R. S. Westfall, *Never at Rest: A Biography of Isaac Newton* (Cambridge: Cambridge University Press, 1980), pp. 411–420.

[19] Newton, *Opticks*, p. 134.

[20] Ibid., p. 132.

Body I call everything tangible in which there is a resistance to tangible things. . . . It is indeed in this sense that the common people always accept the word. . . . Vacuum I call every place in which a body is able to move without resistance. For thus the common people are wont to use the term. If anyone should contend that there are bodies which by touch are neither felt nor cause resistance, he would be disputing the grammatical sense of the word describing as "bodies" what ordinary people do not call bodies. And I would prefer to side with ordinary people, since they have assuredly the ability to give things names.[21]

But in the case of the most important definitions of all, those that preface the first book of the *Principia,* he did not commit himself entirely to common usage. After listing them, he concluded:

Hitherto I have laid down the definitions of such words as are less known, and explained the sense in which I would have them to be understood in the following discourse. I do not define time, space, place, and motion, as being well known to all.[22]

He recognized, therefore, an element of the *prescriptive* in these definitions. In the terminology popularized by Copi,[23] they were proposed as "precising" definitions, that is, they conform broadly to lexical usage but stipulate a more exact sense. Thus their warrant is partly what the "ordinary people" said, and partly Newton's own authority. Did this make the science built on them in any way provisional? Newton obviously did not think so. He insisted that he was "deducing" directly from the phenomena; that was, it seems no element of the "hypothetical" about the definitions he was using. Yet he labored with them through many drafts and rejected so many variant possibilities. How could he be so secure that he had "got them right"? And what did getting them "right" mean for him?

---

[21] Draft material for a set of definitions to be set before the Rules of Reasoning in the third edition of the *Principia,* but never actually incorporated. (The last phrase is lined out in the manuscript.) Printed as an Appendix to McGuire, "Newton's De mundi systemate" [see note 14], pp. 245–6. Translation mine. Original in the University Library Cambridge, Add. 3965, fol. 437$^v$. See also I. B. Cohen, *Introduction to Newton's Principia* (Cambridge, Mass.: Harvard University Press, 1971), pp. 37–8.

[22] *Principia,* Motte–Cajori translation (Berkeley: University of California Press, 1962), p. 6. He later distinguished between the absolute and the relative senses of these latter terms, and noted that "if the meaning of terms is to be determined by their use," it is in the *relative* sense they are to be understood. It would "do violence to language, which ought to be kept precise," to take these words in the absolute sense proper to mathematics only (p. 11).

[23] *Introduction to Logic,* 3rd ed. (New York: Macmillan, 1968), chap. 4, sec. 3.

Matter, motion, and force are the three generic concepts he worked with. He clearly assumed that they were drawn straight-forwardly from ordinary experience. They designated real features of the world to which one could appeal with confidence. What he contributed is the extra "quantity of . . ." qualification, which transforms the definitions into something the "mathematician" can use.

Matter is still "stuff-in-general." But it drops entirely out of sight as a working concept in the *Principia,* and is replaced by "quantity of matter" or mass.[24] Instead of equating matter with extension, as Descartes had persisted in doing despite the obvious difficulties into which this led him, Newton went back to the standard pre-Cartesian way of understanding it as proportional to both extension *and* density. There are obviously "matters" of different density, i.e., which have more or less "stuff" in the same volume. How do we *know* that there is more or less "stuff"? Perhaps by noting that one could sometimes compress the same stuff into smaller and smaller volumes (a popular theme in medieval natural philosophy), and that the *mechanical* behavior of the compressed matter remains the same even when volume changes.

The first Definition of the *Principia* was thus the product of a long tradition of reflection on our everyday experience with bodies. It was grounded in this experience and could plausibly be thought of as an induction or generalization from it. If someone had questioned it, Newton's response would probably *not* have been to say that he was stipulating a special sense for the purposes of the work he was writing. He would very likely have adduced various features of our experience with bodies as evidence in support of the plausibility of taking the "stuff" of bodies to be both an invariant and to be quantifiable in the simple direct proportionality he proposed.

The second Definition might have seemed more in need of justification. From Aristotle's time onward, there had been periodic discussion as to how motion ought to be rendered quantitatively. Aristotle defined local motion as change of place, and represented it as a ratio of space covered to time taken. This gradually became the concept of velocity, understood as a single quantity, space divided by time. But in later medieval natural philosophy, the notion of motion as something communicated from one body to

---

[24] See the Introduction to McMullin, *The Concept of Matter in Modern Philosophy* [note 7], pp. 50–5.

another eventually suggested that another more complex concept would be needed. Descartes thought of "motion" as an invariant of the created universe. God gave motion to the matter of the world at its first creation and conserved this motion ever since. There is no *new* motion, despite appearances to the contrary. Clearly, the concept of velocity will not suffice once these considerations of invariance are introduced. Descartes thus took "motion" to be proportional both to velocity and to "size," and it was this notion that Newton took over ånd clarified.[25]

What makes this Definition more than a nominal one of a quantity (i.e., momentum) that is to be computed in a certain way from two other quantities, is the implicit assumption that it is an invariant of contact action. But this invariant is now rather removed from the ordinary perception of motion. It could hardly be regarded as an abstraction from such individual perceptions. But because it is set equal to the product of volume, density, and velocity, it may still seem innocuous enough in the eyes of the inductivist.

But this is no longer the case with the third notion, force. Newton defined three sets of force: inherent, impressed, and centripetal. The first, *vis insita:*

is a power of resisting by which every body, as much as in it lies, continues in its present state, whether it be of rest or of moving uniformly in a straight line. This force is proportional to the body and differs from the inertia of the mass only in the manner of conceiving it. . . . A body exerts this force only when another force impressed upon it changes its condition, and the exercise of this force may be considered as both resistance and impulse.[26]

*Vis insita* is a notoriously confused concept, and a great deal of Newtonian scholarship has gone into the effort to untangle it. Our

---

[25] Since Descartes equated "size" with quantity of matter, and also used the phrase "force of a body to act" interchangeably with "motion," much clarification was needed. His stress on invariance *did,* however, lead him to one major discovery, the principle of inertia "that every body which moves tends to continue its motion in a straight line," where it is not only momentum but *velocity* that remains the same (since the body remains the same in "size"). See R. S. Westfall, *Force in Newton's Physics: The Science of Dynamics in the Seventeenth Century* (New York: Elsevier, 1971), pp. 57–72.

[26] My translation. The Motte–Cajori translation is faulty here in several respects, notably in having *vis insita* act only when another force *endeavors to* change the body's condition. The phrase "endeavors to" does not occur in the Latin. For a discussion, see *"Vis inertiae,"* chap. 2.3 of my *Newton on Matter and Activity* [note 4].

concern here is only with the status of the Definition itself as a definition. It makes an empirical-sounding claim about the world, that a body tends to continue in the "state" in which it is. This had been denied by Aristotle and had only gradually begun to make its way as an hypothesis in later medieval natural philosophy, prompted mainly by some ingenious thought-experiments. It requires one to abstract from the everyday conditions of nature, in which bodies do *not* tend to continue in their state of motion. The fact that they do not must be capable of being explained in such a way that the "Definition" still holds good. Thus, there is an implicit reference to the frictions and resistances ("impressed" forces), which are held to explain away the actual nonuniform motions of bodies.

Can this sort of counterfactual claim be thought of as inductive? It could be said that Newton was generalizing from our experience of "impediments" to uniform motion. But will this motion, for example, be rectilinear or circular? Galileo had much difficulty with this question and never quite resolved it. And what precisely is to count as an "impediment"? This is no longer a matter of simple generalization. The idealization practiced by Galileo and Newton seems to require a different and much more complex form of inference.

The epistemic status of the Definition becomes even more problematic when one considers the explanation it offers for continuance in the state of motion or rest. The continuance is said to be due to an internal "force," proportional to the mass of the body. How do we know such a "force" operates? Not by observing it, but by postulating that something like it must be operating to account for "resistance" to change of motion.

Newton was in a quandary here. In an addendum to the Definition which he planned for the third edition of the *Principia,* he wrote:

I do not mean Kepler's force of inertia by which bodies tend toward rest, but the force of remaining in the same state whether of resting or of moving.[27]

But why should remaining in the same state require a force? And what in the end is meant by calling it the "force of inertia"? If

---

[27] In Newton's own interleaved copy of the second edition. See Cohen, *Newton's Papers and Letters* [note 15], Introduction, pp. 27–9.

matter is by its nature inert (as Newton insisted it is), how can a force be *inherent* in it, even one that only operates when an outside force affects the body? Later Newtonians eventually cut this cord that still bound Newton to Aristotle; they simply eliminated *vis insita* from their mechanics. But Newton himself was still struggling with the need to explain why inertial motion continues, and thus he postulated this phantom cause.

The third Definition is obviously not just a definition. It proposes a certain sort of motion as the norm, and it proposes to explain why such motion occurs as it does. As has often been noted, the first Law is already implicit in it. And when Newton said that *vis inertiae* is exerted only when external forces act to change a body's motion, and that this exercise can be considered as impulse insofar as "the body, by not easily giving way to the impressed force of another, endeavors to change the state of that other," one can immediately see that both the second and third Laws are also implicated.

The epistemic distinction between the Definitions and the Laws is at best, therefore, a hazy one. The former presuppose the latter just as much as the latter do the former. The former stipulate word-usage, but then so do the latter. The latter draw in a very general way on our experience of bodily motions, but so do the former. The Definitions cannot be singly derived by inductive generalization from observation (in the way in which Boyle's Law, say, might be said to be). But neither can the three Laws of Motion. The inverse-square law of gravitational force (if interpreted as a *description* of gravitational motion and not as an *explanation* of it) comes closest to being an empirical law after the Boyle model.

When force is taken to be an *explanation* of change of motion (as it clearly is in the fourth Definition and the second Law), we have left the realms of inductive science well behind, as Newton's critics were not slow to point out. Newton wanted it both ways. The fifth Definition speaks of bodies being "drawn" by centripetal forces, and later Propositions (e.g., LVII) speak of bodies "attracting" one another. The implication is that such forces and attractions are something more than a redescription of the motions they bring about, that they are real *causes* of the motions. In the eighth Definition, Newton backed away from this: "I here design only to give a mathematical notion of those forces, without considering their physical causes and seats." He went on to enjoin the reader not to assume that because he spoke of centers as "attracting," that he attributes forces "in a true and physical sense" to them.

But if the sun cannot, in the light of the Definitions, be said to attract the earth in a "true and physical sense," has the motion of the earth been explained? If (as the fourth Definition suggests), force "consists in the action only," does this compel us to the further question: What is the cause of the action? Or has the causal question in some sense been answered by specifying the "force"? Newton hoped to deflect these questions by insisting that the *Principia* is restricted to "mathematical notions" only. But does not such a restriction reduce the second law to a *definition* of force, making it nothing more than an abbreviation for the product of mass and correlative acceleration in certain sorts of motion? What exactly is this "mathematical notion" of force anyway?

I am not proposing to address these questions here; they have been the starting point for much recent Newton scholarship.[28] It is sufficient to raise them, however, in order to underline the enigmatic character of Newton's concept of force. As it appears in the Definitions and Laws, it cannot be understood either by reference to the usage of "ordinary people" or to our everyday experience of the behavior of bodies. Nor is it a product of generalization; rather, it is a construct of an ambiguous and puzzling sort.

### THE NETWORK EFFECT

How, then, is the meaning of the key terms in the mechanics of the *Principia* being specified? Through a complex network of interconnections that draws them all together in a single web. Mass is related to density and volume. How is density obtained? Ordinarily, through a knowledge of mass and volume. Force is measured by the acceleration it produces in a given mass, so mass can be determined in this way also. Weight can be measured by the extension of a spring, and the ratio of mass to standard mass can be obtained. And so on. These threads that link each mechanical term to all the others are familiar to every student of physics to-

[28] See, for instance, Westfall, *Force in Newton's Physics,* chap. 7; McMullin, *Newton on Matter and Activity* [note 4], chap. 4; Anita Pampusch, "Isaac Newton's Notion of Scientific Explanation," Ph.D. dissertation, Ann Arbor Microfilms, 1971).

day. And it was in the *Principia* that they were first specified sufficiently exactly to enable the system as a whole to be applied to concrete cases.

The specification was not *entirely* exact, of course. We have already seen something of the confusions that surrounded *vis insita*. And there was the fact that Newton in the second Law made force proportional to change of motion (instead of to rate of change of motion), thus making impulse rather than force the key explanatory concept. And there were the "absolutes" of time, space, and motion, which would leave later Newtonians with some intractable puzzles. Nonetheless, Newton built well, and his successors had for the most part only to adjust the network slightly here and there, to bring the mechanics of the *Principia* close to the formal idea of a conceptual system in which every syntactic relation is precisely specified.

The terms Newton presented in the Definitions prefacing his work are not "observational terms" in the sense in which empiricists, early and late, attempted to define these. They may sound like them. And Newton may have thought that they *did* satisfy the empiricist criteria (though he might have admitted to doubts about "force"!). But there can be no question about their tight and total interconnection. They simply cannot be related one by one to specific "properties" or "observables," somehow discriminable in advance.

Once this be admitted, a second difficulty arises for the strict empiricist. The system of the *Principia* involves all sorts of theoretical presuppositions, some of them very general (e.g., as to how time and length should be measured), some quite specific (e.g., regarding the nature of inertial motion). Because the application of individual terms to the behavior of bodies implicitly involves the network as a whole, it must also implicitly involve the presuppositions underlying the system viewed as a physical theory.

When the system is taken formally (or "mathematically," as Newton would put it), there are no presuppositions; everything is on the same level. The syntactic relations specifying the system may, for instance, have active gravitational mass equal to passive gravitational mass. This then becomes one of the "givens," and from it computations can be made. It is only when the system is considered as a physical theory, as an explanation of real motions, that this way of relating the two masses becomes a *presupposition,* something assumed, for which the evidence offered is less than conclusive. To restrict consideration to the formal system, to the mathematically expressible syntax, as Newton continually tried to do in the *Principia,* has the effect of eliminating the crucial epistemic distinctions between definitions, empirical laws, and theoretical presuppositions. It also allows one to lay aside (as Newton well knew) the really difficult semantic issues involved in interpreting one's causal terms physically. But these issues are crucial to the success of the theory as an explanatory account of real motions.

Fifty years ago, the logical positivists attempted to reconstruct an empiricist theory of science, and in doing so relied heavily on a distinction between "observation terms" and "theoretical terms." Since the former were assumed to be related directly and unproblematically to a class of operations or observations, the distinction served to provide (or so it was hoped) a firm starting point for scientific inference. The later collapse of this distinction in the face of the criticisms of Quine and others is well known.[29] Quine based his criticisms on a general account of language, not on an analysis of a specific scientific theory. But he could well have focused on mechanics, the paradigm science for the positivists, since (as Wittgenstein had already hinted in the *Tractatus*) it exemplifies the network model more clearly than does any other theory.

### THE WARRANT OF THE PRINCIPIA

There is one further way in which the *Principia* serves effectively to undermine the classical empiricist ideal of science, despite appearances to the contrary. We have seen that the linking of words to world in the *Principia* is warranted, not in a one-to-one way by means of notions like observation or abstraction, but through the conceptual system taken as a whole. This is, of course, at odds with the empiricist view that the scientist begins by singling out the appropriate observable qualities, and then proceeds to build up inductive generalizations relating them one to another. In this view, the evidence for a generalization such as Hooke's Law would be a

---

[29] Quine's "Two dogmas of empiricism" (*Philosophical Review, 60,* [1951], 20–43) was the most influential criticism, and launched the "network" model of meaning which is now widely accepted in accounts of physical theory. Many features of this model can already be found in the works of earlier writers, such as Duhem and Collingwood.

specific set of observations. One would then build the science, generalization by generalization (as Bacon advised) until the first principles (that is, the highest generalizations) were ultimately discovered. This is not what Newton did, and the warrant for the *Principia* (and hence for the use of each technical term in it) must be recognized as being of quite a different sort.

Did Newton recognize this? It does not seem so, and indeed it would have been quite remarkable if he had. He struggles to adapt his thought to the two principal models of science of his day, but there are hints of a third model also. The few passages in which he discusses method are so well known that an apology is needed for quoting some of them again. My intent here is not to deal with the very difficult issue of Newton's method in the detail this would require, but only to make two limited points, that the actual warrant[30] for the *Principia* is not what Newton appears to have thought it to be, and second, that it is at odds with the empiricist ideals of science, both classical and contemporary.[31]

Let me begin with Newton's best-known declaration. When Cotes was preparing the second edition of the *Principia,* he queried Newton regarding the epistemic status of the Laws of Motion. Newton's response was that the Laws "are deduced from phenomena, and made general by induction which is the highest evidence that a proposition can have in this philosophy."[32] And he added this passage to the General Scholium with which the *Principia* concludes:

---

[30] Obviously, the notion of "actual warrant" raises many sensitive philosophical issues in regard to scientific rationality and its historical dimension. Some of these will be touched on briefly below, but for the most part, they cannot be dealt with in the space at my disposal here. See McMullin, "The Rational and the Social in the History of Science," in *Scientific Rationality: The Sociological Turn,* ed. J. R. Brown (Dordrecht: Reidel, 1984), pp. 127–63.

[31] It may be noted that these theses are flatly contrary to those defended by R. M. Blake in "Newton and Hypothetico-deductive Method," in *Theories of Scientific Method,* ed. R. M. Blake, C. J. Ducasse, and E. H. Madden (Seattle: University of Washington Press, 1960), pp. 118–43. Blake argued that Newton had grasped the essentials of scientific method as we understand them today, that he exemplified these consistently in the construction of the *Principia,* and that the method he both proposed and used is "inductive," or "hypothetico-deductive" (he equated these two labels). Blake published this piece at a time when the hold of logical positivism was still strong. It is both instruction and warning to see him read Newton so neatly into this perspective.

[32] *The Correspondence of Isaac Newton,* ed. A. R. Hall and L. Tilling (Cambridge: Cambridge University Press, 1975), 5, 386–7.

Hitherto I have not been able to discover the cause of those properties of gravity from phenomena, and I frame (feign) no hypotheses. For whatever is not deduced from the phenomena is to be called an hypothesis, and hypotheses, whether metaphysical or physical, whether of occult qualities or mechanical, have no place in experimental philosophy. In this philosophy, particular propositions are inferred from the phenomena, and afterwards rendered general by induction. Thus it was that the impenetrability, the mobility, and the impulsive force of bodies, and the laws of motion and of gravitation, were discovered.[33]

Before commenting on this puzzling text, it is worth noting that in the very next paragraph Newton spoke of "a certain most subtle spirit which pervades and lies hid in all gross bodies, by the force and action of which the particles of bodies attract one another"; it also accounts for cohesion, electrical attractions and repulsions, the behavior of light, and the operations of the senses. He allowed that we are not yet "furnished with that sufficiency of experiments which is required to be an accurate determination and demonstration of the laws by which this electric and elastic spirit operates." Newton apparently saw no difficulty about juxtaposing his condemnation of "hypothesis" with this flat assertion of the existence of the "subtle spirit," so that we must be careful (as many recent commentators on Newton have reminded us)[34] not to interpret the *"hypotheses non fingo"* as a flat exclusion of hypothesis from science.

Nonetheless, Newton did insist, many times over, that in the "experimental philosophy" he espoused, basic propositions like the Laws of Motion must be "deduced" from the phenomena and then generalized by induction. Presumably, he meant "deduced" to be taken in the weaker sense of "derived" (since one obviously cannot *deduce* a general law from phenomena). But can the Laws be properly regarded as the product of induction/generalization? In his comment to the first Law, Newton mentioned projectiles, tops, planets, comets, and the effects of resistance on their motions, implicitly suggesting an "induction" of sorts. But the Law refers to a counterfactual state where a body is acted upon by no

---

[33] General Scholium to book III of the *Principia*, p. 547.
[34] See, for example, I. B. Cohen, "Hypotheses in Newton's Philosophy," *Physics, 8* (1966), 163–84; A. Koyré, "L'hypothèse et l'experience chez Newton," *Bulletin société française de philosophie, 50* (1956), 59–79.

forces. Can such a claim be regarded as properly inductive? And it mentions impressed forces – but *their* measure requires one to turn to the *second* Law. So that the two cannot be separately evaluated.

In his comment to the third Law, Newton reminded us that "if a horse draws a stone tied to a rope, the horse will be equally drawn back to the stone."[35] Is this observation really derived straightforwardly from the phenomena? And can it be generalized to yield the third Law? Surely it could have been argued (and indeed it *was* argued by some of Newton's predecessors) that action and reaction cannot be equal in cases such as these, for if they were, no motion would occur. When Newton used this everyday experience of horse and stone in the way in which he did here, he relied on a mature mechanical insight – and implicitly assumed a knowledge of the other two Laws. The simple observation of such occurrences as horses drawing stones will never furnish sufficient grounds for the third Law, expressed as it is.

It is surely significant that Newton did not attempt to substantiate each Law in the normal inductive way (as Boyle, for example, did his claim about the inverse relationship of pressure and volume of a mass of gas), other than by citing vague generalities about projectiles and tops. There is *some* element of induction here, of course, a plausible generalization from mechanical experience which makes it sound antecedently likely that mass, for example, should be measured by the product of density and volume. But it is hard to see how the basic insights relating force to *rate of change in motion,* rather than to the more intuitive *motion* of the Aristotelian tradition, could be justified by such an appeal alone.

It seems fair to conclude, then, that the warrant for the *Principia* cannot be taken to be a strictly inductive one in which the system is built up, proposition by proposition, on the basis of generalizations from experience. The Newtonian system stands or falls as a unit; whatever notion of warrant we develop must incorporate this insight.

There is a second strain in Newton's thinking that might seem more promising in this regard. In the Preface to the *Principia,* he wrote:

[35] *Principia*, p. 14.

I offer this work as the mathematical principles of philosophy, for the whole burden of philosophy seems to consist in this: from the phenomena of motions to investigate the forces of nature, and then from these forces to demonstrate the phenomena.[36]

He described the propositions of the first two Books as "mathematically demonstrated." They enabled him (he told us) in the third Book to:

derive from the celestial phenomena the forces of gravity with which bodies tend to the sun and the several planets. Then from these forces, by other propositions which are also mathematical, I deduce the motions of the planets.[37]

The deductivist tone is unmistakable. One might easily come to think of the *Principia* as a work of applied mathematics, to be evaluated by the correctness of its deductions and their utility for practical ends, as well, of course, as by its elegance and clarity. And the *Principia* indeed was, as we know, regarded in this way in the tradition of rational mechanics that derived from it in the eighteenth century.

Newton had, as we have already seen, one special reason for wanting the *Principia* to be taken mathematically rather than physically:

I here design only to give a mathematical notion of those forces, without considering their physical causes and seats. . . . I use the words "attraction," "impulse" or "propensity of any sort to a centre," promiscuously and indifferently, one for another, considering those forces not physically but mathematically, wherefore the reader is not to imagine that I anywhere take upon me to define the kind or the manner of any action, the causes or the physical reason thereof . . .[38]

Though *force* was his central explanatory concept, Newton was never comfortable with it. His critics were quick to urge that to explain in terms of a force of attraction was not to explain at all, because it was not at all clear what forces are or where they are to be located. In addition, to postulate a force of attraction inherent in all matter would make matter essentially active, and this ran

[36] *Principia*, pp. xvii–xviii.    [37] *Principia*, p. xviii.    [38] Definition VIII, *Principia*, pp. 5–6.

counter to one of Newton's most fundamental metaphysico-theo-logical principles, namely, that matter is by its nature inert.[39] In the abundant draft-material from Newton's later years, we find him returning again and again to this problem, but he was never able to find a satisfactory solution to it. His "official" position, as it appears in the *Principia,* is thus as guarded as he could make it. To treat forces "mathematically" is to restrict oneself to their effects, the accelerations they produce, and to leave all other "physical" issues aside. So when he said: "To us it is enough that gravity really exists and acts according to the laws we have explained,"[40] the existence of gravity appears to reduce to the claim that bodies accelerate in the way the inverse-square law prescribes.

Much more could be said about the several ways Newton related the "mathematical" and the "physical," but our interest in the topic here is only due to the emphasis he was led to give to the "mathematical" character of the *Principia* in consequence. The empirical hardly intrudes in the first two Books, except for an occasional scholium, and, of course, the claim that the Definitions and Laws are inductively grounded.

There is, however, the fourth Rule of Reasoning in Book III, which allows that later exceptions *may* be found to inductive generalizations. But the intent of the Rule is to exclude idle hypotheses that rely on the bare possibility of the generalizations' being inaccurate; thus its emphasis is rather on the assertion that inductions are to be taken as "accurately or very nearly true" unless we have reason not to do so.

A more significant reservation is expressed in the *Opticks:*

Although the arguing from experiments and observations by induction be no demonstration of general conclusions, yet it is the best way of arguing which the nature of things admits of, and may be looked on as so much the stronger, by how much the induction is more general. And if no exception occur from phenomena, the conclusion may be pronounced generally. But if at any time afterwards any exception shall occur from experiments, it may then begin to be pronounced with such exceptions as occur.[41]

---

[39] See McMullin, *Newton on Matter and Activity* [note 4], chap. 2.
[40] *Principia,* General Scholium to book III, p. 547.
[41] *Opticks,* p. 404.

So induction does not amount to demonstration. But a very general induction is claimed to be a close approximation to the truth; even if empirical anomalies are found, the original generalization may still be asserted, only that the exceptions now have to be mentioned. Obviously, Newton had no thought that the inductive claim could be overthrown, only that it might have to be qualified in some way. Thus, the fallibilism here is of a limited sort; it is that appropriate to an empirical law, like Boyle's Law. Is it appropriate to the Laws of Motion? Hardly, because the notion of progressive approximation does not apply so well here. Newton himself gave us a clue in this regard. In a Scholium regarding centripetal forces in Book I, he remarked:

In mathematics, we are to investigate the quantities of forces with their proportions consequent upon any conditions imposed; then when we enter upon physics, we compare those proportions with the phenomena of Nature that we may know what conditions of those forces answer to the several kinds of attractive bodies.[42]

What happens in "physics," then, is that the laws of force are adjusted to the observed motions. The *provisional* aspect of mechanics would, then, be restricted to the various laws of force attributed to the "several kinds of attractive bodies." The inverse-square "law" would thus have a rather different epistemic status from the three "Laws" of Motion. Newton never explicitly discussed the possibility that the more basic Laws might also have to be adjusted. They are presented as conceptual relationships of an intuitively satisfactory sort. Did he think them to be definitive in their formulation (as his later followers often tended to do)? It is hard to know, one way or the other. But he certainly gave no indication in the *Principia* that he thought them open to further revision.

So far we have examined two possible types of warrant for the mechanics of the *Principia,* inductivist and deductivist. I mentioned earlier that there are hints of a third possibility in the *Principia* itself. In his Preface to the second edition, he mentioned that the theory of fluid resistance in Book II was "confirmed by new experiments," and that the theory of comets in Book III "was

[42] *Principia*, p. 192.

confirmed by more examples." The notion of empirical confirmation suggested here is a clue worth following.

When the mechanics of the *Principia* is taught today in an elementary physics course, it is usually said that its main warrant was the successful deduction of Kepler's Laws in Book III and of the motion of the pendulum in Book II. Did Newton think this? If he did, he gave little sign of it. It is instructive in this regard to look at the six "Phenomena" with which he opened Book III, "The system of the world." These had been listed as "Hypotheses" in the first edition, and mingled with the Rules of Reasoning; they were evidently to be taken as additional propositions needed in order to extend the work of Books I and II to the system of the world. The "Phenomena" list a sample of the major known astronomical regularities: Kepler's Laws and the periods of the satellites of Jupiter and Saturn. (Others, e.g., in regard to the nodes of the moon, are introduced later in the book.)

Proposition II, for instance, asserts that the forces acting on the planets obey an inverse-square law of distance. The text simply says that this is "manifest" from Phenomenon IV (Kepler's third Law) and Corollary VI to Proposition IV in Book I. It adds that this conclusion is also "demonstrable with great accuracy" from the fact that the aphelion points of the orbits do not move; even a very slight departure from the inverse-square law would bring about a large and perceptible motion of these points. What he was trying to establish, then, was precisely *which* law of force applies to our planetary system. He did not take it for granted that it would be inversely as the square of the distance; this was regarded as contingent and needing empirical determination. But there is no indication that the axiomatic section of Book I, constituting his mechanics proper, is in any need of confirmation.

Proposition XIII asserts that the planets move in ellipses, with the sun in one focus. "Now that we know the principles on which [these phenomena] depend, from them we deduce the motions of the heavens *a priori*."[43] The principles are assumed to be *known*, though the exact form of the laws of force in given cases has to be ascertained empirically. Newton went on to say that the planetary orbits are not *exactly* elliptical. "But the actions of the planets upon one another are so very small that they may be neglected." He

[43] *Principia*, p. 420. The phrase used in Latin is "a priori." Its use here ought not be taken to suggest an "a priori" status for the principles in the later Kantian sense of this phrase.

was thus implicitly *correcting* Kepler's empirical results on the basis of his theory, although of course what he was stressing was the *agreement* (within the limits of observational accuracy of the time) of the calculated and the observed orbital figures.

The most detailed calculations of Book III are devoted to the lunar motions and to cometary orbits. No one before Newton had managed to explain either of these phenomena in terms of a broader theory. He concluded his treatment of the first: "By these computations of the lunar motions I was desirous of showing that by the theory of gravity the motions of the moon could be calculated from their physical causes."[44] Here, for once, he allowed that he was deducing from physical causes. But although some of the results he obtained "exactly agree with the phenomena of the heavens,"[45] his calculations were in general much more fine-grained than the astronomer's observations: "The mean motion of the moon and of its apogee are not yet obtained with sufficient accuracy." Thus more long-term observation was needed, for "the theory of the moon ought to be examined and proved from the phenomena."[46] "Proving the theory from the phenomena" here presumably means testing the applicability to the lunar motions of the inverse-square law, as well as the adequacy of the analysis of the different forces acting on the moon.

Newton's treatment of comets is especially interesting in this regard. After a series of propositions and lemmas detailing how the orbits of comets are to be calculated, he headed the next section: "Example," and went on "Let the comet of the year 1680 be proposed."[47] There follows page after page of observational detail gleaned from astronomers all over Europe. He ended with a table comparing the *observed* positions of the comet over five months with the *computed* positions (based on the assumption that the comet's orbit is a highly elongated elliptical one) and found a satisfying degree of agreement. He ended on a triumphant note:

The observations of this comet from the beginning to the end agree as perfectly with the motion of the comet in the orbit just now described as the motions of the planets do with the theories from whence they are calculated, and by this agreement plainly evince that it was one and the same comet that appeared all that time, and also that the orbit of that comet is here rightly defined. . . . And the theory which justly corre-

[44] *Principia*, p. 473.   [45] *Principia*, p. 463.   [46] *Principia*, p. 477.   [47] *Principia*, p. 507.

sponds with a motion so unequable, and through so great a part of the heavens, which observes the same laws with the theory of the planets, and which accurately agrees with accurate astronomical observations, cannot be otherwise then true.[48]

Here, quite clearly a warrant was being claimed of a different sort from the inductivist and deductivist ones. What made the theory of the comet true in Newton's eyes was that it predicts correctly over a wide variety of dynamic contexts and that it coheres with the successful theory of planetary motions. This is *not* an inductive warrant, because there is no question of working upward from generalizations about cometary motions. This is a properly *retroductive* warrant.[49] The observed results are explained by a theory that introduces hypothetical conceptual elements (like *force*); the theory is confirmed by the range of accurate predictions it makes possible.[50]

Newton was speaking only of the theory of the comet here, not of his theory of mechanics as a whole. He did not, it would appear, claim a specifically retroductive warrant for the broader dynamic theory, perhaps because he did not think it was needed. But the remarkable agreement between theoretical prediction and observed regularities, which is apparent in the treatment of the pendulum in Book II[51] and in the astronomical analysis of Book III, *does* constitute a retroductive warrant of a convincing kind. Newton seemed to treat them as "problems to be worked" rather than as potential confirmations. It is as though he already *knew* that the dynamics was correct, and now he was going to *apply* this knowledge.

What one takes to be the "real warrant" for the *Principia* depends, of course, on one's own theory of science. There is much

---

[48] *Principia*, pp. 515–19.

[49] Peirce's term. I prefer it to "hypothetico–deductive," which was used by the logical posivitists in a similar but rather more restricted way.

[50] As well as by the novelty of the predictions it makes possible, and by its fertility over time. The notion of retroductive warrant is much more complex than simply "getting the predictions right." See, for example, McMullin, "The criterion of fertility and the unit for appraisal in science," *Boston Studies in the Philosophy of Science*, 39 (1976), 395–432.

[51] Where his only remark on the evidential value of his results is the laconic: "and by experiments made with the greatest accuracy, I have always found the quantity of matter in bodies to be proportional to their weight" (Corollary VII to Proposition XXIV, p. 304). Did this constitute in his own mind a *test* of the second Law?

induction in Newton's science, there is a persuasive appeal to simple mechanical experience, there is the coherence of a tight mathematical system. But it is, in the end, the fact that the dynamics of the *Principia* predicted so variously, so well, and for so long, that would today count most in its favor. No theory before it came even close to illustrating the power of retroduction so well.

And now I can return finally to the main theme, and draw a particular moral from the discussion above of the warrant on which the *Principia* rests. Retroduction works for the theory as a *whole*. It is the dynamics as a *whole* that is confirmed by the results of Book II and Book III. Because of the interconnection of the elements of the theory, all of them are implicit in each deduction, in principle at least.

We are now very far from the classical empiricism which Newton inherited, and within which he struggled to clarify his procedures. Not only is the meaning of terms like "mass" and "force" determined by the theory as a whole rather than by abstraction from our experience of matter or of effort, but this meaning is "justified" by the success of the theory of which it is a part. We are not to seek separate inductive support for each of the three Laws; we are not to worry about the supposed circularity of defining mass partly in terms of density. The dynamics functions as an explanatory network, each node playing its part. But it is the network as a whole that is tested against experience by means of prediction, not the separate elements of it.

One further consequence of this is worth noting. If the retroductive warrant of a theory begins to fail, if anomalies multiply, what may happen is not the improvement of approximation (as the inductivist tradition would have had it), but the *replacement* of one theory by another. The earlier theory may not just be corrected; it may be refuted. The reason, note, lies in the difference between the inductive and the retroductive type of warrant.

There is much difference between philosophers of science as to whether (and if so, in what sense) one should say that Newtonian mechanics has been refuted. It is still in general use, so that there is an understandable temptation to hold that it is a limiting case (for low velocity, low mass, etc.) of relativistic mechanics, thus holding true as a sort of approximation. But this will not do (as Feyerabend, in particular, has stressed) because some of the basic assumptions of the older mechanics turn out to be not approximate but wrong. Space and time, for instance, turn out not to be related as Newton thought they were. As a *predictive* system, Newtonian mechanics is still perfectly adequate for most purposes. But as an *explanatory* account, it is no longer acceptable. And the notion of "approximation" does not apply to explanations in the way it does to empirical laws.

CONCLUSION

The kind of revolution that relativistic physics represents is the final challenge to the older empiricism. It is, indeed, a challenge to epistemology at a wider level, because basic notions like objectivity and truth come under pressure and a realistic account of knowledge is imperiled.

It would be silly to suggest that all of this was implicit in the original *Principia*. And Newton's conservative inductivism ("pronounced with such exceptions as may occur") would hardly have prepared him for what happened in 1905. My point is that the *Principia* was the starting point for a set of developments that moved epistemology permanently away from the empiricist maxims that were taken for granted when the *Principia* itself was composed.

There had, of course, been physical theories before Newton's. And the notion of language as a network, to be understood and to be evaluated in a holistic way, is not restricted to physics; it can be nicely illustrated by such philosophical disputes as the medieval one about the unity of substantial form (where the apparently "observational" claims turn out to be quite theory-laden). But the *Principia* holds a special position in the development of this epistemological insight, nevertheless, because it tied the conceptual elements together in a more explicit way than had been done before, a way so explicit and so precise in fact that the network could be almost completely formalized. Meaning and assessment were thus shifted to the system as a whole; the older empiricist notions of abstraction and induction (which related to single terms and single generalizations) no longer would suffice.

It is a commonplace that great works contain more than their creators know. This is true in science as elsewhere. And it is true in science, not only in the sense that the great work contains much that only later scientific research will make explicit, but also in the sense that the creator may well be unaware of the significance of *how* he is proceeding. He intuitively "does it right," but when he tries to tell *how* he did it, or how more generally it *should* be done, he falls far short of the real achievement.

# 3

## The defective diagram as an analytical device in Newton's *Principia*

J. BRUCE BRACKENRIDGE

After the initial limited printing of the first edition of the *Principia,* its acceptance as the paradigm of mathematical science led to multiple editions in Latin, to numerous translations into other languages, and to a large number of English excerpts with multiple editions used largely as teaching volumes. Given such a profusion of derivative editions, it is clear that there exist many variations of the original work. A variorum edition of the three "Newtonian" Latin editions has been published, and there is a projected attempt to trace the dependence of the English "revisions" upon the original Motte translation. But a detailed comparison of all translations and editions is a task of such magnitude and linguistic demands as to be out of the question.

*Tribute:* Contemporary writers on the *Principia* will find many opportunities to acknowledge the insights provided by the work of Professor Westfall, and I am no exception. I knew his work long before I met him at the Newton Symposium at Churchill College in Cambridge University in 1977. Since then I have been privileged on a number of occasions to enjoy the warmth of his personality as well as the stimulation of his intellect. I spent the summer of 1979 on the Indiana University campus and came to better understand the close relationship between him and his colleagues and students. It was this close teacher–student relationship, in fact, that provided the original impetus for this volume. It is therefore fitting in this volume generated by former students of Professor Westfall that my contribution should focus on a subject that arose in my interaction with a former student of mine. In an undergraduate seminar that I conducted at the Lawrence University Center in London, one of my students, Gene Peterson, and I were working on selected propositions in the *Principia* and were thus sensitive to such items as flaws in diagrams. It was his discovery of a variant set of drawings in the Motte translation of the *Principia* in the Senate House Library of the University of London that prompted my interest in this subject. Dr. Peterson now has a medical degree and a Ph.D. in biophysics, but when we meet it is to the memories of our collaboration on the *Principia* that we are drawn. It is a situation, I am sure, that is repeated many times over with Professor Westfall and his former students.

It is possible, however, to use the diagrams as an analytic device to test the care and utility of the editions. Given the number of diagrams in each of the editions, however, even the employment of this device presents a major challenge. It is surprising, however, how much can be learned from even one key diagram in such a comparison. The Keplerian Proposition on elliptical orbits, Proposition XI in Section III of Book I, provides such a key diagram, one which Newton distinguished by referring to "the dignity of the problem." It is a proposition that Newton's successors, with proper historical hindsight, have consistently selected as one of the most important propositions in the *Principia*.

It is of interest, therefore, to note that a critical portion of the diagram for this proposition undergoes a gradual deterioration from its correct form in the first edition to a questionable form in the third edition. It then appears in either a clearly incorrect or clearly correct form in the multitude of subsequent editions, depending upon whether the editor or translator "cleaned up" or "restored" the unclear diagram from the definitive third edition.

## PROPOSITION VI AND THE "CRITICAL RATIO"

In a note to Richard Bentley in 1691, Newton advised that "at the first perusal of my Book it's enough if you understand the propositions with some of the demonstrations which are easier than the rest." He then suggested that Bentley read the first sixty pages and then pass on "to such Propositions as you shall have a desire to know."[1] These first sixty pages include an opening set of definitions, the famous three "Laws of Motion" with a number of corollaries, and the first three sections of Book One: Section I, which uses the method of first and last ratio of quantities to demonstrate a series of eleven lemmas; Section II, which provides examples of the determination of centripetal forces; and Section III, which is concerned with the motion of bodies in eccentric conic sections. It is in this final section that the famous "Keplerian Proposition" resides: Proposition XI, Section III, Book I, "If a body

---

[1] Newton to Bentley, July 1691, *Correspondence of Sir Isaac Newton,* ed. H. W. Turnbull, J. F. Scott, A. R. Hall, and Laura Tilling, 7 vols. (Cambridge: Cambridge University Press, 1957–77), *3*, 367.

revolves in an ellipse; it is required to find the law of centripetal force tending to the focus of the ellipse."[2]

Newton has clearly structured these first 60 pages of the *Principia* as a pedagogical device. Not only does he present the basic definitions and mathematical relationships necessary for what is to follow, but in Section II he provides a series of problems as examples. These introductory problems at the close of Section II thus prepare the reader for the "dignified" Keplerian problem with which Section III opens. These problems all seek to find the mathematical nature of a force necessary to maintain a particle in a given orbit with the force directed toward a given point. The Keplerian problem has an elliptical orbit with a force directed toward the focus of the ellipse. The answer that Newton finds is that the force must vary as the inverse square of the distance of the particle from the focus of the ellipse. The problem that immediately precedes the Keplerian problem also has an elliptical orbit, but the force is directed toward the center of the ellipse instead of toward the focus. The answer that Newton finds for this preliminary problem is that the force must vary directly as the distance of the particle from the center of the ellipse.

In all of these examples, the final step in the solution is the same: The force is found from the critical ratio in Corollary V in Proposition VI.

Hence if any curvilinear figure APQ is given; and therein a point S is also given to which a centripetal force is perpetually directed; that law of centripetal force may be found, by which the body P will be continually drawn back from a rectilinear course, and being detained in the perimeter of that figure, will describe the same by a perpetual revolution. That is, we are to find by computation, . . . the solid $(SP^2 \times QT^2)/QR$ . . . , reciprocally proportional to this force. Examples of this we shall give in the following problems.[3]

The important role that this critical ratio, $(SP^2 \times QT^2)/QR$, plays in Newton's solution to problems in the *Principia* has long been clear within the context of the work and has been a subject of interest to commentators on the work. Recently, Whiteside has

[2] Sir Isaac Newton, *The Mathematical Principles of Natural Philosophy*, trans. Andrew Motte (London: Benjamin Motte, 1729), I, 79 [hereafter cited as *Principia*, trans. Motte].

[3] Newton, *Principia*, trans. Motte, I, 70.

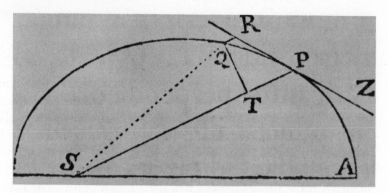

Figure 3.1.    The diagram for Proposition VI from the first edition of the *Principia*. (All diagrams from the first edition were reproduced with permission from the facsimile edition published by William Dawson and Sons.)

demonstrated that the ratio reduces to the orbital equation employed by contemporary physicists.[4]

Figure 3.1 is the diagram that accompanies Proposition VI in the first edition of the *Principia* and very clearly sets forth the core of Newton's attack on the problems to follow in this opening section. The line *RZ* is tangent to the curve *APQ* at the point *P*, the line *QR* is parallel to the line *SP*, and the line *QT* is perpendicular to the line *SP*. In the absence of any forces, the particle *P* would move in a straight line from *Z* to *P* to *R*. Under the influence of a centripetal force directed toward the point *S*, the particle would instead follow the curved line to the point *Q* rather than the tangent line to the point *R*. The line *QR* thus represents the displace-

---

[4] For a discussion of this ratio, see D. T. Whiteside, "The Mathematical Principles Underlying Newton's *Principia Mathematica,*" *Journal for the History of Astronomy, i* (1970), 120. Also Whiteside, ed., *The Mathematical Papers of Isaac Newton,* 8 vols. (Cambridge: Cambridge University Press, 1967–80), *VI,* 42–3, n. 30. For a central force *F(r)*, i.e., a force acting along a line between the particle and a given force center, this critical ratio reduces to what a contemporary physics textbook calls the "orbital equation." This equation is obtained by writing the modern form of Newton's second law in polar coordinates and then eliminating time as a parameter. The result is $F(u) = k\,[u^3 + u^2(d^2u/d\theta^2)]$, where $k$ is a constant and $u = 1/r$, i.e., the reciprocal of the distance $r$ between the particle and the force center. If one knows the polar equation of the path, i.e., $u = u(\theta)$, then the force is simply determined. For example, in Proposition IX, Newton seeks to find the force required if the path is the logarithmic spiral, which is given by $u = e^{-\theta}$, where $\theta$ is the polar angle. Thus, the second derivative is given by $u$, and the force $F(u)$ is proportional to $u$ cubed or the reciprocal of $r$ cubed, which is the result obtained by Newton in the *Principia* employing geometric relationships and the critical ratio.

ment produced by the force and is an important element in Newton's analysis.

Of particular interest are the geometric relationships that hold as the point Q approaches the point P, that is, as the angle QSP becomes very small. In this limit, Newton treats the motion as if it were under the action of a constant force and thus is able to calculate the displacement QR quite simply. Thus, as Westfall and others have pointed out, although the appearance is that of classical geometry, its essence is that of the limiting process of differential calculus.

The strategy required that he implicitly redefine the characteristics of curves in terms of ultimate ratios about the points Q and P as they approach each other. The *Principia* clothed itself in the propositions of classical geometry. The propositions themselves, however, were the ultimate ratios of quantities approaching zero. Euclid would not have recognized his offspring.[5]

Consider, for example, the versine TP of the angle QSP in Figure 3.1.[6] In the limit as Q approaches P, then the displacement QR approaches the versine TP. And it is this displacement QR, read as "versine" QR, that is to figure prominently in the derivation of the critical expression for the force.

In the opening of Proposition VI, Newton refers the reader to Corollary 4 of Proposition I for proof that the versine QR is proportional to the force for a given time and to Corollaries 2 and 3 of Lemma XI for proof that the versine QR is proportional to the inverse square of the time for a given force. And anyone who seeks to follow Bentley's lead in attempting to understand the opening sections of the *Principia* should take Newton's advice and master these corollaries. It is possible, however, to set the background for the discussion of the diagrams in this paper without pursuing the details by which Newton defends (or fails to defend) the assumptions underlying this critical relationship.[7]

[5] Richard S. Westfall, *Never at Rest: A Biography of Isaac Newton* (Cambridge: Cambridge University Press, 1980), pp. 425–6.

[6] A modern text would define the versine (or versed sine) as $1 - \text{cosine}$. Thus, if the line SP were the radius of a unit circle, the line QT would be the sine, the line ST would be the cosine, and the line TP would be the versine. Newton defines the versine TP given above, however, as the "versine of twice the angle." Since the value of the versine is the line TP in either case, it makes no difference which definition is employed.

[7] Whiteside, ed., *Mathematical Papers*. For a discussion of Proposition I, see *VI*, 35–37, n. 19; for Lemma XI, see *VI*, 117, n. 54.

Put quite simply, in the absence of any force a particle at point *P* would move to point *R*. Because of the force directed toward *S*, however, the particle moves instead to point *Q*. In the limit as *Q* approaches *P*, the force that acts for a vanishingly small time is assumed to be a constant in both magnitude and direction. The displacement *QR* is thus given by Galileo's relationship for motion with a constant acceleration, which is discussed by Newton in the Scholium of the opening section of the *Principia*. In contemporary notation, one would write the following:

displacement *(QR)* = ½ (acceleration) (time squared)

or for a given unit mass *m*, the acceleration is proportional to the force, thus in the limit as the displacement *QR* becomes the versine *QR*, the following relationship holds true:

versine *(QR)* is proportional to the (force) × (time squared)

Moreover, Newton has demonstrated Kepler's Area Law in Proposition I, that is, a particle moving in an ellipse under the action of a central force will sweep out equal areas in equal times. Thus, in Figure 3.1 the area of the triangle *SQP* is proportional to the time and is given by ½ base *(SP)* multiplied by the height *(QT)*. Thus, in the limit as *Q* approaches *P*, the measure of the central force *F* directed to the force center *S* is given by the following critical ratio:[8]

force is proportional to (versine)/(time squared) =
$$QR/(SP^2 \times QT^2)$$

In the statement of Proposition VI, Newton promises in what follows to provide problems as examples of how the critical ratio can be employed to determine the nature of forces if one is given the path of the particle *APQ* and the location of the force center *S*. His immediate goal is the Keplerian Proposition, which is the problem of a particle moving in an elliptical path with the force center at a focus. This solution is distinguished by being given the first position in a new section to follow the section that introduces the critical ratio and its immediate examples. But first he describes and presents solutions for the following preliminary examples: (1)

[8] Westfall, *Never at Rest*, p. 425.

various circular orbits and force centers, (2) a logarithimic spiral with its force center at the center of the spiral (see note 4), and (3) an elliptical orbit with its force center at the center of the ellipse.[9] The interest in circular paths arises from the important historical problem of uniform circular motion, although Newton presents a solution to a much more general and complex problem. The spiral would appear to be related to the problem of circular motion with a force center on the circle itself. The final problem of this opening section, the elliptical path with a force at the center, is clearly an example intended to prepare the reader for the "distinguished" problem to follow: to determine the elliptical path with the force at the focus.

### PLANETARY ELLIPTICAL ORBITS

Of all the golden nuggets that lie in the mathematical mire of the *Principia,* none has received more attention than Proposition XI, Section III, Book I: the "Keplerian Proposition." In the square at Grantham, the town where Newton attended grammar school, stands an impressive statue of Newton pointing with his right hand to a scroll on which the diagram for Proposition XI is displayed. A century later, when the Bank of England produced the "Isaac," the new Newton pound note, it clearly bore a diagram that was intended to represent this Keplerian Proposition. It may be argued that Newton himself may not have thought this proposition to be the mathematical highlight of his *Principia;* nevertheless, posterity has so considered it.[10]

The *Principia* has on occasion been referred to as "a work more revered than read." But whether for its prestige or for its content, copies of it have been much in demand by the learned and academic establishments. There are some sixteen Latin editions and

[9] Problem III in Proposition VIII is not strictly a central force problem. The particle moves on a semicircular path with a force center so remote that the force lines are parallel.

[10] Whiteside, "The Mathematical Principles Underlying Newton's *Principia Mathematica,*" 130. Whiteside points out that many contemporary portraits of Newton exist in which he sits before a half-open *Principia,* and that it is reasonable that the choice of section would be Newton's. Unfortunately, in only one portrait is the diagram clear, the one by Enoch Seeman now in the National Portrait Gallery in London. There Newton points to Proposition LXXXI of Book I in which he must in effect solve a triple integral to evaluate the external gravitational effect of a uniform spherical distribution of mass.

reprints and nine translations into other languages with some twenty-one various English revisions and reprints.[11] Substantive changes have been made under Newton's direction during the revisions of the first three Latin editions. Later Latin editions are faithful to the third Latin edition but vary from the so-called Jesuit edition of LeSeur and Jacquier with their extensive commentaries (often more comments than text on a page) to the Glasgow edition of Thomson and Blackburn, who had "been induced to reprint Newton's last (third) edition without note or comment."

In the first edition of the *Principia*, the same diagram is used for three different propositions that are concerned with planetary motion in an ellipse: Propositions X and XI, both of which seek "to find the law of centripetal force" about different force centers, and Proposition XVI, which seeks to find the law of velocity dependence. The decision to employ the same diagram for both the central force center, Proposition X, and the focal force center, Proposition XI, was dictated by economy of costs and not by ease of exposition because the line *QR*, the critical "versine to be" from Proposition VI, should be in the direction of the force, which differs in each case. In Proposition X, it should be parallel to a line directed toward the center of the ellipse, and in Proposition XI it should be parallel to a line directed toward the focus of the ellipse. Close examination of the diagram employed in Propositions X and XI, however, reveals that the identical woodcut was used in both diagrams. Thus, the direction of the critical versine *QR* must be wrong for at least one of the propositions.[12]

Figure 3.2A is the diagram for Propositions X and XI as it appears in the first edition of the *Principia*. It is important to identify the major elements of the critical relationship from Proposition VI in it: first, the "arbitary" curve passing through the points *A*, *P*, and *Q* in Figure 3.1 has now grown into the full ellipse passing

[11] For an extended bibliography of the *Principia*, see *Isaac Newton's Philosophiae Naturalis Principia Mathematica, the Third Edition with Variant Readings*, ed. Alexandre Koyré and I. Bernard Cohen, 2 vols. (Cambridge, Mass.: Harvard University Press, 1972), *II*, 851–83 [hereafter cited as variant *Principia*].

[12] That the woodcuts are identical and not just similar can be attested to by counting the dots and breaks in a given small area. These "fingerprints" of the woodcut clearly indicate the identity. Further, Propositions X and XI occur in different signatures in the first edition. Thus, it is possible to employ the same woodcut for different figures. See Koyré and Cohen, variant *Principia*, *I*, xxvi for an extended discussion of the diagram of the planetary ellipse of the *Principia*. Also see J. A. Lohne, "The Increasing Corruption of Newton's Diagrams," *History of Science*, 6 (1967), 81–3.

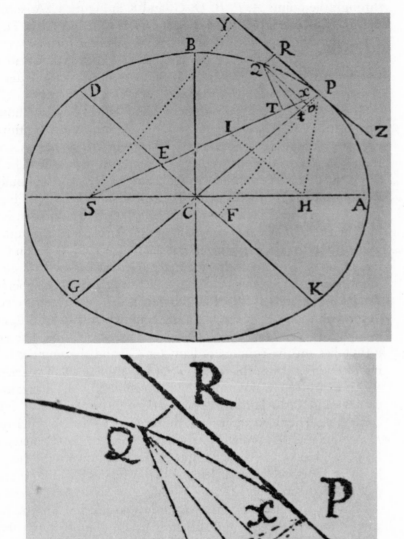

(A)

(B)

Figure 3.2. (A) The common diagram for Propositions X and XI from the first edition of the *Principia*. (B) An enlargement of the critical area of diagram A. The ordinate *Qv* crosses the focal line *TP* and terminates on the conjugate diameter *tP*.

through the points $A$, $P$, $B$, $D$, $G$, and $K$ in Figure 3.2A; second, the line $QT$ in Figure 3.1 now appears as $QT$ and $Qt$ in Figure 3.2A, the latter serving Proposition X and the former serving Proposition XI; and third, the tangent line $ZPR$ and the all-important versine $QR$ remains unchanged except that the latter has changed its inclination slightly.

The diagram also has a number of lines added beyond those in Proposition VI. The lines $AC$ and $AB$ are simply the semimajor and semiminor axes of the ellipse. The line $DK$ is a diameter of the ellipse that is drawn parallel to the tangent line $ZPR$. The line $GP$ is the diameter conjugate to the diameter $DK$, and it should be noted that $GP$ is not, in general, perpendicular to $DK$.[13] Figure 3.2A has four final additions: the lines $PF$, $Qv$, $PH$, and $IH$. First, the line $PF$ is a normal to the diameter $DK$ that passes through the point $P$ and thus is constructed perpendicular to $DK$. Second, the line $Qv$ is an ordinate to the diameter $DK$ that passes through the point $Q$ and thus is constructed parallel to $DK$ and hence perpendicular to the normal $PF$. Note, however, that Newton has elected to terminate the ordinate $Qv$ on the conjugate diameter $GP$, even though it is constructed perpendicular to the normal $PF$. The point $v$ thus lies on the conjugate diameter $GP$ and the point $x$ is the intersection of the ordinate $Qv$ with the focal line $PS$. The triangle $Pxv$ is employed in the proof of Proposition XI. The diagram becomes quite crowded in this area and will be the site of future confusion. The third addition is the line from $P$ to the other focus $H$ and the line $HI$, which is another ordinate to the diameter $DK$ (as is $Qv$) but extended beyond to its intersection $I$ with the focal line $PS$. Thus, the tangent $RP$ and the ordinates $Qv$ and $HI$ are all parallel to the diameter $DK$.

In what follows, however, attention is directed for the most part to the versine $QR$ and the ordinate $Qv$ (Figure 3.2B). From Proposition VI, the force is proportional to the versine, and hence they should be in the same direction. It is clear that in the first edition the versine $QR$ is parallel to the conjugate diameter $GP$ and is thus correct for a force directed toward the center $C$ of the

---

[13] A diameter of an ellipse is any line through the center $C$, such as $DK$; a conjugate diameter, such as $GP$, is the locus of midpoints of cords parallel to the diameter $DK$. Thus, in general, the conjugate diameter $GP$ is not perpendicular to the diameter $DK$. An error in the direction of the conjugate diameter can lead to considerable confusion when the geometry is explored. See Lohne, "Newton's Diagrams," 83.

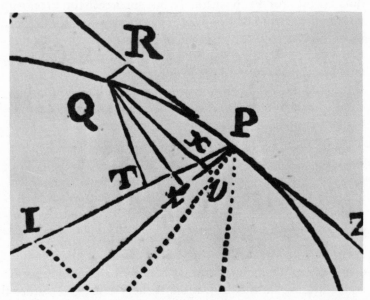

Figure 3.3. The critical area of the common diagram for Propositions X and XI from the second edition of the *Principia*. The versine *QR* is inclined at an angle greater than the focal line *TP* (Prop. XI) but less than the conjugate diameter *tP* (Prop. X). (Source: University of London Library.)

ellipse, as in Proposition X. But it is therefore incorrect for the force directed to the focus *S* of the ellipse, as in Proposition XI. Moreover, the ordinate *Qv* clearly cuts the focal line *PS* at *x* and terminates correctly on the conjugate diameter *GP* at *v*.

The second edition was published in 1713 under the editorship of the able young mathematician Roger Cotes, and the figures reflect his ability. The diagrams for Propositions VI, X, and XI are slightly revised. Here in the second edition, as in the first edition, the identical woodcut is used for both Propositions X and XI.[14] Thus, it is again impossible for the direction of the versine *QR* to be in the correct direction for both Propositions X and XI. Careful inspection of Figure 3.3, which is an enlargement of the critical section of the diagram, will demonstrate, however, that Cotes has arrived at a compromise. Whereas the first edition had the versine *QR* in the direction of the conjugate diameter *GP*, and

[14] Again, a close examination of details on the diagram indicates the identity. Now the two figures appear in the same signature but on opposite sides of the same page.

thus correct for Proposition X but incorrect for Proposition XI, now in the second edition Cotes has the versine $QR$ in the approximate direction of the bisector of the angle $SPC$. This direction is midway between that required for Proposition X, toward the center $C$, and that required for Proposition XI, toward the focus $S$; it appears to be a deliberate compromise.[15] The significance of this compromise has not always been recognized, even when the change has been noted, as in the following commentary.

In the first edition there is an error in the figure, since the line QR, if extended, would intersect the major axis slightly to the right of S, that is, between S and C, which is not possible if the figure is accurate to illustrate Prop XI. For the second edition the figure has been redrawn; the inclination of QR is altered slightly so that the intersection with the principal diameter occurs slightly to the left of S, rather than to the right.[16]

Although it is true that these projections of the versine $QR$ do change, it is the inclination that is of fundamental importance. By concentrating on the projection of the versine $QR$, rather than on its absolute inclination, the commentator misses the significance of the compromise.

Another area in which Cotes's concern for detail can be found is in the construction of the ordinate $Qv$. Here, as in the first edition, $Qv$ terminates correctly on the conjugate diameter $GP$. But Cotes has introduced a dot between $GP$ and $PF$ to indicate that it is constructed normal to $PF$ or perhaps to aid in locating the point $v$. The editors of the variant *Principia* note that there is no existing correspondence between Cotes and Newton concerning the changes in this diagram.[17] They assume, therefore, that the changes were made by Newton, since he would feel free to do so without consulting Cotes whereas the reverse might not be so. In the third edition, separate diagrams are employed and thus the question of inclination is eliminated; however, the utility of the dot remains. Nevertheless, the dot is eliminated in this next edition. If, on the one hand, Newton had inserted it, then it is difficult to understand why he eliminated it. If, on the other hand, Cotes had inserted it without consultation, then perhaps Newton removed it in the following edition. In any event, the dot continues to appear and disappear in the eighteenth- and nineteenth-century editions. Thus,

[15] Lohne, "Newton's Diagrams," p. 82.
[16] Koyré and Cohen, variant *Principia*, I, xxvi.     [17] Ibid. I, xxvi (see n2).

both the "compromise" and the "dot" are clear indicators of concern for and control over the critical details of the diagram.

The third edition appeared in 1726 under the editorship of Henry Pemberton. In this edition, different individual diagrams are provided for each of the propositions. In the diagram for Proposition X, the versine QR is correctly inclined in the direction of the conjugate diameter GP, which is directed toward the force center at C, and the ordinate Qv clearly terminates on the conjugate diameter GP. There is no question concerning the location of the point v.

The diagram for Proposition XI, by necessity, is slightly more complicated. Figure 3.4A is the diagram as it appears in the third edition and Figure 3.4B is an enlargement of the critical section. The versine QR no longer need be a compromise as in the earlier editions and it is clearly in the direction of the focal line PS, which is directed toward the force center at the focus S. The ordinate Qv is not, however, as clearly resolved. Careful inspection of the diagram indicates that it is intended that the ordinate Qv terminate correctly on the conjugate diameter GP. The woodcut, however, has not produced a clear line and it fades after crossing the focal line PS and disappears before reaching the conjugate diameter GP.[18] Thus, it is not evident, out of context of the solution itself, where the point v lies. In this rendering of the diagram, in fact, v almost appears to be part of the normal PF. The solution requires the use of the triangle Pvx, however, and it is clear from the properties of this triangle that v lies on the conjugate diameter GP and that x lies on the focal line SP. It is not clear from the diagram in all printings, however, that the termination of the ordinate Qv is the conjugate diameter GP. This fact is particularly significant when

---

[18] Ibid. *I*, xxvi. The editors state that "in the third edition the line from Q to v through x stops midway between x and v, whereas in the second edition it goes on to v, as it does in the first edition." It would appear that "fades" rather than "stops" would be a better description. A comparision of three different impressions of this edition, identified by different page sizes, indicates a variation in the degree of fading of the line. In the three different impressions held in the Royal Society, it would appear that the largest page size contains the best reproduction of the diagram. Since this large impression was intended as presentation copies and since it contained the fewest in number, one might think that it was the first impression. The editors of the variant edition quoted above argue otherwise, however. Basing their ordering on other examples of apparent progressive deterioration, they conclude that the presentation copies were printed last. Examining all their examples and employing all their criteria, I arrive at just the opposite conclusion. In any event, in all impressions various degrees of fading do occur.

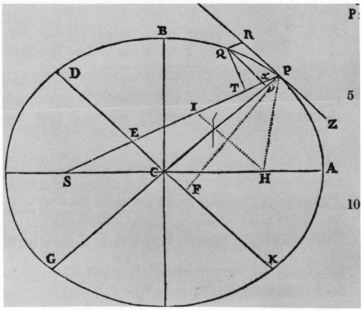

(A)

(B)

Figure 3.4.   (A) The separate diagram for Proposition XI from the third edition of the *Principia*. (Source: Variant edition of the *Principia*, Harvard University Press.) (B) An enlargement of the critical area of diagram in A. The ordinate Qv begins to fade after crossing the focal line *TP*.

one notes that it is this third edition, the final revision done during Newton's lifetime, that is to become the definitive edition for all to follow.

The Latin editions that follow the definitive third edition by Pemberton can be divided into two groups. The first group is centered on the 1739–42 Jesuit editions of LeSeur and Jacquier published in Geneva with their extensive commentaries. This Geneva edition is followed by two Latin editions that contain the LeSeur and Jacquier commentaries: one published in Prague in 1780–5, which is edited by Tessanek, and the other published in Glasgow in 1822, which is edited by Wright. The second group consists of reprints of the third Latin edition by Pemberton with minimal corrections or comments. Into this group fall the reprinting edited by Horsely and published in London in 1779–82 as part of a five-volume set of Newton's collected works and the reprinting edited by Thomson and Blackburn in Glasgow in 1871.

The 1738 Geneva edition by LeSeur and Jacquier correctly produces the diagram for Proposition XI, as would be expected from such detailed analysts. The proposition is accompanied by extensive comments, and both the inclination of the versine $QR$ and the termination of the ordinate $Qv$ are correct. In a second reprinting in 1760, the location of the symbol $v$ is slightly shifted, but the diagram is otherwise identical to the 1738 edition. The 1780 Prague edition by Tessanek does not fare so well, however. The versine $QR$ is inclined correctly but the terminal point $v$ of the ordinate $Qv$ is located so close to the planetary point $P$ that one cannot tell if the ordinate $Qv$ terminates on the conjugate diameter $GP$ or on the focal line $PS$. One cannot use this diagram in a proof of the proposition. The third and final member of the Jesuit group is the 1822 Glasgow edition by Wright, who is clearly a working editor, being also the author of an extensive English commentary on the *Principia*. The diagram for Proposition XI has the inclination of the versine $QR$ correct, but there is a variation in the drawing of the ordinate $Qv$. The solid line from $Q$ terminates on the focal line $SP$, and a single dot is placed between $x$ on $SP$ and $v$ on $GP$. This dot therefore represents one side of the triangle $Pxv$, in contrast to Cotes's dot, which simply located the point $v$ and was located beyond $GP$.

The second group is represented by the editors Horsely and Thomson/Blackburn. In the Horsely edition of 1779, the inclination of the versine $QR$ is correct, but the ordinate $Qv$ is represented between the focal line $SP$ and the conjugate $GP$ by a single dot, as in the 1822 Wright edition. Since the Horsely edition is earlier, it may well be the source of Wright's later revision. The Thomson/Blackburn edition has the correct inclination of the versine $QR$, but the ordinate $Qv$ clearly terminates incorrectly on the focal line $SP$. Since this edition claims to be taken directly from the third Latin edition, the fading line between $x$ and $v$ has been incorrectly shortened rather than extended.

The bibliographical essay in the variant *Principia* lists five Latin excerpts: two edited by William Whiston in 1710 and 1726, one with a group editorship of Jebb, Thorp, and Wollaston in 1765, one edited by Wright in 1831, and one edited by Whewell in 1846. The diagram for the joint effort of 1765 is correct. The letter $v$ is moved so that it lies between the focal line $SP$ and the conjugate diameter $GP$ to make clear that the ordinate $Qv$ terminates on $GP$. The Wright excerpt of 1831, on the other hand, clearly allows the ordinate $Qv$ to terminate incorrectly on the focal line $PS$. The dot that appeared in Wright's earlier full Latin edition is not shown. Whewell's excerpt of 1846 has both indicators correct, but the ordinate $Qv$ is beginning to fade like the original Pemberton diagram.

The first English translation appeared in 1729, closely following the third Latin edition of 1726 and Newton's death in 1727. The translator was Andrew Motte, of whom little is known, and who may also have served as the editor because the volume was published by his brother Benjamin Motte. A second independent English translation of Book I of the *Principia* by Robert Thorp was published in 1777 with a promise of a translation of Books II and III to follow. In 1802, a reprint of the translation of Book I appeared, but there was no trace of the promised translation of Books II and III. The Motte translation was reprinted with "corrections and explanations" by three editors: William Davis in London in 1803 and 1819, Nathaniel Chittenden in New York in 1848 and 1850, and Florian Cajori in California in 1934 with multiple reprints including a 1971 paperback edition in two volumes.

In the first English translation, Motte is, in general, faithful to the diagrams as they appear in the third Latin edition. The figures in the Motte edition appear on separate foldout pages rather than as internal figures on the printed page as in the Latin editions. The publisher Dawson of Pall Mall has recently reprinted an edition of the Motte translation in which the foldout diagrams are photographically reproduced from the original Motte diagrams but are bound as internal, unnumbered pages.[19] Figure 3.5A is Plate IV, which contains the diagrams for Propositions X and XI as it appears in the Dawson reprint, and Figure 3.5B is an enlargement of the critical section of Proposition XI. The critical inclination of the versine $QR$ is correct in both diagrams, being parallel to the conjugate diameter $GP$ in the upper diagram and being parallel to the focal line $SP$ in the lower diagram. The letter $T$ remains in the upper diagram but is missing from the lower diagram, and the letter $x$ has been moved from the right side of the ordinate $Qv$ to the left side, which makes it somewhat unclear that $x$ represents the intersection of $Qv$ and $PS$. Moreover, the normal $QT$ appears in the proof as part of the critical ratio; with $T$ missing, this important line is undefined in the lower diagram. The ordinate $Qv$, however, is clearly extended to and terminates correctly on the conjugate diameter $GP$, so that in the context of the proof, the triangle $Pvx$ is clearly delineated, as may be seen in Figure 3.5B.

A comparison of the diagrams from the Dawson reprint with the diagrams from the first Motte edition held by the University of London at the Senate House Library, however, reveals a variant edition. Figure 3.6A is Plate IV as it appears in this variant reprint, and Figure 3.6B is an enlargement of the critical section of Proposition XI. In the variant edition the critical inclination of $QR$ is correct in both figures. In the lower diagram for Proposition XI, however, the ordinate $Qv$ now terminates incorrectly on the focal line $SP$, rather than on the conjugate diameter $GP$, as it does in the first printing. Thus, the triangle $Pvx$, which is required for the proof, is no longer uniquely determined from the diagram.

The most obvious difference between the two sets of plates is the capital $F$ in the heading "Fig.": in one set it is a block letter, and in the other it is a script letter. This variation runs through all the figures in both editions. Further, it is also clear that the printed pages are the same although the diagrams differ. For example, page 85, which appears in Section III, is mislabeled "SECT II" in

---

[19] Sir Isaac Newton, *The Mathematical Principles of Natural Philosophy*, trans. Andrew Motte (London: Dawsons of Pall Mall, 1968). See the introduction by Bernard Cohen to this edition for a discussion of the birth and growth of this edition.

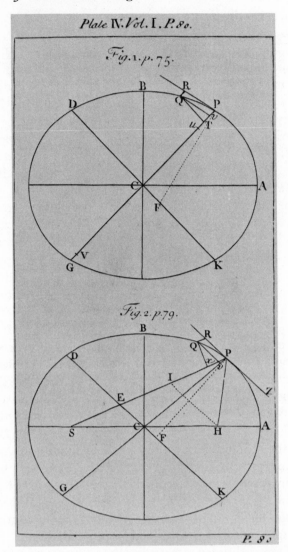

*Plate* IV. *Vol.* I. *P.* 80.

Figure 3.5.   (A) The separate diagrams for Propositions X and XI from the Dawson reprint of the Motte edition of the *Principia*. (Reproduced with permission of the publishers Dawsons of Pall Mall.)

both the Dawson and variant editions. The variant edition may well be a second but unacknowledged printing. This supposition is supported in part by a comparison of the higher quality of paper used in the original edition with the poorer quality in the variant edition. In the first printing, it is likely that the numbered printed

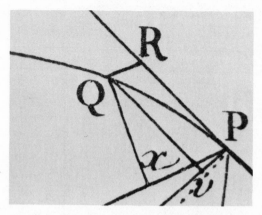

(B)

Figure 3.5 *(cont'd.)*. (B) An enlargement of the critical section of the lower diagram for Proposition XI from part A. The ordinate *Qv* clearly defines the triangle *Pxv*.

pages contained the text and the unnumbered foldout pages containing the diagrams were produced separately and brought together for collating and binding. Hence when the decision was made to run off more copies of the text, it would appear that the original woodcuts were missing or worn and had to be replaced.[20]

An examination of fifteen copies of Motte's first edition revealed that only three contained the variant diagrams: the copy in the Senate House Library in London, the copy in the Library of Congress in Washington, D.C., and one of the three copies held in the library of the University of Edinburgh. It would appear, therefore, that the number printed of the variant edition was much smaller than the number printed of the original edition. The six first editions listed in the bibliographical essay in the variant *Principia* are all of the first printing, which may explain why the variant edition discussed here was not listed.

It is of interest, therefore, to follow the three major revisions of the Motte translation that took place in the nineteenth and twentieth centuries. The first was done in London by William Davis in 1803 and in 1819; the second was done in New York by Nathaniel Chittenden in 1848 and in 1850; and the third was done in Califor-

---

[20] This scenario developed from a conversation at Trinity College Library, Cambridge, with Dr. Philip Gaskell, author of *A New Introduction to Bibliography* (Oxford: Oxford University Press [Clarendon Press], 1972). The point of interest is, of course, the existence of such a variant printing and not how it came about.

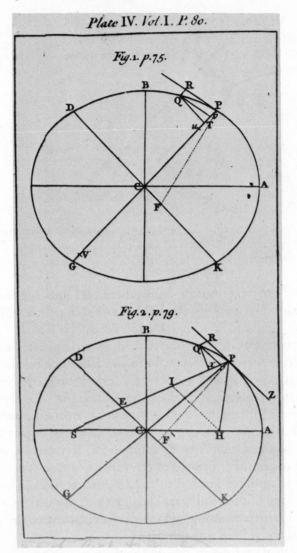

Figure 3.6.  (A) The separate diagrams for Propositions X and XI from the Senate House Library copy of the Motte edition. (Source: University of London Library.)

nia by Florian Cajori in 1934. All claim to have done the revisions with "corrections" and "explanations." But not everyone agrees that they were successful. Of the 1848 New York editions, Lohne says, "Chittenden's edition of Motte's translation of the *Principia*

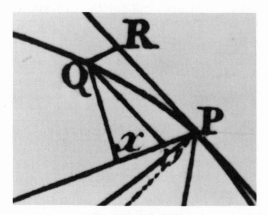

(B)

Figure 3.6 *(cont'd.)*.   (B) An enlargement of the critical area of the lower diagram for Proposition XI from part A. The ordinate *Qv* fails to define the triangle *Pxv*.

is not always so 'carefully revised and corrected' as it pretends to be."[21] Of the 1934 California edition, Cohen says, Cajori chiefly replaced "obsolete mathematical expressions and notations by modern forms, but he . . . did not take any pains to make Newton's mathematical equations easily intelligible, as by adding parentheses and grouping terms together. . . . The result is neither Newton's original version nor a truly modern text."[22] In all three editions, the versine *QR* has the correct inclination. The ordinate *Qv*, however, varies from edition to edition.

The editor of the London edition, William Davis, was described as a "mathematician and a member of the Mathematical and Philosophical Society."[23] He was a bookseller who also edited works on "Fluxions" by Maclaurin (1801), Simpson (1805), and Rowe (1809). In both the 1803 and 1819 London editions of the *Principia*, a new variation on the drawing of the ordinate *Qv* appears. The line changes thickness dramatically as it crosses the focal line *SP*, and it becomes much thinner. Moreover, it continues on beyond the conjugate diameter *GP* and terminates instead upon the normal *PF*. Only the triangle *Pxv* is required in the proof of the prop-

---

[21] Lohne, "Newton's Diagrams," 80.
[22] Newton, *Principia*, Dawson reprinted [see note 19], xii–xiii. For a detailed discussion of the Cajori edition, see I. Bernard Cohen, "Pemberton's translation of Newton's *Principia*," *Isis*, 54 (1963), 341–348.
[23] *Dictionary of National Biography* (London: Oxford University Press, 1917), *V*, 623.

osition, but the construction does call for the ordinate $Qv$ to be perpendicular to the normal $PF$. Thus, the "overshoot" has a logical function and may not simply be accidental. A comparison of a number of copies confirms that the line was in the original plate and was not simply added by a reader. The letter $T$, and hence the critical normal $QT$, is missing in this edition, as it was in the Motte edition.

The editor of the New York edition, Nathaniel W. Chittenden, was an 1837 graduate of Columbia University of New York. This edition is dedicated to the "teachers of the normal school of the state of New-York" and opens with the observation that "a stirring freshness in the air, and ruddy streaks upon the horizon of the moral world betoken the grateful dawning of a new era" and concludes with the identification of the *Principia* as "the greatest work of the greatest Teacher."[24] Despite the editor's claim that this first American edition is "carefully revised and corrected," the ordinate $Qv$ clearly terminates incorrectly on the focal line $SP$ in the diagram for Proposition XI, as in the variant Motte edition.

The editor of the California edition, Florian Cajori, was a professor of physics with an extended interest in the history of physics. The 1934 printing by the University of California Press was released with a variant edition that was published by the Cambridge University Press for release in England. Six photoreprints have been made since the first release in 1934, the final one being a paperback in 1962. Despite the editor's qualifications, however, the diagram for Proposition XI presents difficulties. In all editions, the ordinate $Qv$ clearly terminates incorrectly on the focal line $PS$. The letter $T$, which is missing from the diagram for Proposition *XI* in both of the Motte printings but which did appear in the third Latin edition, has been restored in the Cajori editions. Moreover, the location of the point $x$ is as in the Latin edition and not as in the Motte edition. Thus, it appears that the diagram in the Cajori edition may have been taken from the Latin edition and not from Motte's English translation.

[24] Sir Isaac Newton, *Newton's Principia, the Mathematical Principles of Natural Philosophy* (New York: Daniel Adee, 1846), pp. 3–4. There were four different printings in 1848 and a fifth in 1850. They differ in title page, portrait frontispiece, and quality of paper, but all appear to be printed from the same plates. See Koyré and Cohen, variant *Principia*, pp. 863–4, for a discussion of the bibliographical problems created by this edition.

There exists only one eighteenth-century English excerpt of note, that by John Clarke in 1730. He was the younger brother of Samuel Clarke, who had a long association with Newton and was, following the death of John Locke in 1704, England's leading metaphysician. Early in his career, Samuel had published a "fresh" Latin translation of Rohault's very popular Cartesian textbook, *Traité de physique*. Clarke added extensive "Newtonian" footnotes, however, that were designed to refute the Cartesian text. This work became a standard Cambridge textbook, and the fourth edition of 1718 was said to be still in use at Samuel's death in 1730. John, who was distinguished as a mathematician and who resided for much of his life in Cambridge, published in 1723 an English translation of his brother's Latin translation of Rohault's textbook. Therefore, John Clarke was extremely well qualified to produce an English translation of the *Principia*.[25]

A comparison of the Clarke and Motte translations demonstrates on the one hand that Motte has supplied a literal translation with few additions except those necessary to render the Latin into acceptable English. On the other hand, Clarke has added full sentences and even paragraphs to the translation of the original Latin text without explicitly designating them as additions. The astronomical data from Book III is now interspersed in Book I where appropriate. The periods and radii for Kepler's third law are now given in Book I, but interestingly do not appear in Proposition XVI, which gives the third law. Instead, it follows Proposition IV, which in effect discusses a circular centripetal force and its dependence upon the square of the velocity and the reciprocal of the radius, from which in turn the third law can be derived. The dangers of such hidden and presumably unauthorized additions are obvious, particularly in sections that are not strictly mathematical in nature. The advantages are equally obvious, particularly in the mathematical proofs, and it is with such proofs that Clarke is largely concerned in the limited sections he has elected to translate.[26]

---

[25] John Clarke, *A Demonstration of Some of the Principal Sections of Sir Isaac Newton's Principles of Natural Philosophy* (New York: Johnson Reprint, 1972). See also *Dictionary of National Biography IV*, 433.

[26] The first three sections (as per Newton's advice to Bentley), Section X on the pendulum and its use in determining the shape of the earth, Section XI on centers of gravity, and Section XII on attractive forces between spherical bodies.

Clarke has also revised diagrams where, presumably, he thought they were confusing, such as the critical diagram for Lemma XI, which is an important underpinning for the Keplerian Proposition.[27] The diagrams for Propositions VI, X, and XI are correct, including the ordinate $Qv$, which terminates correctly on the conjugate diameter $GP$. Clarke has introduced, however, an interesting variation on the inclination of the versine $QR$ in the diagram for Proposition XI. The point $R$ on the tangent line $RPZ$ is now determined by extending a straight line from $S$ through $Q$ to $R$. Thus, the versine $QR$ is in the direction of $SQ$ rather than in the direction of $SP$. The two directions coincide in the limit as $Q$ approaches $P$, of course, but it is a distinct departure from Newton's original construction.

The *Principia* and its unique quasi-geometric version of the calculus continued to be studied in the English universities well into the nineteenth century. The nineteenth-century excerpts are largely compiled by editors seeking to serve that student market by providing commentaries on the *Principia* to be used as study guides. The editors range from established educators such as William Whewell, master of Trinity College, Cambridge, to free-lance educators such as J. M. F. Wright.[28] As would be expected, such editors present the diagrams correctly, and there are no errors in the inclination of the versine $QR$ or the length of the ordinate $Qv$, in any of the twenty-eight various editions or reprints. The form of the drawings varies from editor to editor, however, and this variation often indicates the source of the original version of the drawing. In the 1825 version of the Carr excerpt, there appears a dot in the ordinate $Qv$ between the focal line $SP$ and the conjugate diameter $GP$, in the fashion of the 1779 Horsely Latin edition. In the 1855 version of the Wright edition, the ordinate $Qv$ "over-

---

[27] Lemma XI is concerned in part with the radius of curvature to a general curve at a given point. Newton's proof requires two circles that are tangential to the curve at the given point: a large circle and a smaller circle, which in the limit is the "osculation" circle whose radius is the radius of curvature at that point. His diagram, however, is confusing in that it shows only a portion of one circle. Clarke has not only revised the diagram to show the two circles but has also inserted the clause "now if we suppose two circles to pass through the points AGB, Abg" for the single word "but," which appears in the Motte translation. The help that such an extension provides for the first (or second) reading of the lemma is immense.

[28] J. M. F. Wright, *A Commentary on Newton's Principia* (New York: Johnson Reprint Corp., 1972). See the "Introduction to the Reprint Edition" by I. Bernard Cohen for a discussion of Wright and his contributions.

shoots" the conjugate diameter *GP* and terminates on the normal *FP,* in the manner of the 1803 Davis English edition.

One of the most interesting of the English excerpts is that by Percival Frost, Fellow and Mathematical Lecturer of Kings College. It appeared in six editions between 1854 and 1900. Frost sets forth his charge in the preface:

In publishing the following work my principal intention has been to explain difficulties which may be encountered by the student on first reading the *Principia,* and to illustrate the advantages of a careful study of the methods employed by Newton, by showing the extent to which they may be applied in the solution of problems.[29]

Frost presents only the first three sections of Book I; he does not include the opening sections of the three laws of motion, claiming that to do so would simply be to repeat material that is already available to students. Moreover, he has interspersed throughout the work sets of problems selected from the papers set in the Mathematical Tripos and in the course of the College Examinations, much in the form of a modern physics textbook. In fact, Frost's work is a fascinating metamorphosis of Newton's opening pedagogical sections of the *Principia* into a modern textbook, even to the extent of providing answers and solutions at the end of the book, although he does so only after overcoming some initial reservations.

At the end of the work I have given hints for the solution, and in many cases complete solutions, of the problems; and in doing so I am acting in direct opposition to my previously expressed opinion, but additional experience of fifteen years has shown me that it is a satisfaction to a student to see a solution of it; and, even when he has been successful, to compare his solution with that of an older hand.[30]

The diagram for Proposition XI is sharp and clear and is correct in every respect.

There exist translations of the *Principia* in eight other languages: in the eighteenth century, a French translation (1756); in the nineteenth century, a German translation (1872) and in the twentieth century, full or partial translations in Russian (1915), Swedish (1927),

---

[29] Percival Frost, *Newton's Principia, First Book, and Sections I, II, III* (London: Macmillan Press, 1883), v.
[30] Ibid., vii.

Japanese (1930), Dutch (1932), Romanian (1956), and Italian (1966).[31] Here again, the key diagrams serve as universal indicators of the care and concern of the editor. The figures range from those that are clear and concise to those that are so jumbled as to be useless in analysis.

The French edition of 1756 was produced by Madame la Marquise Du Chastellet while she was living with Voltaire, who himself wrote the historical preface for the translation. The final revision and commentary, however, appears to be in part the work of the great French mathematician Clairaut.

Voltaire admitted that many of the ideas in the commentary were Clairaut's, that – although Mme Du Chastellet had made the calculations by herself – Clairaut went over each chapter as she finished it and corrected it.[32]

In the diagram for Proposition VI, which contains the critical ratio, the illustration of the versine $QR$ has been revised. The third Latin edition gives it as a solid straight line between $R$ and $Q$ with the same slope as the focal line $PS$, and then shows a dashed line from $Q$ to $S$. Thus, the dashed line $QS$ must be in a different direction from the solid versine $AR$. The French edition, however, generates the point $R$ by continuing the dashed line directly from $S$ through $Q$ to the tangent $RP$. Thus, the versine $QR$ is directed toward the force center $S$ from the point $Q$, and not from the point $P$, although in the limit $Q$ will ultimately approach $P$. This variation is essentially the same revision as employed earlier by Clarke in his diagram for Proposition XI (note 25).

In the diagram for Proposition X, the versine $QR$ is not directed as in Proposition VI, but is correctly shown as a straight line parallel to the conjugate axis $GP$. The diagram, which is dedicated to Proposition X, is unnecessarily crowded as it contains lines such as the focal line $PS$ and the ordinate $HI$, which are needed only in Proposition XI. Comparison with the diagram for Proposition XI, however, displays that the two diagrams are distinctly different in many other respects and there is no intent, as in the first two Latin editions, to reuse the same diagram for different propositions.

---

[31] For bibliographical details, see Koyré and Cohen, variant *Principia, II,* 875–83.

[32] I. B. Cohen, "The French Translation of Isaac Newton's Philosophiae Naturalis Principia Mathematica (1756, 1759, 1966)," *Archives internationales d'histoire des sciences, 21* (1968), 265–6.

In the diagram for Proposition XI, the ordinate $Qv$ is correct, clearly continuing to and terminating on the conjugate diameter $GP$. The versine $QR$, however, displays an interesting variation; it is slightly curved. As a planet moves along the ellipse from the point $P$ to the point $Q$, the direction of the force, and hence the direction of the versine changes. But because the segment $QR$ was intended by Newton to represent the displacement due to the force over a very small time interval, it was represented by a straight line, even though in any finite time it would be a curved line, because it was to be evaluated in the limit as $Q$ approached $P$. Du Chastellet (Clairaut?) elects, however, to show it as a curved line, and further, it is curved in the correct direction, that is, concave with respect to $P$.

The German edition of 1872 was produced by Jakob Philipp Wolfers, a German mathematician and astronomer who also produced an annotated edition of Euler's mechanics with commentaries.[33] In this translation, the editor has abandoned the traditional numbering of propositions that appears in the Latin editions and introduces a numbering system of his own. Moreover, he has modernized the mathematical notation and formulation by presenting relationships as equations set on separate lines, as in a standard nineteenth-century physics text. But the figures are faithful copies of the third Latin edition. The versine $QR$ is correctly inclined in Propositions VI and X, but it is incorrect in Proposition XI, where it remains parallel to the conjugate diameter $GP$, as in Proposition X, rather than being inclined parallel to the focal line $SP$. The ordinate $Qv$ does terminate correctly on the conjugate diameter $GP$.

The Russian edition of 1915 was produced by Lieutenant-General de Marine and Professor Emeritus A. N. Krylov. The copy held by the Royal Society was received during World War II and was presented "To the Royal Society of London on the 300th anniversary of the birth of its great president Sir Isaac Newton from the Academy of Sciences of the U.S.S.R." There appear to be no revisions of the diagrams and they are faithfully reproduced from the third Latin edition. The versine $QR$ is correctly inclined in Propositions VI, X, and XI, and the ordinate $Qv$ in Proposition XI terminates correctly on the conjugate diameter $GP$. In this edi-

---

[33] *Allgemeine Deutsche Biographie* (Leipzig, 1898), *44*, 6–7.

tion, the ordinate $Qv$ is not the usual solid line but is dashed in the same fashion as the ordinate $IH$.

The Swedish edition of 1927 was produced by C. V. L. Charlier. The translator has taken great liberties with the text, condensing the proofs and highlighting the pertinent results. This procedure is particularly noticeable in Proposition VI, where there is extensive interpolation of letters from the diagram into the text and where the critical ratio appears explicitly highlighted in the boldface summary that precedes the proposition. The versine $QR$ is correctly inclined in all three propositions, and the ordinate $Qv$ clearly terminates correctly on the conjugate diameter $GP$.

The Romanian edition of 1956 was produced by Professor Victor Marian. In contrast to the German and Swedish translations, this one appears to be a literal translation of the third Latin edition without a revised number system or the insertion of contemporary mathematical symbols. The figures also appear to be faithfully reproduced without additions or revisions. The versine $QR$ has the correct inclination in all three propositions, and the ordinate $Qv$ is very sharp and dark in Proposition XI and terminates clearly and correctly on the conjugate diameter $GP$.

The three remaining editions all contain errors, either in the inclination of the versine $QR$ or in the termination of the ordinate $Qv$. The Japanese edition of 1930 contains the complete text but the figures are very poorly done. It would be difficult to use them in actual calculations of the propositions. In the Italian edition of 1966, the ordinate $Qv$ clearly terminates incorrectly on the conjugate diameter $GP$ in Proposition XI. In Proposition X, the point $v$ is almost at the planetary point $P$, as in the 1780 Tessanek Latin edition published in Prague, and consequently it is very difficult to use the diagram. The Dutch edition is a summary and paraphrase with some explanatory comments and scattered excerpts in Dutch. It has the termination of the ordinate $Qv$ correct, but the inclination for the versine $QR$ is incorrect.

What may one then conclude concerning the utility of this diagram as a simple analytic device to test the care and utility of the editions? On the one hand, editors such as Roger Cotes and John Clarke, who not only correctly reproduced but also creatively revised diagrams, display their qualifications as mathematicians and careful editors. This care is substantiated by a more extensive examination of their editions. On the other hand, editions such as the Japanese translation of 1930, prepared for a volume entitled *Collections of Great Thoughts of the World,* display almost no concern for rigor in the diagrams. The figure for Proposition XI is so crowded and distorted as to be useless in working the proof of the proposition. Such a translation stands in stark contrast to the Swedish and German translations, which are clearly the work of mathematicians and careful editors. But between these two extremes there lie a wide variety of editions: the Motte edition with its two sets of drawings, one for which the diagram of Proposition XI is correct and one for which it is not; the Cajori edition, in which the diagram for Proposition XI has been taken from the third Latin edition and the fading ordinate $Qv$ has been incorrectly shortened, even though the editor is a physicist and a historian who should have been alert to this critical portion of the diagram. Perhaps the most that can be claimed is that when one finds the details of the diagrams correct, then it is probable that the other elements of the text are also carefully done. But when the diagram is faulty, then there may or may not be trouble elsewhere.

### POSTSCRIPT: THE BRITISH POUND NOTE

But the tale of the ordinate $Qv,$ whose tip $xv$ fades and reappears in the best Cheshire cat tradition, does not finish with these twentieth-century translations. Somewhere deep within the mysterious confines of "the old lady of Threadneedle Street," the Bank of England decided to produce a new design for the pound note that would honor England's premier scientist: Sir Isaac Newton. The date of first issue was the ninth of February, 1978, and it suffered through newspaper articles with leads such as "Newton becomes man of note" and letters that purported to confuse his triangular prism with a bar of Tobleröne chocolate. There were, of course, many more criticisms than congratulations, but overall the design of the note is sound, given that the purpose of the design of any banknote is primarily to be secure from forgery and only secondarily to be attractive or historically accurate. And with the exception of one major glaring error, the note as it is currently printed fulfills both its primary and secondary goals.

Figure 3.7 is the "Newton side" of the new note as it was first issued. The portrait is a composite of three parts: the head is from

Figure 3.7.   The "Newton side" of the 1978 issue of the pound note by the Bank of England. (Permission to reproduce this photograph and the following photographs was received from the Bank of England.)

a portrait done in 1702 by G. Kneller, which now hangs in the National Portrait Gallery in London; the body is from another portrait done by Kneller in 1720, which is now located in Petworth House, Surrey[34]; and the somewhat oversized hands are the imaginative reconstruction of the artist engaged by the Bank of England. The tripartite portrait thus serves the primary purpose of security without serious damage to the secondary purpose of accuracy. On the table beside Newton sits his reflecting telescope and his triangular prism, both meaningful representations of major areas of his scientific interests. Around his head are what appear to be apple blossoms that presumably are "Flower of Kent," which tradition has it was the apple tree that graced the yard at Woolsthorpe. In his lap he holds a copy of the first edition of the *Principia*, open to pages 50–1, which contain the diagram and text for Proposition XI, the Keplerian ellipse.

---

[34] A mirror-image copy of the original portrait appears in the frontispiece of vol. 7 of the Newton *Correspondence*.

The large diagram that occupies the left half of the note is also from Proposition XI, but evidently from the Cajori edition of the Motte translation of the Pemberton third Latin edition. Figure 3.8A is an enlargement of the critical section of the diagram as it appeared in the 1978 issue of the pound note. The diagram is not from the Pemberton edition, for there the ordinate $Qv$ fades between $x$ and $v$. It is not from either of the original Motte editions because the note contains the letter $T$, which is missing from both of Motte diagrams. It therefore appears to be from the 1934 Cajori–Motte–Pemberton edition, in which the $T$ was restored and in which the ordinate $Qv$ was incorrectly terminated on the conjugate diameter $GP$. In March of 1981, however, a new and unheralded revision of the "Isaac" was quietly issued by the Bank of England. The occasion appears to have been the appointment of a new chief cashier, for the signature on the side of the note containing the portrait of the Queen changes from J. B. Page to D. H. F. Somerset. The original 1978 note was printed with two colors, but the revised 1981 note appears in three colors. But of greater interest to the historian of science is that the troublesome ordinate $Qv$ now terminates correctly on the conjugate diameter $GP$. Figure 3.8B is an enlargement of the upper section of the diagram as it appears in the 1981 revised note. Now at last the diagram is correct: no faded ordinate as in the Pemberton edition, no missing $T$ as in the Motte edition, and no incorrect ordinate termination as in the Cajori edition.

There remains, however, the one glaring error that cannot be corrected with a simple added line, as with the ordinate $Qv$. To correct this error, the entire plate would have to be reworked and here the primary objective of security must take precedence over the secondary objective of historical accuracy. Figure 3.9 is an enlargement of the entire diagram of the revised note, and in it one notices that the artist has placed in the center a cowl of flames around the point that presumably represents the Sun. In Proposition XI, the Sun is located at the focus of the ellipse, labeled $S$ (for Sun). Unfortunately, the artist has elected to locate his Sun at the center of the ellipse, which is labeled $C$ (for center). Thus, one is faced with the embarrassment of a misplaced Sun and thus an undermining of Kepler's great achievement. There are only three possible resolutions of the difficulty: first, to assume that the designer had in mind Proposition X, in which the Sun is at the center

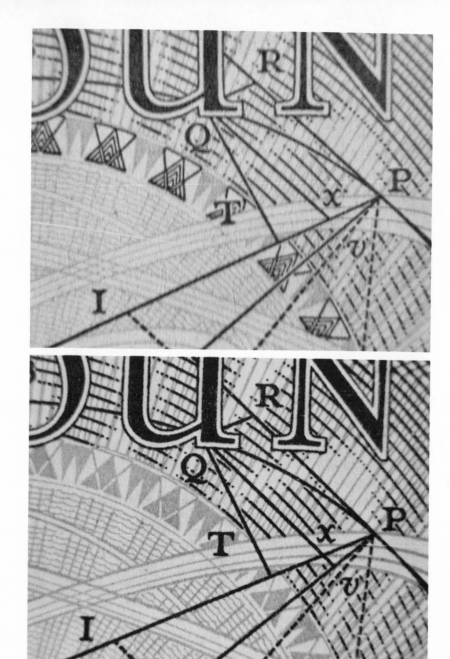

Figure 3.8. (A) An enlargement of the critical section of the diagram for Proposition XI from the 1978 issue of the pound note. The ordinate $Qv$ fails to define the triangle $Pxv$. (B) An enlargement of the critical section of the diagram for Proposition XI from the 1981 issue of the pound note. The ordinate $Qv$ clearly defines the triangle $Pxv$.

Figure 3.9.   The diagram for Proposition XI from the 1981 issue of the pound note. The cowl of flames representing the Sun surrounds the center *C* of the ellipse rather than the focus *S*.

of the ellipse with the gravitational force directly proportional to the distance instead of to the inverse square of the distance, but then Newton would be unhappy; second, to assume that the designer is suggesting a return to Copernican celestial centricity and that in the best modified Aristotelian tradition the Sun occupies that center, but then Kepler would be unhappy; or third, to admit that this diagram, as critical and as clear as it appears, is now, and always has been, a treacherous trap for all but the most careful of designers.

# 4

## Force, electricity, and the powers of living matter in Newton's mature philosophy of nature

R. W. HOME

One of Newton's most widely quoted methodological pronouncements appears in the preface he prepared for the first (1687) edition of his *Principia*. "The whole burden of philosophy seems to consist in this," Newton there wrote: "from the phenomena of motions to investigate the forces of nature, and then from these forces to demonstrate the other phenomena."[1] He made the same point in his other great work, his *Opticks*, towards the end of the long, final Query that he added to the Latin edition published in 1706:

To tell us that every Species of Things is endow'd with an occult specifick Quality by which it acts and produces manifest Effects, is to tell us nothing: But to derive two or three general Principles of Motion from Phaenomena, and afterwards to tell us how the Properties and Actions of all corporeal Things follow from those manifest Principles, would be a very great step in Philosophy, though the Causes of those Principles were not yet discover'd.[2]

The procedure thus outlined has been seen as encapsulating the very essence of the "Newtonian revolution" in science. Though Newton's world was composed, like that of mechanical philosophers such as Descartes and Gassendi, of material corpuscles in motion, no longer was Newton's science constrained, as the earlier mechanical philosophy had been, to reducing all natural phenomena to empirically inaccessible, and hence merely imagined,

[1] Isaac Newton, *Mathematical Principles of Natural Philosophy*, trans. Andrew Motte, rev. Florian Cajori (Berkeley: University of California Press, 1934), pp. vii–viii.
[2] Newton, *Opticks, or a Treatise of the Reflections, Refractions, Inflections and Colours of Light* (New York: Dover, 1952), pp. 401–2.

motions and impacts of particles. On the contrary, as the passages just quoted reveal, Newton admitted a further explanatory principle into his science, namely force. For Newton, armed as he was with a clearly understood principle of inertia, force became that which changed (or tended to change) the motions of bodies; individual forces came to be measured in terms of the changes of motion they produced; and the chief object of science became the discovery of the various forces acting in the world.[3] As Newton himself demonstrated so convincingly in *Principia* with respect to gravity, such forces were at least in some cases empirically determinable in a way that corpuscular mechanisms were not. Elsewhere in *Principia,* and even more so in *Opticks,* Newton held out the hope that many other kinds of force might eventually become equally as well characterized as gravity now was.

At least as important, in the eyes of many Newtonian scholars, is the fact that, in Newton's hands, "force" became a quantifiable concept. His approach thus held out the prospect of a truly mathematical physics in which various natural effects would be shown to follow in a rigorously demonstrative and quantitatively exact manner from mathematically expressed force laws. This is, of course, what Newton himself achieved in *Principia* in relation to gravity. Though success on a similar scale in other areas of physical inquiry proved elusive, Newton, by focusing on the concept of force, had dramatically enlarged man's expectations regarding the degree of precision possible in science.

On this much, historians of science are generally agreed. In other respects, however, and especially on the question of what Newton's real attitude was to the various forces he invoked, opinion is divided. Were passages such as those quoted above merely methodological in import, or did they carry with them, as well, an ontological commitment on Newton's part? In other words, when Newton so ostentatiously refrained from offering mechanical explanations of the kind demanded by his critics for the forces of which he wrote, did he in fact do so not only because he believed that such explanations were unnecessary within the context of his inquiry, but also because he believed that these forces had a real

---

[3] The evolution of the concept of force during the seventeenth century is described in R. S. Westfall, *Force in Newton's Physics: The Science of Dynamics in the Seventeenth Century* (London: MacDonald, 1971).

existence of their own as true actions at a distance, independent of all mechanical explanation?

Westfall is one of those who has argued most strongly for the latter view. More than this, he has offered a reconstruction of the development of Newton's thought from which such an attitude on Newton's part toward his forces emerges as a natural consequence.[4]

Newton, Westfall has shown, was in his early years profoundly influenced in his thinking about the physical world by the ideas of the seventeenth-century mechanical philosophers – of men such as Descartes, Gassendi, Gassendi's English mouthpiece Walter Charleton, and Robert Boyle.[5] In line with this, we find Newton as an undergraduate embracing (though with a hint of occasional reservations perhaps inspired by Henry More) the fundamental credo of the mechanical philosophy, the reducibility of natural events to the motions and impacts of particles of matter. Similarly, a decade and more later, we find him trying to account for a wide range of natural phenomena in typically mechanistic style in terms of interactions between particles of matter and an all-encompassing material aether (or aethers).[6]

It is now, however, generally recognized that by the early 1680s, Newton had abandoned his former belief in a universal dense aether and had begun to speak instead in what we now see as characteristically "Newtonian" style of forces acting between particles at a distance.[7] Westfall claims to have found an explanation for this change of outlook on Newton's part in the alchemical investiga-

---

[4] Richard S. Westfall, *Never at Rest: A Biography of Isaac Newton* (Cambridge: Cambridge University Press, 1980), chaps. 7–9.

[5] Richard S. Westfall, "The Foundations of Newton's Philosophy of Nature," *British Journal for the History of Science*, 1 (1962), 171–82; idem, *Never at Rest*, pp. 83–93. Cf. also A. R. Hall, "Sir Isaac Newton's Notebook, 1661–65," *Cambridge Historical Journal*, 9 (1948), 239–50; and J. E. McGuire and Martin Tamny, eds., *Certain Philosophical Questions: Newton's Trinity Notebook* (Cambridge: Cambridge University Press, 1983).

[6] Newton, "An Hypothesis explaining the Properties of Light," in H. W. Turnbull et al., eds., *The Correspondence of Isaac Newton* (Cambridge: Cambridge University Press, 1959–77), I, 362–86; "De Aere et Aethere," in A. R. and M. B. Hall, eds., *Unpublished Scientific Papers of Isaac Newton* (Cambridge: Cambridge University Press, 1962), pp. 214–20 (English trans., pp. 221–8); Newton to Robert Boyle, 28 February 1679, in *Correspondence*, II, 288–95.

[7] Henry Guerlac, "Newton's Optical Aether: His Draft of a Proposed Addition to His *Opticks*," *Notes and Records of the Royal Society of London*, 22 (1967), 45–57; J. E. McGuire, "Force, Active Principles, and Newton's Invisible Realm," *Ambix*, 15 (1968), 154–208.

tions that engaged so much of Newton's attention at this period. In alchemy, Westfall has argued, Newton found an idea "that refused to be reconciled with the mechanical philosophy. Where that philosophy insisted on the inertness of matter, such that mechanical necessity alone determines its motion, alchemy asserted the existence of active principles in matter as the primary agents of natural phenomena."[8] Westfall finds hints of the idea even in the generally mechanistic "Hypothesis explaining the Properties of Light" of 1675, for example in Newton's willingness to ascribe the immiscibility of oil and water to "some secret principle of unsociablenes" between them. However, he dates the full-blown transformation in Newton's thinking to a few years later when, armed with this insight drawn from alchemy, Newton turned anew to the analysis of the planetary motions:

As it appears to me, Newton's philosophy of nature underwent a profound conversion in 1679–80 under the combined influence of alchemy and the cosmic problem of orbital mechanics, two unlikely partners which made common cause on the issue of action at a distance. . . . Henceforth, the ultimate agent of nature would be for him a force acting between particles rather than a moving particle itself.[9]

Just as widely recognized among Newtonian scholars as the change in Newton's approach in the early 1680s is the fact that, beginning in about 1707 (that is, shortly after the publication of the Latin Optice) Newton once more began actively to consider possible causes for many of the forces he had invoked. In particular, Newton, fascinated, it seems, by various electrical experiments devised by Francis Hauksbee, began to explore the notion that many natural powers were manifestations of the activity of a subtle electric spirit that pervaded gross bodies. Hints of these speculations appear in the final paragraph of the "General Scholium" that Newton appended to the second (1713) edition of Principia, and in certain passages in the Queries published at the end of the

[8] Westfall, Never at Rest, p. 299.
[9] Ibid., p. 390. Cf. also Westfall, Force in Newton's Physics, chap. 7. McGuire has likewise seen a definite ontological commitment in Newton's attitude towards his forces, arguing, indeed, that these become the chief components of the universe as Newton conceived it in the years immediately after the first publication of Principia: "Not only did Newton enrich his ontology by including forces, but in the nineties they became essential to his natural philosophy. By 1706, he seemed to consider them, rather than matter, to be the primordials of nature" (McGuire, "Force, Active Principles, and Newton's Invisible Realm," p. 161).

1717 edition of *Opticks*. They survive in more ample form in extensive unpublished drafts intended for these editions of Newton's two great books. By 1717, however, Newton's thinking had evolved still further, and he now attributed many of the phenomena previously accounted for in terms of the electric spirit to a new form of universal aether, the activity of which might even, he suggested, be sufficient to explain the force of gravity.[10]

Historians such as Westfall and McGuire, who have argued strongly for an ontological rather than a merely methodological commitment to the doctrine of forces on Newton's part in the years after 1680, have found these later vacillations something of an embarrassment. McGuire ignores the many indications that Newton provides that he sees the electric spirit as a material agency, and treats it instead as a generalized force, "an electrical *arche* connecting mind with matter"; and he dismisses the 1717 aether on the ground that it does not conform to his general view of Newton's philosophy of nature:

it is difficult to suppose that Newton took it seriously. It was subject to obvious conceptual inconsistencies; it was a flagrant example of the sort of intermediate entity which Newton had always tended to reject; and more significantly it repudiated his basic metaphysics of God in an empty universe.[11]

Westfall, too, treats these later views of Newton's as aberrations, arising, he suggests, from a "growing philosophical caution" that "reflected at once the impact of Hauksbee's electrical experiments on Newton and perhaps also the effect of unrelenting criticism by mechanical philosophers on an aging man no longer able to sustain a revolutionary position in the face of general opposition."[12] He sees Newton's invocation of an electric spirit as a retreat toward a

---

[10] Newton, *Opticks*, Queries 17–24 (pp. 347–54). The "classic" papers on Newton's flirtation with the electric spirit are by Henry Guerlac: "Francis Hauksbee: expérimentateur au profit de Newton," *Archives internationales d'histoire des sciences, 16* (1963), 113–28; "Sir Isaac and the Ingenious Mr. Hauksbee," in I. B. Cohen and R. Taton, eds., *Mélanges Alexandre Koyré* (Paris, 1964), I, 228–253; Guerlac, "Newton's Optical Aether" (note 7). See also Joan L. Hawes, "Newton and the 'Electrical Attraction Unexcited,' " *Annals of Science, 24* (1968), 121–130 and "Newton's Two Electricities," *Annals of Science, 27* (1971), 95–103; and, more recently, R. W. Home, "Newton on Electricity and the Aether," in Z. Bechler, ed., *Contemporary Newtonian Research* (Dordrecht: D. Reidel, 1982), pp. 191–213.

[11] McGuire, "Force, Active Principles, and Newton's Invisible Realm" [see note 7], pp. 176, 187.

[12] Westfall, *Never at Rest*, pp. 644, 793.

mode of explanation "which at least appeared acceptable to conventional mechanical philosophers," and concludes that "there is no satisfactory explanation of his return to such fluids. . . . It is hard to imagine Newton acquiescing in such a retreat at an earlier, more vigorous age."[13] Consistent with this view, the preparation of the new edition of *Opticks,* with its famous Queries about the aether, is allotted a scant two pages in Westfall's 874-page biography, and these are consigned to a chapter entitled "Years of Decline."[14]

To be sure, there are aspects of the "aether" Queries of the 1717 *Opticks* that give the historian pause. Years before, in composing the concluding sections of Book II of *Principia,* Newton had marshaled overwhelming arguments against the possibility of any dense Cartesian-style aether filling all space. The motions of the planets in accordance with Kepler's laws were, he showed, inconsistent with the existence of such a medium. Hence the new aether of 1717 was, of necessity, of a very different character. It was both exceedingly rare, in order that it might offer no perceptible resistance to the motions of the planets through it, and exceedingly elastic, in order that it might transmit vibrations with the speed required if it were to fulfill the role in the theory of light that Newton wished to ascribe to it. Newton also supposed, however, that it was composed of particles. It followed inevitably, on Newtonian principles, that those particles exerted powerful forces of repulsion upon each other.[15] Hence, even if the 1717 aether offered a satisfactory basis for explaining gravity (and there is some evidence that Newton remained less than fully convinced on this score),[16] it by no means met the fundamental objection leveled by mechanical philosophers against Newton's physics: its reliance on unexplained forces acting at a distance. All the evidence we have suggests that Newton's "aether" Queries do indeed represent a defensive response of an old man to sustained criticism.

There are even indications that Newton subsequently repented their publication. Though historians have often failed to notice the fact, the new Queries were strictly inconsistent with the final paragraph of the concluding General Scholium of the 1713 *Principia,*

[13] Ibid., pp. 793, 747.    [14] Ibid., pp. 792–4.    [15] Newton, *Opticks,* p. 352.
[16] J. T. Desaguliers to Sir Hans Sloane, 4 March 1730/31, quoted by Guerlac, "Newton's Optical Aether" [see note 7], p. 51.

because various phenomena – the reflection, refraction, inflection, and heating effects of light, and the transmission of sensations through the nerves to the brain and commands of the will from the brain to the muscles – which were there attributed to the action of "a certain most subtle spirit which pervades and lies hid in all gross bodies," were in the 1717 Queries ascribed to a universally disseminated aether. This latter was no mere transmogrification of the earlier subtle spirit, but was a second and quite distinct subtle matter. Even after reintroducing the aether, Newton continued to believe, as well, in the existence of the subtle spirit associated with matter, and to ascribe certain natural effects such as electricity to it;[17] but now he transferred some of the other actions previously ascribed to the subtle matter to the aether. At the very least, therefore, some modification of the General Scholium was called for. In fact, Newton envisaged deleting the entire final paragraph dealing with the subtle spirit from the next edition of *Principia*.[18] When that edition finally saw the light of day, however, in 1726, he did not do so; instead, the subtle matter continued to be invoked to explain the same effects as in 1713. This suggests that, at the very end of his life, Newton may have changed his mind yet again about the aether, and reverted to his previous mode of explanation of the various optical and physiological effects in terms of a subtle spirit associated with bodies.

If we may thus accept Westfall's judgment as to the relative insignificance of the 1717 aether for our general understanding of Newton's thinking about forces, the situation is, I believe, very different with respect to his writings on the subtle spirit. In the various published hints and unpublished drafts on this subject, we surely have before us no last-minute concessions by a declining 74-year-old, as the aether Queries appear to be, but the fruit of at least ten years of speculation and inquiry by a Newton still capable of prodigious feats of intellectual activity. Although much of what he had to say on the matter was consistent with views he had held at an earlier period of his career, these were now given a new lease of life by Hauksbee's remarkable experiments. There is ample evi-

---

[17] Home, "Newton on Electricity and the Aether" [see note 10], passim.
[18] A. Koyré and I. B. Cohen, eds., *Isaac Newton's Philosophiae Naturalis Principia Mathematica: The Third Edition (1726) with Variant Readings* (Cambridge: Cambridge University Press, 1972), *II*, 764.

dence of the renewed interest and enthusiasm with which Newton pursued the work.[19] Furthermore, this interest pre-dates the main Continental onslaught on his doctrine of forces, and so cannot be dismissed as a mere defensive response to that criticism. Rather, it should be seen as yet another constructive phase – perhaps, indeed, the last major constructive phase – in his thinking about matter and its powers. If this be accepted, it then becomes of considerable interest to discover precisely what views Newton arrived at concerning the subtle spirit and, more particularly, what implications these views might have for our broader understanding of his philosophy of nature.

I have argued elsewhere that, even at this late period of his life, Newton's ideas regarding the various modes of action of the subtle spirit depended much less than has generally been allowed on his invoking unexplained forces acting at a distance. I have suggested, indeed, that in many respects his ideas concerning the subtle spirit even then more closely resembled those of that arch-mechanist, Descartes, than they did those of a programmatic "Newtonian," systematically reducing each class of natural phenomena, in turn, to the action of some force or other.

Consider, first of all, the case of electricity, which lies at the heart of Newton's discussions of the subtle spirit. Newton sets out his ideas on this subject at length in a number of surviving documents. One of these is a long, unnumbered draft Query that predates the introduction of the aether hypothesis of 1717. "Do not all bodies abound," Newton here asks rhetorically, "with a very subtil active vibrating spirit by w$^{ch}$ . . . the small particles of bodies cohære when contiguous, agitate one another at small distances & regulate almost all their motions amongst themselves as the great bodies of the Universe regulate theirs by the power of gravity?" In particular, Newton believes that diffusion of this subtle matter into the space surrounding a rubbed body such as glass is what brings about electrical attraction, "for electric bodies could not act at a distance without a spirit reaching to that distance." Hauksbee's experiments provide, according to Newton, ample evidence that this diffusion occurs:

---

[19] Guerlac, "Francis Hauksbee: expérimentateur au profit de Newton" [see note 10], passim.

[B]y several experiments shewn by M$^r$ Hawsksby before y$^e$ R. Society it appears that a cylindrical rod of glass or hard wax strongly rubbed emitts an electric spirit or vapour w$^{ch}$ pushes against the hand or face so as to be felt, & upon application of the finger to y$^e$ electric body crackles & flashes, & that the electric spirit reaches to y$^e$ distances of half a foot or a foot from the glass or above . . . ; & that if a globe of glass be nimbly turned round upon an axis & in turning rub upon a man's hand to excite its electric virtue, . . . the glass emitts an electric vapour or spirit w$^{ch}$ may be felt by the hand & w$^{ch}$ in dashing upon the hand or upon white paper or a handkerchief at the distance of a quarter of an inch or half an inch from the glass or above, illuminates the hand or paper or handkerchief with a white light while the glass continues in motion, the spirit by striking upon those bodies being agitated so as to emit the light. . . . There is therefore an electric spirit by w$^{ch}$ bodies are in some cases attracted in others repelled & this spirit is so subtile as to pervade & pass through the solid body of glass very freely . . . , & is capable of contraction & dilatation expanding itself to great distances from the electric body by friction. & thefore *[sic]* is elastic & susceptible of a vibrating motion like that of air whereby sounds are propagated. & this motion is exceeding quick so that the electric spirit can thereby emit light.[20]

The introduction of the aether hypothesis did not in any way affect Newton's thinking about what lay behind the various phenomena of electricity. This we discover from a passage which Newton at one point numbered "Qu. 18B," intending it to be inserted between two of the early "aether" Queries of the new edition of *Opticks:*

Do not electric bodies by friction emit a subtile exhalation or spirit by which they perform their attractions? And is not this spirit of a very active nature & capable of emitting light by its agitations? And may not all bodies abound with such a spirit & shine by the agitations of this spirit within them when sufficiently heated? ffor if a long cylindrical piece of Ambar be rubbed nimbly it will shine in the dark & if when it is well rubbed the finger of a man be held neare it so as almost to touch it, the electric spirit will rush out of the Ambar with a soft crackling noise like that of green leaves of trees thrown into a fire, & in rushing out it will also push against the finger so as to be felt like the ends of hairs of a fine brush touching the finger. And the like happens in glass. If a long hollow tube of flint glass about a inch be rubbed nimbly with a paper held in the hand . . . , the electric spirit which is excited by the friction will rush out

---

[20] University Library, Cambridge, Add. MS. 3970, fols. 241$^v$–241$^r$.

of the glass with a cracking noise & push against the skin so as to be felt, & in pushing emit light so as to make the skin shine like rotten wood or a glow worm. And if the glass was held neare pieces of leaf brass scattered upon a table the electric spirit w$^{ch}$ issued out of the glass would stir them at the distance of 6, 8 or 10 inches or a foot, & put them into various brisk motions, making them sometimes leap towards the glass & stick to it, sometimes leap from it with great force, sometimes move towards it & from it several times with reciprocal motion, sometimes move in lines parall [*sic*] to the tube, sometimes remain suspended in the air, & sometimes move in various curve lines. Which motions shew that this spirit is agitated in various manners like a wind.[21]

The thoroughly mechanistic flavor of these passages is reinforced by yet another draft Query, numbered "Qu.23" and apparently also intended to accompany the "aether" Queries of the published edition. Whereas in the passages quoted above, Newton focuses chiefly on the new electrical phenomena discovered by Hauksbee, he here discusses, in detail, the cause of the electrical forces of attraction and repulsion:

Qu.23. Is not electrical attraction and repuls; performed by an exhalation which is raised out of the electrick body by friction expanded to great distances & variously agitated like a turbulent wind, & w$^{ch}$ carries light bodies along with it & agitates them in various manners according to its own motions, making them go sometimes towards the electric body, sometimes from it & sometimes move with various other motions? And when this spirit looses its turbulent motions & begins to be recondensed & by condensation to return into the electrick body doth it not carry light bodies along with it towards the Electrick body & cause them to stick to it without further motion till they drop off?[22]

The import of passages such as these is, in my view, unmistakable. Newton does not regard electricity as a true action at a distance. On the contrary, the various electrical effects are, in his view, brought about in a straightforwardly mechanical way following the agitation, by friction, of subtle spirit residing in the pores of electrifiable bodies such as amber and glass. (As the argument proceeds, Newton in fact concludes that the spirit resides in the pores of all bodies whatsoever.) Set in motion by the friction "like a turbulent wind," the subtle spirit simply sweeps along any light objects it encounters until, as it slows, it recondenses and

[21] Ibid., fol. 295.    [22] Ibid., fol. 293$^{v}$.

carries such objects with it towards the electrified body. If it strikes against larger bodies, it may be set vibrating, and it then emits light in just the same way as a vibrating column of air generates sound. This spirit is no generalized immaterial force, but a material agency. To be sure, Newton usually referred to it as a "spirit," but on occasion he called it a "vapour" or "exhalation," terms which in Newton's vocabulary certainly implied materiality. Likewise, the word "spirit" could be used in this way, as he himself made clear in a draft intended for *Principia:*

Vapours and exhalations on account of their rarity lose almost all perceptible resistance, and in the common acceptance often lose even the name of bodies and are called spirits. And yet they can be called bodies in so far as they are the effluvia of bodies and have a resistance proportional to density.[23]

As I have argued elsewhere, Newton's ideas about electricity, as set out in passages such as those quoted above, were in almost every respect typical of his day. The notion that electrical effects were caused by subtle matter, originally contained in the rubbed body, being excited by the friction, spreading out as effluvia into the surrounding air and sweeping along any light objects lying in the way, had been widely held at least since the time of William Gilbert. It continued to be generally espoused until the 1750s, when it began to be displaced by Benjamin Franklin's notion that the phenomena in question were brought about by static accumulations or "charges" of subtle electric fluid transferred by friction from one body to another.[24] In Newton's day, the main point of contention was not the correctness of the effluvial picture in general terms but the nature of the subtle matter or matters involved. In particular, whereas some maintained that the effluvia were composed of the matter of the rubbed body itself ("the finer parts of the attrahent," as Boyle put it),[25] others, including both Descartes and Newton, held that the stuff involved was a distinct sub-

---

[23] University Library Cambridge, Add. MS. 3965.13, fol. 437ᵛ; quoted by J. E. McGuire, "Body and Void in Newton's De Mundi Systemate: Some New Sources," *Archive for History of Exact Sciences, 3* (1966), 206–48; p. 219 (Latin original, p. 245).

[24] Home, "Newton on Electricity and the Aether" [see note 10], pp. 204–7. Cf. idem, *The Effluvial Theory of Electricity* (New York: Arno Press, 1981), and J. L. Heilbron, *Electricity in the 17th and 18th Centuries: A Study of Early Modern Physics* (Berkeley: University of California Press, 1979).

[25] Robert Boyle, *Experiments and Notes about the Mechanical Origine or Production of Electricity* (London, 1675; reprinted Oxford University Press, 1927), p. 6.

tle matter – whether the Cartesian "first element" or Newton's "electric and elastic spirit" – common to all bodies. Others again, most notably Niccolò Cabeo and his Jesuit followers, and also Hauksbee, thought that the air, too, was involved in bringing about the attraction.[26]

Magnetism was a power that Newton often cited, together with electricity, as an example of a force known to act in the world. He saw it, however, as a power that was limited in its operation, being confined, so he thought, to iron and some of its ores. Accordingly, only rarely did he mention it in connection with the electric spirit; in particular, magnetism is strikingly absent from the list of effects ascribed in the 1713 General Scholium to the action of this spirit, or in the 1717 "aether" Queries to the action of the aether.[27] For Newton, magnetism seems to have been a separate power, subject to its own laws. Here too, however, such evidence as we have suggests that Newton did not see the magnetic force as a true action at a distance. On the contrary, it appears that, as in the case of electricity, he assumed the existence of an underlying mechanism of unmistakably Cartesian provenance.

Descartes's invention of a mechanism that could account for the archetypically occult power of the magnet had been one of the major early triumphs of the mechanical philosophy. According to his scheme, associated with any magnet were circulating streams of subtle material effluvia of a distinctive kind. These streams passed axially through the magnet, emerging from one pole and returning to the other through the surrounding air before resuming their course. The various known magnetic effects were then explained in terms of interactions between such streams of matter, or between a single stream and pieces of iron that came in its way.

This general picture quickly won wide acceptance and, indeed, remained predominant for a century and more, even though various details propounded by Descartes were challenged and in some cases largely rejected.[28] Newton's early adherence to at least the

---

[26] On Hauksbee's ideas, see R. W. Home, "Francis Hauksbee's Theory of Electricity," *Archive for History of Exact Sciences, 4* (1967), 203–17.

[27] On at least one occasion, however, Newton does inexplicably include magnetism in such a list (University Library Cambridge, Add. MS. 3970, fol. 241; quoted by Joan L. Hawes, "Newton's Two Electricities" [see note 10], p. 97). I find it difficult to understand this except as a slip of the pen.

[28] R. W. Home, introduction to *Aepinus's Essay on the Theory of Electricity and Magnetism* (Princeton, N.J.: Princeton University Press, 1979), chap. 4, "Magnetism."

broad outlines of the theory is clearly displayed in an unpublished manuscript dating from 1666 or 1667, in which he actually sketches the likely patterns of flow of the subtle matter in various circumstances.[29] In another paper, which Westfall dates to the same period, Newton discusses magnetism at greater length than anywhere else. Throughout, he assumes the existence of peculiar "streams" in the vicinity of a magnet; indeed, he assumes (though in terms somewhat different from those used by Descartes) two separate and "unsociable" streams entering a magnet at its two opposite poles and passing through it in opposite directions.[30]

It is tempting to see these papers as characteristic of Newton's youthful flirtation with the mechanical philosophy, and to suppose that he would later have abandoned such unrestrained hypothesizing, as he did in relation to the cause of gravity. The evidence, however, suggests otherwise, for although it is nowhere specific enough to show that Newton remained faithful to the details of his early explanatory scheme, it does unmistakably reveal that even after he had abandoned the search for a mechanical explanation for gravity, he continued to believe that magnetic effects were brought about by the action of subtle effluvia of some kind. Even in the 1690s, at the height of Newton's disillusionment with mechanism, David Gregory recorded, while visiting him, his host's view that the magnetic virtue "seems to be produced by mechanical means."[31] A casual allusion to "magnetick Effluvia" in the first edition of *Opticks,* published ten years later, tells the same story.[32] Finally, from the period with which we are chiefly concerned in this chapter, in one of the "aether" Queries of 1717, Newton uses the existence and activity of the magnetic effluvia, which he again takes as beyond dispute, to justify by analogy assumptions he is making about the nature of the aether: "If any one would ask how a Medium can be so rare," he suggests, "let him tell me . . . how the Effluvia of a Magnet can be so rare and subtile, as to pass through a Plate of Glass without any Resistance or

[29] University Library Cambridge, Add. MS. 3974, fols. 1–3.
[30] University Library Cambridge, Add. MS. 3970, fols. 473–4. Cf. Westfall, *Force in Newton's Physics* [note 3], p. 332.
[31] Newton, *Correspondence, III,* 335 (English trans., p. 338). The entry in Gregory's diary is dated 5–7 May 1694.
[32] Newton, *Opticks* (1st ed., 1704), Book II, p. 69. Cf. reprint ed. (New York, 1952), p. 267.

Diminution of their Force, and yet so potent as to turn a magnetick Needle beyond the Glass."[33]

Newton frequently cited electricity and magnetism as other instances, besides gravity, of forces known to act in the world, to
support the suggestion that there were many other forces at work
as well. Whenever he did so, however, he surrounded the argument with disclaimers, of which the one near the beginning of
Query 31 is the most famous:

How these Attractions may be perform'd, I do not here consider. What
I call Attraction may be perform'd by impulse, or by some other means
unknown to me. I use that Word here to signify only in general any
Force by which Bodies attract one another, and what are the Laws and
Properties of the Attraction, before we enquire the Cause by which the
Attraction is perform'd.[34]

Historians have tended to place little weight on such remarks, and
to dismiss them as little more than symptoms of Newton's habit-
ual caution in expressing his views publicly. In the light of the
evidence just presented, however, they take on a new significance.
Whatever his attitude toward gravity might have been, Newton,
we discover, did not at all suppose that either electricity or magnetism was an irredeemably nonmechanical power. On the contrary, he thought both that there was an underlying mechanism in
each case, and that he knew what this was. The disclaimer in the
passage just quoted must be taken seriously; Newton believes that
some of the attractions of which he speaks *are* "perform'd by impulse." In other words, what is being proposed here is not a general philosophy of nature in which forces have ontological primacy, but rather a methodology only.

According to this methodology, in studying magnetism, for example, one ought above all to try, as Newton did,[35] to discover
the mathematical law governing the magnetic force. Such a methodological priority is, however, in no way inconsistent with one's
simultaneously having views about the existence and nature of a
mechanism that might bring about the attraction in accordance
with that law. Nor, for that matter, is having views about the

[33] Newton, *Opticks* (reprint ed.), p. 353. For further discussion of Newtonian views on
magnetism, see R. W. Home, " 'Newtonianism' and the Theory of the Magnet," *History of Science*, 15 (1977), 252–66.
[34] *Opticks*, p. 376.
[35] Newton, *Mathematical Principles of Natural Philosophy*, p. 414.

existence of such a mechanism inconsistent with allowing that this may perhaps itself ultimately be found to rest upon the behavior of a fluid the particles of which exert unexplained forces on each other at a distance. Whether that be so, or whether they act on each other only by contact, remains undetermined, in the case of magnetism, at the point reached by Newton's inquiry. In the case of electricity, on the other hand, he seems rather closer to the view that the particles of the subtle matter that provides the mechanism do indeed exert forces on each other at a distance.[36]

Conclusions such as these are further reinforced when we turn to what Newton had to say on another subject, of which the views he expressed have yet to attract systematic study: namely, the powers of living matter. Here, too, Newton's slowness to ascribe an ontological significance to the forces he discovers at work in the world becomes apparent. Once again, he appears much more inclined to assume fairly orthodox mechanistic explanations than one might have expected of a committed "Newtonian" bent on reducing natural events to the actions of independently existing forces. This is not to say, however, that Newton is here bowing to the conceptual constraints imposed by the mechanical philosophy. On the contrary, after reducing a variety of life processes to mechanisms of one kind or another, he eventually concludes that there is associated with living matter a peculiar power for which he offers no hint of an explanation, a power which appears to lie behind the various mechanisms he has described, but which itself appears to be intrinsically nonmechanical.

Newton's best-known statements about living processes comprise, first, anatomical descriptions of the optic system that offer no additional insights into his philosophy of matter[37] and, second, various comments about the operation of the nerves in transmitting sensations from the "organs of sense" to the brain and commands of the will from the brain to the muscles. These latter comments appear in writings from all periods of his career, the best known, however, being his late-in-life remarks in the final paragraph of the 1713 General Scholium and in the last two of the

[36] Home, "Newton on Electricity and the Aether" [see note 10], p. 209.

[37] *Opticks*, pp. 15–17. Cf. also three letters of Newton's to William Briggs, 1682–5, in his *Correspondence*, II, 377–8, 381–5, 417–9; and his "Description of the Optic Nerves and their Juncture in the Brain," published in D. Brewster, *Memoirs of the Life, Writings and Discoveries of Sir Isaac Newton* (Edinburgh, 1855), I, 432–6.

"aether" Queries of 1717 (Queries 23 and 24). In 1713, sensation and the commands of the will were held to be transmitted by vibrations of the subtle electric and elastic spirit "mutually propagated along the solid filaments of the nerves, from the outward organs of sense to the brain, and from the brain into the muscles." By 1717, the vibrations were held to occur not in the subtle spirit but in the newly introduced "Æthereal Medium": In all other respects, however, the account remained as before.[38] In both cases, the explanations were conceived in wholly mechanistic terms, and bore obvious affinities to the opinion, commonly held at the time, that the nerves functioned as conduits for "animal spirits" flowing between the brain and the various sensory and motor parts of the body.[39] Insofar as they offered scope for characteristically "Newtonian" interpretations, they did so only by admitting the possibility, noted earlier, that the fluid in question might derive the elasticity requisite to sustain the vibrations to which Newton referred, from a repulsive force operating between its particles.[40]

Much more interesting from the point of view of this paper are, first, a manuscript entitled "De vita et morte vegetabile" and, second, sections of some associated draft Queries, prepared by Newton for a new edition of *Opticks* at about the time he composed the concluding General Scholium for the second edition of *Principia,* but never in fact published. In these papers, Newton confronts directly, albeit briefly, the nature of the distinction between living and nonliving matter. In doing so, he reveals yet again the extent to which his physical ideas were in tune with those held more generally in his day. At the same time, the discussion sheds new light on the place of unexplained forces in his mature philosophy of nature.

The passages in question are, in fact, linked with those discussed earlier in which Newton set out at length his ideas on electricity; indeed, the most important of them occur in the very same draft of the Queries from which most of the passages quoted above

---

[38] Newton, *Mathematical Principles of Natural Philosophy,* p. 547; *Opticks,* pp. 353–4.

[39] Cf. René Descartes, *Treatise of Man,* trans. T. S. Hall (Cambridge, Mass.: Harvard University Press, 1972), pp. 21 ff.

[40] In Newton's earliest surviving discussion of the operation of the nerves, in the "Hypothesis explaining the Properties of Light" of 1675, he invokes the additional nonmechanical concepts of sociableness and unsociableness to explain how impulses could be confined to the nerves and how, once arrived at a muscle, they could cause this to contract (Newton, *Correspondence, I,* 368–9).

concerning electricity were drawn. Here as nowhere else in Newton's papers we see revealed the full extent to which his powerful
imagination had been captivated by Hauksbee's astonishing experiments. Both the traditional electrical attraction of light objects
and the striking new effects studied by Hauksbee arose, Newton
had concluded, from the agitation of a very active subtle spirit
associated with the particles of bodies. The role of friction in
bringing about the phenomena was not, he thought, to generate
the "virtue," but only to expand it, the subtle spirit under normal
circumstances "reach[ing] not to any sensible distance from the
particles."[41] Now, however, he goes on to consider the possibility
that there may be other causes, too, by which "the electric vertue
is invigorated." One such cause, he says, may be the power of life:

The vegetable life may also consist in the power of this spirit supposing
that this power in substances w$^{ch}$ have a vegetable life is stronger then in
others & reaches to a greater distance from the particles. ffor as the electric vertue is invigorated by friction so it may be by some other causes.
And by being stronger in the particles of living substances then in others
it may preserve them from corruption & act upon the nourishment to
make it of like form & vertue w$^{th}$ the living particles as a magnet turns
iron to a magnet & fire turns its nourishment to fire & leaven turns past
to leaven.[42]

As Newton himself makes plain, he is concerned here not with
all properties of living things but only with those that had, since
ancient times, been technically distinguished as constituting the
vegetative aspects of life. These amounted to the ability of living

---

[41] University Library Cambridge, Add. MS. 3970, fol. 235; quoted by Westfall, *Force in
Newton's Physics* [see note 3], p. 394, and Home, "Newton on Electricity and the Aether"
[see note 10], p. 197.

[42] University Library Cambridge, Add. MS. 3970, fol. 241; quoted by Home, "Newton
on Electricity and the Aether" [see note 10], p. 199. Folios 235 and 241 are the two
halves of a single, folded sheet. A long unnumbered Query that begins on fol. 241$^v$ was
drafted first. This carries over to fol. 241$^r$ and then, briefly, to fol. 235$^v$. At this stage,
however, Newton evidently turned the sheet over and started again on fol. 235$^r$, writing
out new Queries which he numbered 24 and 25, intending them to be inserted into the
middle of what is now Q. 31 but in the Latin *Optice* of 1706 was Q. 23. A series of
dashes in the proposed Q. 25 indicate that the long first paragraph of the unnumbered
Query on fol. 241$^v$ is to be taken in at this point. The manuscript headed "De vita et
morte vegetabile" is now marked as fol. 237 in the same bundle, with drafts on fols. 238
and 240. It deals with the same subject matter but, as we shall see, in a rather different
way. The first few paragraphs and the concluding sentences of this paper have been
published by Westfall, *Force in Newton's Physics* (note 3), pp. 417–8.

matter, plant or animal, to maintain its organized form and to organize other matter likewise, whether in the processes of nutrition and growth, whereby existing tissues or organs are renewed or enlarged through the accretion of new matter, or in reproduction, where new living individuals are generated. In these passages, Newton does not discuss the additional powers that distinguish animate from mere vegetative life, or rational animals (i.e., human kind) from the beasts. For Newton, the subtle spirit, the agitation of which through friction gives rise to the phenomenon of electricity, also sustains the organic character of living matter by activity brought about in another way. When, at death, that activity ceases, putrefaction sets in; that is, the organized structures that characterized the matter in question during life now fall into decay.

In order to uphold this general position, Newton in these documents sets out at some length his general views on nutrition, growth, and generation, and contrasts these essentially constructive processes with the destruction and dissolution that constitute putrefaction. The ideas he displays are, in fact, very typical ones for his day.

So far as the reproduction of living forms is concerned, Newton reveals himself as an "ovist," that is, an upholder of that version of the preformation theory that held that the primordium or embryo of the young was present, already formed, in the female egg, and that the role of the male parent was merely to provide an indirect stimulus for its growth.[43] On this view, the process of generation reduces to one of growth of the embryo from suitable nutriment and, at an appropriate stage, a budding off of the embryo from the mother's body. Indeed, Newton draws the analogy with budding explicitly:

Generation is nothing else then separating a branch from the tree and giving it better nourishment. If a separated branch takes root in the earth or a separated twigg or bud by grafting or inoculation is nourished from y$^e$ root of a young stock, it grows into a new tree as big as the tree from which it was separated being better nourished from a young root than from an old one. The seed of a tree has the nature of a branch or twig or

---

[43] For a lively account of the preformation theory and its intellectual milieu, see Elizabeth Gasking, *Investigations into Generation: 1651–1828* (London: Hutchinson, 1967), Chapters 2–5. Also see Jacques Roger, *Les sciences de la vie dans la pensée française du XVIII$^e$ siècle* (Paris: Armand Colin, 1963).

bud. While it grows upon y$^e$ tree it is a part of the tree: but if separated and set in the earth to be better nourished, the embryo or young tree contained in it takes root & grows into a new tree. [In] like manner the egg of a female w$^{th}$ the embryo formed inside while it grows in the ovarium is a branch of y$^e$ mothers body & partakes of her life, yet the Embryo is as capable of being separated from the mother & growing great by due nourishment as a branch or twigg or bud or seed of a tree is of being separated from the tree & growing into a new tree.[44]

Newton's adherence to the preformation theory, and to the ovist version in particular, is made plain by his reference to "the egg of a female w$^{th}$ the embryo formed inside," growing in the ovarium prior to the intervention of the male parent. This theory was widely held during the last decades of the seventeenth century, though it was being challenged at the end of the period by the animalculist alternative according to which the preformed individual derived not from the female but from the male parent, being present in the spermatozoa newly discovered in the male semen by Leeuwenhoek. Newton does not mention the spermatozoa, which suggests that he adhered to the fairly common opinion that, far from being the essential part of the semen, they were in fact parasites. For him, the male semen played an entirely different role, namely the essentially chemical one of reacting with the female juices to produce the nourishment required by the embryo if it were to grow:

by the act of generation nothing more is done then to ferment the sperm of the female by y$^e$ sperm of y$^e$ male that it may thereby become fit nourishment for y$^e$ Embryo. ffor y$^e$ nourishment of all animals is prepared by ferment & the ferment is taken from animals of the same kind, & makes the nourishment subtile & spiritual. In adult animals the nourishm$^t$ is fermented by the choler and pancreatic juice both w$^{ch}$ come from the blood. The Embryo not being able to ferment its own nourishment w$^{ch}$ comes from the mothers blood has it fermented by the sperm w$^{ch}$ comes from y$^e$ fathers blood, & by this nourishment it swells, drops off from y$^e$ Ovarium & begins to grow with a life distinct from that of y$^e$ mother.[45]

According to this line of thinking, nutrition and growth become *the* fundamental biological phenomena. Nutrition, in turn, was for Newton a two-stage affair. First, a fermentation occurs,

---

[44] University Library Cambridge, Add. MS. 3970, fol. 235.     [45] Ibid.

brought about by the digestive juices or, in the act of generation, by the male semen. This chemical process breaks down the nutriment into minute, extremely subtle parts:

Now in all fermentation w$^{ch}$ generates spirits, the ferment abounds w$^{th}$ a supprest acid w$^{ch}$ being more attracted by the other body forsakes its own to rush upon & dissolve y$^e$ other & by the violence of the action breaks both its own particles & the particles of y$^e$ other body into smaller particles & these by their great subtilty volatility & continual digestion resolve y$^e$ whole mass into as subtile parts as it can be resolved by putrefaction.

Then, these separated parts reassemble under the organizing influence of the body being nourished:

And when the nourishment is thus prepared by dissolution & subtillation, the particles of the body to be nourished draw to themselves out of the nourishment the particles of the same density & nature w$^{th}$ themselves. ffor particles of one & the same nature draw one another more strongly then particles of different natures do.[46]

Newton alludes at this point to the formation of mineral deposits in the earth as a known instance in which the "like attracts like" principle operates in nature. Yet, such a principle remains insufficient for his purposes because it does not explain that assembling of the new materials into appropriate structures, which manifestly occurs in the nourishment and growth of living creatures. Newton argues that the growth of crystals provides a model here:

And when many particles of the same kind are drawn together out of y$^e$ nourishment they will be apt to coalesce in such textures as the particles w$^{ch}$ drew them did before because they are of the same nature as we see in the particles of salts w$^{ch}$ if they be of the same kind always crystallize in the same figures.

The analogy is one that was commonly drawn upon by mechanical philosophers. Newton, however, is intellectually rigorous enough to recognize that it remains but an analogy, and does not in any sense constitute an explanation. It is precisely at this point, in order to provide an explanation for the remarkable power that living forms possess of organizing new matter into existing structures, that he invokes the force of electricity:

[46] Ibid., fols. 235–235$^v$.

And for faciliating [*sic*] this assimilation of y<sup>e</sup> nourishment & preserving the nourished bodies from corruption it may be presumed that as electric attraction is excited by friction so it may be invigorated also by some other causes & particularly by some agitation caused in the electric spirit by the vegetable life of the particles of living substances: & the ceasing of this vigour upon death may be the reason why y<sup>e</sup> death of animals is accompanied by putrefaction.[47]

Throughout the different versions of this long draft Query, whenever Newton mentions electricity, he presents it as an effect produced by the agitation of a subtle spirit that pervades the pores of ordinary matter. By contrast, in the document "De vita et morte vegetabile," the electric spirit is not mentioned. Instead, Newton refers repeatedly to the "electrical force" associated with vegetating matter, without any further hint as to its ontological status. It is therefore tempting to date this paper to the period before Hauksbee's spectacular experiments rekindled Newton's belief in the existence of the subtle spirit and his faith in its explanatory possibilities. Other features of the paper seem firmly to link it, however, with the "subtle matter" Queries. Not only are there close parallels between the arguments presented in the two documents, "De vita et morte vegetabile" displays the same striking conviction – which must surely postdate Hauksbee's work – that we find in the "subtle matter" Queries, that electricity is a power universally associated with matter that lies behind many well-known effects not normally thought of as electrical. In this case, the document would stand as a suggestive instance of Newton confining his discussion to forces and offering no explanation for these, even though he actually had an elaborate explanation worked out at the time.

Whether or not this be allowed, however, these discussions by Newton of the vegetative power of living matter are revealing in other ways. In particular, as indicated earlier, they show that here, too, as with (or so I have argued) electricity and magnetism, Newton preferred to adopt a mechanistic (or at least pseudomechanistic) theory current in his day, the preformation theory, rather than invoking yet another category of unexplained forces to explain the effects in each case.

[47] Ibid., fol. 235<sup>v</sup>.

Newton, I would suggest, is in fact much more unwilling than has often been supposed to invoke a specific kind of in-principle-inexplicable force to account for each different category of natural event. Unexplained forces remain full of mystery for him; indeed, his conviction that the most famous of them, gravity, brought him close to the activity of God himself is well known,[48] and there is no reason to suppose that others would have had less profound implications.

Prima facie, then, one might expect that Newton would have recourse to entities of this kind only for the most universal of natural powers; and this does, indeed, seem to be the pattern that emerges. Neither the "amber effect" – the drawing of little light objects to rubbed amber or glass – nor the attraction of iron to a magnet, is an effect of sufficient universality; but a force of repulsion acting between the particles of an all-pervading subtle matter and ultimately responsible not just for the macroscopic electrical attraction but for the wide range of natural phenomena listed in the 1713 General Scholium, might be. Similarly, the various specific forces associated with living forms are unlikely to be the end point of the analysis, whereas a very general power of activating the subtle electric spirit even without friction might be what universally distinguishes living from nonliving matter. Something presumably does; and whatever it be, it is likely to be closely linked to the activity of God as the Creator of life and to be irreducibly nonmechanistic in character.

Meanwhile, Newton's scientific method leads him to invoke forces of all kinds in studying natural processes, and encourages him to focus on discovering the mathematical laws according to which those forces act rather than on explaining their action. In these cases, I believe, as I have indicated above, that we ought to take more seriously than has usually been done Newton's oft-expressed caveat about the causes lying behind the forces of which he spoke. Some of these forces might, he suggested, be "perform'd by impulse."[49] If the interpretation I am suggesting is correct, Newton would in fact have thought that a mechanism was involved in most instances. I have presented evidence, indeed, that he did think so in certain leading cases. He would also not have been surprised, however, if, as these mechanisms were investi-

---

[48] Westfall, *Force in Newton's Physics* [see note 3], pp. 395ff.    [49] *Opticks*, p. 376.

gated, they were found ultimately to depend upon a small number of irreducibly nonmechanical powers of quasi-universal application. At that point, he would have said – but only at that point – we are indeed approaching a knowledge of the first cause, which is God.

# 5

## Concepts of inertia: Newton to Kant

PETER M. HARMAN

In his *Force in Newton's Physics,* Richard S. Westfall described Newton's concept of inertia as paradoxical, for there are striking ambiguities in Newton's designation of the concept of inertia as the "force of inertia *[vis inertiae].*" In the *Principia,* the force of motion of a body is defined as the cause of a change of motion, whereas inertia is held to preserve a body in a state of rest or uniform rectilinear motion. For Newton to link his concepts of force and inertia in denoting the "force of inertia" would therefore seem to be anomalous.

Despite this terminological ambiguity, Newton does not confuse his concepts of impressed force and inertia in his usage in the *Principia;* but the "paradox of Newton's *vis inertiae*"[1] may be seen as the tip of the iceberg of problems concerning matter and force, which Newton carefully excluded from consideration in his treatise on the mathematical principles of motion, but which are nevertheless central issues in his natural philosophy. The problem of force and its relation to inertia remained at the heart of the debates on the foundations of physics in the eighteenth century, and in this essay I shall seek to illuminate the problematic status of *vis inertiae* as perceived by Leibniz, Euler, and Kant, and to review their attempts at conceptual clarification. Although subsequent discourse about the role of forces in natural philosophy was fundamentally shaped by Newton's physics, Newton's treatment of

[1] Richard S. Westfall, *Force in Newton's Physics: The Science of Dynamics in the Seventeenth Century* (London and New York, 1971), p. 450.

the concepts of force and inertia was subjected to searching criticism by his successors.

The problems these natural philosophers were seeking to resolve had their origin in the debates surrounding the mechanical philosophy of the late seventeenth century. These issues concerning the conceptual foundations of physics were not put to rest by Newton's *Principia,* despite the importance of the transformation in the conceptual structure of physics brought about by the establishment of the Newtonian laws of motion and concept of universal gravitation. Debates over the status of the concepts of force and inertia not only continued to flourish but were integral to the development of natural philosophy in the eighteenth century. My intention here is not to discuss this theme systematically but to illustrate some of its salient features by focusing upon the problem of Newton's concept of *vis inertiae.*

It should be appreciated that Leibniz, Euler, and Kant approached these matters with different intentions, though it is not possible here to properly characterize their diverse objectives and worldviews. Leibniz aimed to refute the conceptual framework of the *Principia* and to create his own distinctive science of "dynamics"; Euler sought to elaborate an essentially Newtonian mechanics by clarifying the ambiguities of the Newtonian concept of *vis inertiae;* whereas Kant sought to justify the intelligibility of Newtonian physical laws by reconstructing the metaphysical foundations of Newtonian science.[2]

### FORCE AND INERTIA IN NEWTON'S PHYSICS

Newton based his first law of motion in the *Principia* on a relationship between *force,* as an external action generating change of motion, and *inertia,* conceived as a fundamental property of matter by which bodies resist changes in their state of rest or uniform rectilinear motion. Whereas the separation of his mathematical theory of the motion of bodies from a physical interpretation is fundamental to the strategy of Newton's argument, the laws of motion provide a physical basis for the mathematical theory of motion. On one level of explanation, Newton considered "forces

---

[2] For a broader discussion of these themes, see P. M. Harman, *Metaphysics and Natural Philosophy: The Problem of Substance in Classical Physics* (Brighton, Sussex, and Totowa, N. J., 1982).

not physically but merely mathematically" and maintained that the concept of force, as defined by its usage in the mathematical formalism of the *Principia*, does not imply any "physical cause or reason" for the motions of bodies.[3] But at the same time he emphasized that the treatment of the motion of bodies in the *Principia* is to be given a physical interpretation by means of the laws of motion. The laws of motion thus provide the physical rationale of the mechanics of the *Principia*, and are based on concepts of force and inertia.

Newton's interpretation of force and inertia as physical concepts had its origin in his reading of Descartes's theory of the motion and collision of bodies. In particular, Newton reformulated Descartes's first two laws of motion to establish the basis of the first law of motion of the *Principia*. In expounding his own laws of motion, Descartes explained the tendency of a body to persevere in a state of rest or a state of motion in terms of the "force" of a body to remain at rest, and its "force" to remain in a state of motion; force is therefore conceived as the cause that maintains the existence of a body in any particular state, whether of rest or motion.[4]

In developing his own theory of mechanics, Newton adapted and significantly transformed the Cartesian concept of force. Newton's arguments show significant development from his early writings on the theory of mechanics in the 1660s (which show a pronounced Cartesian influence) to the mature formulation in the *Principia*.[5] The key feature of this transformation is the introduction of two concepts, *impressed force* and *inertia*, to explain the motion of bodies. Newton adopted the term *inertia* from Descartes who had employed the word to denote the sluggishness of bodies. Newton, however, used *inertia* to denote the internal power by which a body tends to persevere in a state of rest or uniform rectilinear motion.[6] In the *Principia*, Newton employed the concept

[3] *Isaac Newton's Philosophiae Naturalis Principia Mathematica: The Third Edition (1726) with Variant Readings*, ed. A. Koyré and I. Bernard Cohen, 2 vols. (Cambridge, Mass., 1972), I, 46.

[4] A. Gabbey, "Force and Inertia in the Seventeenth Century: Descartes and Newton," in *Descartes: Philosophy, Mathematics and Physics*, ed. S. Gaukroger (Brighton, Sussex, and Totowa, N. J., 1980), pp. 230–320.

[5] J. W. Herivel, *The Background to Newton's 'Principia'* (Oxford, 1965); Westfall, *Force in Newton's Physics*, pp. 424–67.

[6] I. Bernard Cohen, *The Newtonian Revolution* (Cambridge, Engl., 1980), pp. 189, 333–4.

of *impressed force* to denote causes generating change of motion or rest, the "action exerted on a body to change its state either of resting still or moving uniformly straight on." The disjunction between the concepts of impressed force and inertia expresses a distinction between an external cause generating changes in the motion of a body, which consists in "action alone, and does not endure in the body after the action is over," and the internal, inherent power of inertia, by which bodies resist changes in their state of rest or uniform rectilinear motion.[7]

Newton's designation of the concept of inertia as the "force of inertia" is ambiguous but reflects his early (Cartesian) conception of force as the principle maintaining a body in a state of rest or motion. To denote the power of matter by which a body persists in its state of rest or uniform rectilinear motion, Newton had used the term *vis insita* (innate force), which in seventeenth-century usage meant natural power or inherent force, ultimately introducing *vis inertiae* and declaring that "the innate force may be called, by a very significant name, the force of inertia."[8] Newton's ambiguous terminology in designating inertia as a *force* reflects contemporary usage, but in his later writings especially, he seeks to differentiate sharply between the status of the concepts of inertia and impressed force.

The concept of inertia is fundamental to the meaning of Newton's first law of motion, providing an explanation of the natural tendency of a body to persevere in the same state of rest or uniform rectilinear motion. He supported his view of the physical significance of the concept of inertia by claiming that inertia is proportional to the mass of a body: Inertia defines the materiality of bodies. In the third rule of philosophizing, appended to the second edition of *Principia* (1713), Newton listed inertia together with extension, hardness, impenetrability, and mobility as the "universal" qualities of matter, the essential properties an entity would have to possess as a necessary and sufficient condition of its materiality.[9] Inertia is thus a defining property of material substances; and in contrast to inertia, impressed force is held to be "not essential to body" and to be external to the material substance of bodies.[10]

---

[7] Newton, *Principia*, I, 40–1.

[8] Newton, *Principia*, I, 40. See Cohen, *Newtonian Revolution*, pp. 190–1, 334.

[9] Newton, *Principia*, II, 552.

[10] Newton, *Principia*, I, 41 (an unpublished annotation to the text of the second edition).

Despite his general disparagement of metaphysical arguments, he supported his statement of the physical significance of inertia by appeal to philosophical principles. In support of this claim that inertia is essential to matter, Newton declared that the universal qualities of bodies, including inertia, are known through experience to be essential properties of matter. He justified their universality, their ascription to the imperceptible indivisible particles of bodies, by appeal to the "analy y of nature," and declared that the ascription of these qualities to all bodies in nature is the "foundation of all philosophy."[11] The universal qualities are conceived as being those that could not change without affecting the materiality of bodies. Unlike nonessential qualities like heat and cold, these qualities do not manifest continuous and successive gradations of intensity: Were inertia to change, then a body would necessarily change its essential nature. The ascription of essential qualities to the fundamental particles of matter is justified by an appeal to the uniformity and homogeneity of nature.[12] The primordial atoms thus differ from observable particles only in size, not in their possession of essential properties such as inertia.

The definition of inertia as an essential property of matter contrasts with Newton's definition of force as external to matter. Newton's denial in his letter to Richard Bentley in 1693 that the force of gravity is "innate inherent and essential" to matter[13] emphasized that gravity was in a different category from inertia, reflecting Newton's distinction between force as external to the material substance of bodies and as acting to bring about change of motion, and inertia as a "universal" or "essential" property of matter. Hence unlike inertia, gravity is not an inherent or essential property of matter.

In the *Opticks,* Newton contrasted inertia as a "passive principle by which bodies persist in their motion or rest" with "active principles, such as are the cause of gravity." Matter in itself does not have the capacity to sustain motion; the motion and activity of bodies in nature is therefore seen in terms of forces and the causal agency of "active principles." Newton referred to active principles

[11] Newton, *Principia, II,* 552–4.

[12] J. E. McGuire, "Atoms and the 'Analogy of Nature': Newton's Third Rule of Philosophising," *Studies in History and Philosophy of Science, 1* (1970), 3–57; E. McMullin, *Newton on Matter and Activity* (Notre Dame and London, 1978), pp. 13–21.

[13] Newton to Bentley, 25 February 1692/3, in *The Correspondence of Isaac Newton,* ed. H. W. Turnbull, J. F. Scott, A. R. Hall, and L. Tilling, 7 vols. (Cambridge, Engl., 1959–77), *3,* 254.

both as being the "cause of gravity" and as being "general laws of nature . . . such as is that of gravity," and it would seem that active principles functioned both as laws of nature (though not "passive" laws of matter) and as the mediating agents by which God conserved motion and gravity in the cosmos.[14] The operations of active principles in nature were not reducible to the passive principles of matter such as inertia.[15]

The introduction of a conceptual disjunction between active and passive principles is clearly intended as a clarification of the status of gravity: Gravity, not being an essential property of matter, could not be explained in terms of passive principles of matter. But the introduction of active principles to explain gravity raised further difficulties. Leibniz ridiculed the appeal to a theory of gravity "performed without any mechanism" or "by a law of God" as tantamount to supposing gravity" an unreasonable occult quality."[16] Newton was accused of reviving the discredited notion of occult qualities, a view of nature vehemently contested by the mechanical philosophers of the seventeenth century.

Newton's introduction of his famous aether theory in the 1717 *Opticks* was probably governed by his attempt to refute Leibniz's charges, developed at length in Leibniz's correspondence with Samuel Clarke in 1715–16. Rejecting the reducibility of gravity to a contact–action model based on the impact of particles, which would have given gravity the status of a "passive" principle, Newton endowed the aether with active properties, which enabled it to function as a cause of gravity. The aether is composed of material particles and would ostensibly appear to fall under the category of passive principles. But Newton sought to distinguish between the passivity and inertia of ordinary matter and the active properties of ether.[17] The use of aether as an "active" principle thus threatened the active–passive dualism of Newton's natural philosophy, just as the ambiguous term "force of inertia" blurred the conceptual distinction between inertia as a passive property of matter and force as an external action generating change.

---

[14] Isaac Newton, *Opticks: or a Treatise of the Reflexions, Refractions, Inflexions and Colours of Light*, 4th ed. (1730; reprint ed., London, 1952), pp. 397–401.

[15] J. E. McGuire, "Force, Active Principles, and Newton's Invisible Realm," *Ambix*, 15 (1968), 154–208.

[16] Quoted in A. Koyré, *Newtonian Studies* (London, 1965), p. 141.

[17] P. M. Heimann [Harman], "Ether and Imponderables," in *Conceptions of Ether: Studies in the History of Ether Theories 1740–1900*, ed. G. N. Cantor and M. J. S. Hodge (Cambridge, Engl., 1981), pp. 61–83.

## LEIBNIZ AND THE CONCEPT OF FORCE

In his *Specimen Dynamicum* (1695), Leibniz sought to resolve the difficulties (as he perceived them) arising from Newton's dualist ontology by elaborating a theory of motion that was not based on a dualism of forces acting to bring about change and matter passively resisting change. By contrast, Leibniz proposed to explain motion in terms of a monism of force. Forces act to bring about change of motion; the resistance to motion, *inertia,* was explained in terms of forces resisting change of motion. Leibnizian dynamics revoked the disjunction between the concepts of force and inertia in Newton's physics. Leibniz rejected Newton's supposition that inertia is an essential property of matter: As a resisting power of matter, inertia does not define the nature of substances but is conceptualized as a "force." By supposing a monism of force, Leibniz aimed to resolve the ambiguities of the Newtonian disjunction between impressed force and inertia (conceived as a defining property of matter).

Leibnizian dynamics rested upon a theory of substance which differed radically from that held by Newton. Leibniz did not envisage a dualism between matter and force: In his view, substances are characterized by fundamental forces, and the forces of empirical physics are conceived as phenomenal analogues of these fundamental forces. The relationship between substances and empirical forces was thus conceived in quite different terms from the ontological dualism proposed by Newton.[18]

Leibnizian dynamics was thus grounded on a distinctive theory of substance, and Leibniz emphasized the importance of explaining the relations between ontological foundations and the laws of empirical physics. Unlike Newton, Leibniz maintained that the formulation of an intelligible natural philosophy requires explicit appeal to metaphysical principles. Leibnizian dynamics, concerned with empirical forces, was thus shown to be intelligible by a metaphysical explication of the relation between substances and empirical forces.

Leibniz criticized the Cartesian theory of substance, the thesis that matter is defined by spatial extension. In Leibniz's view, spatial extension is merely a manifestation of substance, not its defin-

---

[18] For general discussion, see Gerd Buchdahl, *Metaphysics and the Philosophy of Science. The Classical Origins: Descartes to Kant* (Oxford, 1969), pp. 388–469; Westfall, *Force in Newton's Physics,* pp. 283–322.

ing property. The inherent activity of substances, caused by forces, is the most fundamental characteristic of bodies, and force therefore defines the nature of substances. Leibniz termed the fundamental forces that characterize substances "primitive forces," but he denied that these fundamental forces could be invoked to explain the laws of empirical physics. Primitive forces explain the ultimate "sources of things" but not the "reasons for the laws of nature." The laws of empirical physics are explained in terms of "derivative" forces, the forces "by which bodies act on one another or are acted upon by each other." Derivative or empirical forces, which alone provide the basis for physical theorizing, were conceived as phenomenal analogues of the primitive forces that define the nature of substances.[19]

In characterizing the laws of nature, Leibniz distinguished between "active" and "passive" derivative forces, these terms having a distinctive meaning in his dynamics. The derivative active forces, which generate and sustain motion, were termed "living" and "dead" forces. The explanation of their relationship, based on an appeal to the metaphysical principle of the law of continuity, that "all change occurs gradually," forms the core of the Leibnizian theory of dynamics, based on the relationship between infinitesimal and finite quantities.[20] Living force, which measures the force of motion of a body, arises from an infinite number of infinitesimal impulses of dead force – the force that arises in the tendency of a body to achieve its state of motion. The conservation of living force in the mechanical interaction of bodies was for Leibniz a fundamental law expressing the inherent activity of nature and the order and self-sufficiency of natural processes.[21]

The derivative passive forces were the impenetrability of bodies and their resistance to motion, constituting the effort of bodies to remain in their state of rest or motion. For Leibniz, the resistance of bodies to motion (inertia) was not, as with Newton, explained by appeal to the essential nature of matter. Resistance to motion, or "natural inertia," was explicated in terms of the manifestation

[19] G. W. Leibniz, *Specimen dynamicum* [1695], in *Leibnizens Mathematische Schriften*, ed. C. I. Gerhardt, 7 vols. (Berlin and Halle, 1849–63; reprinted Hildesheim, 1960–1), *6*, 233–8.

[20] Leibniz, *Mathematische Schriften*, *6*, 241.

[21] P. M. Heimann [Harman], " 'Geometry and nature': Leibniz and Johann Bernoulli's Theory of Motion," *Centaurus, 21* (1977), 1–26.

of "forces."[22] Impenetrability and "inertia" were not conceived as essential or defining properties of matter, but were explicated as derivative passive forces, conceived as the phenomenal manifestation of primitive forces that define substances. The Newtonian dualism between force and matter was therefore avoided. For Leibniz, *inertia* represented a power of matter: Whereas active force generates motion, passive force resists motion. Bodies therefore resist motion by the derivative passive forces of impenetrability and inertia. In Leibnizian dynamics, the motion of bodies is explained by the interplay between active and passive derivative forces, whereas in Newton's physics, the theory of motion is based on the dualism of force and matter (defined by fundamental properties of inertia and impenetrability).

Living and dead forces, i.e., impenetrability and resistance to motion, were thus explicated as falling under different categories of *force*. Whereas living and dead forces (*active* forces) generate motion, inertia (*passive* force) resists motion. The Newtonian dualism of force and inertia conflates to a monism of force, an interaction of active and passive derivative forces. The relation between active and passive derivative forces was not envisaged as analogous to the Newtonian disjunction between impressed force and inertia; hence Leibniz sought to avoid the problems of the Newtonian *vis inertiae*. The notion of inertia as a resisting power is made intelligible in terms of the Leibnizian category of derivative force; and primitive forces, not inertia, define the essential nature of substances.

### EULER'S CRITIQUE OF "FORCE OF INERTIA"

In an essay "Recherche sur l'origine des forces" (1750), Leonhard Euler's concern lay in analyzing the foundational questions arising from Newton's concept of force. Euler contested the Leibnizian science of dynamics, which explicated empirical forces as phenomenal manifestation of fundamental forces; and he also rejected the view of D'Alembert, who sought to eliminate forces from the science of mechanics on the ground that the concept of force was obscure and metaphysical. In Euler's view, the concept of force was neither redundant nor "primitive" (in the Leibnizian sense);

---

[22] Quoted in Westfall, *Force in Newton's Physics*, p. 318.

adopting a Newtonian theory of force, he nevertheless recognized that "force" required conceptual explication.[23]

Euler perceived that there is an ambiguity in Newton's notion of "force of inertia." He argued that for Newton, force is the cause of a change of motion of a body, whereas inertia is defined in terms of the perseverance of a body in its state of rest or uniform rectilinear motion. Euler emphasized that strictly speaking, the concepts of *force* and *inertia* were therefore "directly contrary to one another." He concluded that to define *inertia* in terms of a *force* is therefore self-contradictory.

Euler nevertheless agreed with Newton's claim that inertia and impenetrability are defining properties of matter. Euler alluded to Newton's argument in the third rule of philosophizing of the *Principia*, that the essential properties of matter, such as impenetrability and inertia, are not subject to alteration of degrees of intensity, for he claimed that "impenetrability does not admit of degrees." Euler thus maintained that impenetrability is an absolute quality of bodies: "If a body is not completely impenetrable, it is penetrable." Because of the "absolute impossibility" of bodies to admit penetration by other bodies, impenetrability is an essential property of matter.[24] For Euler, the property of impenetrability is fundamental, providing a basis for his explication of force as arising from the impenetrability of bodies, manifested when bodies collide.

These arguments were developed further in Euler's *Lettres à une princesse d'Allemagne* (1768–72), in which he asserted that impenetrability is a "necessary property of all bodies" and defined body as an "impenetrable extension." Inertia, too, was conceived as a defining property of matter: "It would be impossible for a body to exist without inertia." Defining inertia as "a repugnance to everything that tends to change the state of bodies," he contrasted inertia with force, the "external cause," which signifies "everything that is capable of changing the state of bodies." Whereas inertia, like impenetrability, is essential to matter, force is not *"inherent* in matter." Thus the term "force of inertia" is an abuse of language, because inertia is "the opposite of a force." By rejecting

---

[23] S. Gaukroger, "The Metaphysics of Impenetrability: Euler's Conception of Force," *British Journal for the History of Science*, 15 (1982), 132–56.

[24] L. Euler, "Recherches sur l'origine des forces" [1750], in *Leonhardi Euleri opera omnia*, 2nd ser., vol. 5, ed. J. O. Fleckenstein (Lausanne, 1957), 112–15.

Newton's ambiguous term *vis inertiae,* Euler sought to clarify the Newtonian concept of inertia.

Euler sought to explain the origin of forces in terms of the fundamental concept of impenetrability. He took collision, contact–action, as the basic mode of change in nature. He suggested that bodies act on one another "no more than is necessary to prevent penetration." Hence collision is the source of "all the changes which occur in the world," and he concluded that the impenetrability of bodies is "the great spring by which nature works all her effects." He argued that forces, which arise from the impenetrability of bodies, are manifested only in resisting penetration. Forces are therefore dispositional, for although the impenetrability of matter gives rise to force, matter is not "endowed with a determinate force" but is "rather in a condition to manifest force" necessary in order to prevent penetration.[25] Thus the "origin of forces is based on the impenetrability of bodies." Euler therefore sought to clarify Newton's arguments, rejecting the term *vis inertiae* as self-contradictory, and affirming the (Newtonian) disjunction between the concepts of force and inertia.

### KANT AND THE CONCEPT OF INERTIA

In his *Metaphysische Anfangsgründe der Naturwissenschaft* (1786), Kant seems to have provided an analysis of the presuppositions about matter and force that underlie Newton's statement of the laws of motion and the concept of universal gravitation. Kant therefore sought to appraise the relationship between Newtonian physical laws and the false metaphysical assumptions that he believed were fundamental to Newton's statement of his physics in the *Principia.* Kant declared that all natural philosophers who sought to create a mathematical physics had in fact "made use of metaphysical principles and were obliged to make use of them, even though they otherwise [like Newton] solemnly protested against any claim of metaphysics on their science."[26]

---

[25] L. Euler, *Lettres à une princesse d'Allemagne sur divers sujets de physique et de philosophie,* 3 vols. (St. Petersbourg, 1768–72), in *Leonhardi Euleri opera omnia,* 3rd ser., vol. 2, ed. A. Speiser (Zürich, 1960), 150–71.

[26] Immanuel Kant, *Metaphysische Anfangsgründe der Naturwissenschaft* [1786], in *Kants gesammelte Schriften. Herausgegeben von der Königlich Preussischen Akademie der Wissenschaften,* 24 vols. (Berlin, 1902–38), 4, 472.

Kant contested Newton's atomistic ontology and the doctrine that impenetrability and inertia are defining properties of matter. By reappraising metaphysical foundations, by defining matter in terms of inherent forces of attraction and repulsion, Kant sought to establish the fundamental status of the concept of force (which he believed was confused in Newton's account) and thus to demonstrate the intelligibility of Newton's concept of the attractive force of gravity. Rejecting Newton's philosophical argument that inertia is an essential property of matter, Kant nevertheless sought to clarify the status of the Newtonian concept of inertia as a physical principle. Indeed, Kant claimed that only by rejecting Newton's metaphysical arguments could the status of inertia as a physical principle be adequately justified.

The argument of Kant's book is elaborate and at times discursive; its unifying theme is the application of the a priori categories of the *Critique of Pure Reason,* which he regarded as being constitutive of all experience, to the concept of matter. Although Kant claimed that there are links between physical theory and the a priori principles of cognition, he did not claim that the validity of Newtonian physics – the law of gravitation and the laws of motion – can be derived from a priori premises. Kant's aim was rather to establish the intelligibility of Newtonian physical concepts by demonstrating links between the categories and Newtonian physical laws.[27] In Kant's view, this enterprise involved a total reappraisal of Newton's dualistic ontology of matter and force: Only then would gravity and inertia be shown to be intelligible as physical concepts. Kant's reappraisal of Newton's dualistic ontology of force and matter was based on a metaphysical argument, which proceeded by the application of the categories of cognition to the concept of matter. The metaphysical argument was intended to establish the possibility of physical concepts and laws, not their actuality, yielding principles of constraint on possible physical hypotheses.

Kant's critique of Newton's theory of matter was developed in the second chapter of his book, on "Dynamics." He rejected the doctrine of essential qualities proposed by Newton and Euler, that impenetrability and inertia were the defining properties of all material substances. Kant maintained that forces define the essence of matter: "Matter fills space, not by its mere existence, but by a

[27] Buchdahl, *Metaphysics and the Philosophy of Science,* p. 678.

special moving force." Matter fills space by repulsive or extensive forces, which form the "basis of its impenetrability." The property of impenetrability is derivative not fundamental and "rests on a physical basis," the repulsive force. Kant went on to assert that "the possibility of matter requires a force of attraction as the second essential fundamental force of matter," a force that limits the extensive effect of the repulsive force.[28]

The dynamic account of matter was formulated with reference to the categories of reality, negation, and limitation, which correspond analogically to the account of matter in terms of fundamental forces of attraction and repulsion, which counteract or limit each other and determine the "degree of the filling of space" by matter. Kant rejected the concept of "absolute impenetrability": the repulsive force has "a degree which can be overcome," and the filling of space by matter is dependent on the "degree of compression" or "relative impenetrability" of matter. The dynamic theory of matter supposes that extension and impenetrability are derivative properties of matter, and that these qualities can change by degrees of intensity; impenetrability is derivative, being grounded on the "fundamental forces" which define the essence of matter.[29]

Kant claimed that in Newton's "mathematico-mechanical" philosophy, "forces were philosophized away," being superadded to matter, which was defined by the "empty concept" of "absolute impenetrability." By contrast, the Kantian "metaphysico-dynamical" theory made forces fundamental, avoiding the problems raised by Newton's dualistic matter–force ontology.[30] Moreover, the Kantian scheme provided an explanation of the intelligibility of the concept of gravity. By its attractive force, matter acts through empty space on other matter: The attractive force is extended through the universe. Thus in establishing that the "possibility of matter requires a force of attraction," Kant claimed that he (unlike Newton) had thereby justified the notion of action at a distance, the concept of gravitational attraction.[31] Kant's argument thus avoided the need to appeal to a physicotheological argument to justify the concept of gravity (as in Newton's reference to "active principles") or to an aether model (as in Newton's anomalous aether concept).

Kant's discussion of Newton's first law of motion in his chapter

---

[28] Kant, *Schriften*, *4*, 497, 502, 508.   [29] Kant, *Schriften*, *4*, 501, 502, 525.
[30] Kant, *Schriften*, *4*, 524–5.   [31] Kant, *Schriften*, *4*, 516–17.

on "Mechanics" sought to clarify the meaning of Newton's concept of inertia. Having established that matter is defined by fundamental forces of attraction and repulsion, and that impenetrability is a relative and derivative concept, Kant aimed to show that inertia should be construed as a law of mechanics, not as a defining property of matter. As before, the argument in the chapter on "Mechanics" is based on the application of the categorical principles to the concept of matter. The principle of causality states that "every change has a cause"; applied to the concept of matter, it yields a metaphysical "law of mechanics" that "every change of matter has an external cause." This metaphysical law of mechanics establishes the possibility of Newton's first law of motion (though not its physical actuality), as Kant indicated by placing brackets around his added statement of Newton's first law of motion. Kant did not seek to prove Newton's first law of motion but to establish that "all change of matter is based on an external cause" and that matter "undergoes no changes except by motion."[32]

Kant described this law of mechanics as the "law of inertia," which denotes the passivity of matter, its inability to "determine itself to motion or rest." Kant therefore denied that matter possesses an "internal principle," an inherent power or "special force of matter under the name of the force of inertia." The supposition that matter possesses an "internal principle" or force would contradict the law of inertia, which proves that matter does not possess an inherent power. Kant was thus in agreement with Euler in rejecting the "designation force of inertia" as contradictory.

Kant's reappraisal of the metaphysical foundations of Newton's physics thus aimed to clarify Newton's *vis inertiae:* Inertia is not an inherent or essential property of matter nor is inertia a "special and entirely peculiar force [of resistance]." The law of inertia merely signifies that all changes of matter require an external cause.[33] Kant's rejection of the notion that matter possesses an inherent principle or force of resistance (which he believed to be implied by the term "force of inertia") also signals his rejection of the conceptual framework of Leibnizian dynamics. Although he disagreed with Euler about the relative status of the concepts of inertia, impenetrability, and force he echoed Euler's critique of the term "force of

---

[32] Kant, *Schriften, 4,* 543. See Buchdahl, *Metaphysics and the Philosophy of Science,* pp. 676–8.

[33] Kant, *Schriften, 4,* 543–4, 549–51.

inertia."[34] Kant's intention was to reconstruct the relationship between Newton's physical laws and the assumptions about matter and force which underlie the physics of *Principia*. The clarification of Newton's concept of inertia was central to this endeavor and required the Kantian critique of Newton's "mathematico-mechanical" philosophy of nature. Kant's analysis and reappraisal of the metaphysical foundations of Newtonian physics thus resolved the paradox of Newton's *vis inertiae*.

[34] P. M. Harman, "Force and Inertia: Euler and Kant's *Metaphysical Foundations of Natural Science*," in *Nature Mathematized*, ed. W. R. Shea (Dordrecht and Boston, 1983), pp. 229–49.

# Part II

SCIENCE AND RELIGION

# 6

## Celestial perfection from the Middle Ages to the late seventeenth century

EDWARD GRANT

Within medieval Aristotelian cosmology, the concept of celestial perfection had two easily distinguishable aspects. One was concerned with the relationship between the celestial and terrestrial regions, those two radically different, but fundamental parts of Aristotle's universe. Until the seventeenth century, Scholastic authors were virtually unanimous in judging the celestial part more noble than the terrestrial. The other aspect focused on differences in perfection that might exist in the heaven itself and what these might be. Here there was disagreement and contention from the thirteenth to the seventeenth centuries. Within the Aristotelian worldview, celestial perfection was an important concept and raised significant issues. Eventually, Christian concepts of animate and inanimate beings along with the Copernican theory would produce dramatic changes in seventeenth-century Scholastic ideas about the perfection and nobility of the celestial region. Because the second aspect of celestial perfection involved by far the greater degree of contention, we shall consider it first.

### ON INTRACELESTIAL DIFFERENCES IN PERFECTION

Because Aristotle had argued that the celestial region was incorruptible,[1] did this also imply that all celestial bodies were equally

---

[1] I have briefly described medieval and early modern Scholastic conceptions of celestial incorruptibility in "Were There Significant Differences Between Medieval and Early Modern Scholastic Natural Philosophy? The Case for Cosmology," *Nous, 18,* no. 1 (March 1984), 5–14. A much longer and fully documented version will appear in the proceedings of a workshop on "The Interrelations Between Physics, Cosmology and Astronomy: Their Tension and its Resolution 1300–1700," held in Israel, April 29–May 2, 1984.

perfect? Although Aristotle wrote relatively little on the subject of intracelestial perfection, there is no doubt of his conviction that celestial bodies differed in degrees of perfection. Despite his presentation of conflicting interpretations, Aristotle's opinions served as the major points of departure for subsequent discussions.

In the first book of *De caelo,* Aristotle linked the perfection of the celestial region directly to its distance from the earth, or the sublunar region as a whole.[2] Presumably a celestial body was more perfect the farther it was removed from the earth. Averroes viewed this as a higher level analogue of the relations between the four sublunar elements, where the nobler the element, the higher its location, as, for example, fire is higher and therefore nobler than earth.[3] Under the influence of Neoplatonists like Macrobius, many would make a body's degree of perfection directly proportional to

[2] "Thus the reasoning from all our premises goes to make us believe that there is some other body separate from those around us here, and of a higher nature in proportion as it is removed from the sublunary world." Aristotle, *De caelo,* 1.2.269b13–18. The translation is by W. K. C. Guthrie, *Aristotle On the Heavens,* Loeb Classical Library (London: William Heinemann Ltd.; Cambridge, Mass.: Harvard University Press, 1960). Here, in the thirteenth-century translation of Michael Scot, is one medieval version of this brief statement: "Et habens rationem potest ex omnibus predictis ratiocinari, quod aliud corpus est praeter ista corpora vicinantia et continentia nos, separatum ab eis, cuius natura nobilior est naturis eorum secundum suam remotionem ab eis et collocationem eius super ipsa." The translation appears in *Aristotelis opera cum Averrois commentariis,* 9 vols. in 11 parts plus 3 suppl. vols. (Venice: Junctas ed., 1562–74), *5, De caelo,* bk. 1, text 16, fol. 11v, col. 2.

[3] Averroes, *Aristotelis opera cum Averrois commentariis, 5: De caelo,* bk. 1, comment 16, fol. 11v, col. 2. Because the earth and moon marked the beginning of the ascent of perfection for the inferior and superior regions, respectively, Averroes observed that in the *Liber de animalibus,* Aristotle declared that "the nature of the moon is similar to that of the earth" ("Unde Arist. in *lib. de Animalibus* dicit quod natura lunae similis est naturae terrae"; Averroes repeated this reference in bk. 2, comment 42, fol. 127r, col. 1). Without mention of Averroes, Copernicus repeated this statement when he declared that "Aristotle says in a work on animals, the moon has the closest kinship with the earth." See *Nicholas Copernicus On the Revolutions,* ed. Jerzy Dobrzycki; trans. and commentary by Edward Rosen (Warsaw: Polish Scientific Printers; London: Macmillan Press, 1978), bk. 1, chap. 10, p. 22. In a note (p. 360), Rosen, who gives Copernicus's source as Averroes, *De substantia orbis,* explains that Averroes and Copernicus were mistaken because "Aristotle's *Generation of Animals,* (4.10.777b.24–27) regards the moon, not as akin to the earth, but as a 'second and lesser sun.' " In that very same passage, Aristotle also says that "the moon is a first principle because of her connexion with the sun and her participation in his light" (trans. Arthur Platt in vol. 5, pt. 3 of the Oxford English translations, ed. J. A. Smith and W. D. Ross). Apparently with this same statement in mind, Averroes again refers to the *Liber de animalibus* in *De caelo,* bk. 2, comments 32 (fol. 115v, col. 2) and 42 (fol. 127r, col. 1), where he says that Aristotle linked the natures of moon and earth because neither is self-luminous.

its proximity to the first heaven.[4] The author of the pseudo-Aristotelian *De mundo,* composed in the first or second century A.D., expressed similar sentiments. Perfection diminishes as things are more remote from God, who occupies the supreme and highest place in the universe. For

the heavenly body which is nighest to him most enjoys his power, and afterwards the next nearest, and so on successively until the regions wherein we dwell are reached. Wherefore the earth and the things upon the earth, being farthest removed from the benefit which proceeds from God, seem feeble and incoherent and full of much confusion.[5]

But in the second book of *De caelo,*[6] Aristotle employed a different criterion for celestial perfection. Here it was the number of motions that a celestial body required to fulfill its cosmic objectives that determined its degree of perfection. From this vantage point, Aristotle would explain that "the first heaven reaches it [i.e., its ultimate end or perfection] immediately by one movement, and the stars that are between the first heaven and the bodies farthest from it reach it indeed, but reach it through a number of movements." Within this overall scheme, the greatest perfection resides in the outermost sphere of the fixed stars, with its single daily motion, and the least perfection in the earth, which moves not at all. Between these two extremes, however, the degree of perfection did not correspond to the ascending or descending order of the planets, although as Aristotle explained, "considering that the primary body [i.e., the sphere of the fixed stars] has only one motion, it would seem natural for the nearest one to it to have a very small number, say two, and the next one three, or some similar proportionate arrangement." In fact, Aristotle argues, "the op-

---

[4] The creative process for Macrobius and all Neoplatonists was one in which there was a continual debasement and materialization as one moved farther "downward" from God, or the One. The celestial spheres were thus hierarchically ordered, those nearest the outermost reaches of the cosmos being more perfect than those nearer the earth. See *Macrobius, Commentary on the Dream of Scipio,* trans. with an introduction and notes by William H. Stahl (New York: Columbia University Press, 1952), pp. 143–144.

[5] *De mundo,* 397$^b$26–32. Trans. E. S. Forster, in *The Works of Aristotle,* trans. into English under the editorship of W. D. Ross, vol. 3 (Oxford: Oxford University Press [Clarendon Press], 1931). At least two Latin translations from the Greek original were made during the Middle Ages. For the Latin text of the passage above, see *Aristoteles Latinus,* vol. 11, 1.2: *De mundo,* ed. W. L. Lorimer (Rome, 1951), pp. 40 (anonymous translation) and p. 69 (translation of Nicholas Siculus).

[6] Aristotle, *De caelo,* 2.12.291$^b$.28–292$^b$.25. The translations given below from this Aristotelian passage are by W. K. C. Guthrie, *On the Heavens* [see note 2].

posite is true, for the sun and moon perform simpler motions than some of the planets, although the planets are farther from the centre and nearer the primary body, as has in certain cases actually been seen. . . ."

Perfection could thus be interpreted according to distance from the earth or sublunar region, so that order of perfection or nobility simply followed the ascending order of the planets; or it might be determined by the number of motions that a celestial body required to move around the heaven, ìn which event the order of perfection did not follow the ascending order of the planets and orbs. These were the major options from Aristotle. Over the centuries they were elaborated and new ones were added. Even passages not directly concerned with the problem of celestial nobility, primarily from Aristotle's *Physics* and *Metaphysics,* were introduced and adapted to the literature on perfection. Celestial perfection would eventually be compared to organic life, indeed even to the very lowest forms of life.

Although concern about celestial perfection became most intense during the course of the sixteenth and seventeenth centuries, there was certainly no lack of interest in the thirteenth and fourteenth centuries. St. Thomas Aquinas repeated, with apparent approval, Aristotle's declaration of ascending celestial nobility, but added a brief confirmation of that doctrine from Aristotle's *Physics* when he explained that "in the universe, containing bodies are related to contained bodies as form to matter and act to potency."[7] In later centuries, this "container argument" would often be cited in defense of ascending celestial nobility, as when Bartholomew Amicus invoked it in declaring that "as the whole is nobler than the part, so [also] is the container [nobler] than the contained because it is related as whole to part."[8]

---

[7] St. Thomas Aquinas, *In Aristotelis libros De caelo et mundo; De generatione et corruptione; Meteorologicorum Expositio,* ed. P. Fr. Raymundi M. Spiazzi, O.P. (Turin/Rome: Marietti, 1952), *De caelo,* bk. 1, lectio 4, p. 23, par. 50. Although Thomas's reference to the fourth book of the *Physics* is to Aristotle's discussion on place, it should be emphasized that Aristotle does not consider whether the container, or place of a thing, is nobler than what is contained.

[8] Bartholomew Amicus, *In Aristotelis libros De caelo et mundo dilucida textus explicatio et disputationes in quibus illustrium scholarum Averrois, D. Thomae, Scoti, et Nominalium sententiae expenduntur earumque tuendarum probabiliores modi afferuntur* (Naples, 1626), Tractatio quinta, questio 4, dubit. 2: "De ordine caelorum," p. 264, col. 2. As did Aquinas, Amicus also cited book 4 of Aristotle's *Physics.* In the fifteenth century, Dominicus de Flandria O.P., following Thomas, accepted the container argument as upholding the ascending

John Buridan (ca. 1295–ca. 1358), who devoted a whole question to the problem in his *Questions on the Metaphysics,* admitted that it was difficult.[9] As Buridan conceived the question, perfection involved a material orb, which might or might not carry a planet; and an immaterial, separated intelligence, which actually moved that orb. Their perfections, however, were identical. Whatever the level of perfection assigned to the orb had also to be assigned to the intelligence. But how was perfection to be measured? Since each orb or sphere was moved by an immaterial, separated intelligence, one might first compare the orbs by the straight lines drawn from their common center and then relate the intelligences moving the orbs in the same way. But this method fails because, as Aristotle argues in the twelfth book of the *Metaphysics,* each intelligence constitutes a separate species and therefore no two intelligences are comparable, no more than a man could be compared to an ass. No increase or multiplication of qualities could enable an ass to equal or exceed a man.

In the absence of direct means of comparative measure, Buridan suggests that perfection be measured by effects. Because each intelligence was actually a motive power causing the motion of the sphere under its control, one might make certain effects the basis of comparison. Other things being equal, one might compare the perfection of mobiles (planets, stars, or orbs); the magnitude of mobiles, where greater magnitude is equated with greater perfection; and the velocity of mobiles, where greater speed is equivalent to greater perfection.[10]

As an illustration, Buridan compared the Prime Mover, God,

order of planetary nobility (see *Questionum super xii libros Methaphisice* [1523; reprinted, Frankfurt: Minerva G.M.B.H., 1967], sig. R3ᵛ, col. 2; the work is unfoliated). Although he did not accept the ascending order of celestial nobility, Illuminatus Oddus, in the seventeenth century, presented it as the second of four arguments usually offered in support of ascending celestial nobility (see *Disputationes De generat. et corrupt. ad mentem Scoti, cuius doctrina adhuc magis elucidatur, defenditur, roboratur. Opera et industria R. P. Illuminati a Collisano Capuccini. Cum resolutione aliquorum dubiorum ad libros De coelo & Meteoris spectantium* . . . [Naples, 1672], p. 35, col.2).

[9] John Buridan, *In Metaphysicen Aristotelis questiones argutissimae Magistri Ioannis Buridani* . . . (Paris: Badio, 1518; reprinted Minerva, 1964), bk. 12, question 12 ("Whether the order of the celestial spheres in position is [also] their order in perfection"), fols. 74ʳ, col. 2 – 75ʳ, col. 1.

[10] "Notandum est ergo quod rationabiliter ex multis arguitur maior perfectio potentie moventis primo ex maiori perfectione mobilis ceteris paribus; secundo ex maiori magnitudine mobilis ceteris paribus; tertio ex maiori velocitate motus ceteris paribus." Ibid., fol. 74ᵛ, col. 1.

who moved the whole heavens, to the intelligence that moved the eighth sphere of the fixed stars. Since the former completed one circulation in one day and the latter moved only one degree in one year, it followed that the ratio of their velocities, and therefore their perfections, would be as 36,000 to 1.[11] The greater perfection of the Prime Mover was also measurable in other ways. It moved the whole heaven simultaneously, whereas any other intelligence moved only one partial sphere – that is, a single sphere within any complex of spheres that together accounted for the motion of a single planet. The Prime Mover also produced the quickest motion, moving the whole heaven in a day as compared to the mover of the moon, which took a month, that of the sun, which took a year, and so on.

If the Prime Mover is left aside, Buridan asks which of the motive intelligences would emerge as most perfect among those that move only single spheres, or at most move a whole complex of spheres for a single planet? Did the order of perfection follow the ascending order of the planets and their orbs? Buridan acknowledged that the sun posed the most serious problem, as indeed it did throughout the history of inquiries about celestial perfection. The sun seemed so much nobler than any other planet, and yet three other planets – Mars, Jupiter, and Saturn – lay beyond it. The sun's apparent superior nobility over all other planets was commonly noted. Buridan presents most of the usual reasons for so believing. It was not only the greatest of the planets in magnitude but it supplied light to all the other stars and planets; it was the most obviously active planet and the cause of generation. As the fourth and middle planet from the earth, the sun was like a king in the middle of his kingdom.[12] Moreover, astronomers re-

---

[11] Of course, Buridan realized that God was of infinite power and therefore could not be compared to the finite power of an intelligence. Only for the sake of the example was God assumed to have finite power (see ibid., fol. 74$^v$, col. 2).

[12] In the fourteenth century, Themon Judaeus also spoke of the sun lying in the middle as a king in his kingdom (for the reference, see my article, "Cosmology," *Science in the Middle Ages,* ed. David C. Lindberg [Chicago: University of Chicago Press, 1978], p. 279). The same figure was rather commonly cited in the sixteenth and seventeenth centuries: for example, the Coimbra Jesuits *(Commentarii Collegii Conimbricensis Societatis Iesu In quatuor libros De coelo Aristotelis Stagiritae,* 2nd ed. [Lyon, 1598], bk. 2, chap. 5, p. 255) who cite Albumasar's *Great Introduction,* Tract 3, differ. 3, as their source (I was unable to locate the passage in the 1506 edition, *Introductorium in astronomiam Albumasaris abalachi octo continens libros partiales,* published in Venice by Jacobus Pentius Leucensis [for more on Albumasar's treatise, see note 21]; Bartholomew Amicus, *De caelo,* p. 264,

lated the motions of the other planets to the sun. And, finally, in a reference to Aristotle's discussion in the second book of *De caelo* (see above), Buridan explains that because the sun moves with fewer motions than Saturn and Jupiter, its motive intelligence must be nobler than theirs.

Other discrepancies, based on widely received astrological properties assigned to the planets, also threatened the ascending order theory of perfection. Thus Venus and Jupiter were called "fortunate" planets, whereas Mars and Saturn were considered unfortunate. Similarly Jupiter and Venus were characterized as better and more benevolent planets than Saturn, from which it might be inferred that their intelligences are more perfect than Saturn's.

Despite these apparent discrepancies, Buridan eventually supported the ascending order theory of celestial perfection. He rightly declared that Aristotle had not provided the specific perfection with respect to which the spheres were to be ordered positionally. But the commentator, Averroes, had provided a proper response that took into account the apparent discrepancies described above. Although the Prime Mover, God, exceeded all intelligences with regard to perfectional properties, other intelligences varied in their perfections, exceeding in some and exceeded in others. This is evident in terrestrial things, as, for example, that "a man is absolutely more perfect than a horse and yet a horse exceeds him in magnitude, speed, and in strength. And so, although the three superior planets are absolutely nobler than the sun, yet it is not absurd that the sun should exceed them in some properties."[13]

Although Aristotle had not decided on the specific perfections by which the planetary spheres and their intelligences could be arranged in ascending order, Buridan believed that Thomas Aquinas had already proposed reasonable criteria for such an arrangement. Thomas had decided that an intelligence moving a superior sphere was more perfect than one moving an inferior sphere, a conviction based on the widely accepted container theory, which

col. 1 [for his statement, see below]; and Bartholomew Mastrius and Bonaventura Bellutus, *Philosophiae ad mentem Scoti cursus integer . . . Editio novissima a mendis expurgata,* 5 vols. (Venice, 1727; earlier editions in 1678, 1688, and 1708), 3: *Continens disputationes ad mentem Scoti in Aristotelis Stagiritae libros: De anima, De generatione et corruptione, De coelo, et Metheoris,* p. 499, col. 1, par. 108 (in their commentary on *De coelo*).
13 Buridan, *Metaphysics* [see note 9], fol. 75<sup>r</sup>, col. 1.

assumed that because a superior sphere contained an inferior sphere, the former was prior and nobler than the latter. And although lower spheres are quicker in their movements, superior spheres are greater in magnitude and have to traverse vastly greater distances on their circumferences. Thus the superior sphere retains its overall and absolute superiority because, as Buridan put it, "having posited that an inferior is moved quicker than a superior, nevertheless it is not moved so much quicker than the superior sphere is greater [in magnitude]."[14] Also of significance for Buridan was his belief that in governing the inferior world, the three superior planets, and the sun as well, governed the being and permanence of things, whereas the three planets below the sun governed the motions and mutations of things.[15] But because "it is nobler to have a property concerning the being and permanence of things than [a property] concerning the motions and mutations of things,"[16] it followed, on the basis of greater prefection in a number of key properties, that the three superior planets were nobler than the sun and the three inferior planets.

But how did Buridan, and others, reconcile the defense of ascending nobility with Aristotle's order of perfection by the number of motions required to complete a single revolution for each celestial body? As we saw above, the sun and moon had fewer motions than some of the planets. Were they therefore nobler than Saturn and Jupiter, thus making impossible an ascending order of nobility for celestial bodies? Buridan considered this problem in his *Questions on De caelo* where he inquired "whether the sun and

---

[14] Ibid., fols. 74ᵛ, col. 2 – 75ʳ, col. 1.

[15] Buridan says the same thing in his *Questions on De caelo*, bk. 2, quest. 21 (*Iohannis Buridani Quaestiones super libris quattuor De caelo et mundo*, ed. Ernest A. Moody [Cambridge, Mass.: The Mediaeval Academy of America, 1942], p. 225). In his *Questions on De caelo*, bk. 2, quest. 17, Albert of Saxony, who derived much of his material directly from Buridan's *De caelo*, repeats the same statement (*Questiones et decisiones physicales insignium virorum: Alberti de Saxonia in octo libros Physicorum; tres libros De celo et mundo; duos lib. De generatione et corruptione; Thimonis in quatuor libros Meteororum; tres lib. De anima; Buridani in lib. De sensu et sensato; . . . Aristotelis. Recognitae rursus et emendatae summa accuratione et iudicio Magistri Georgii Lokert quo sunt Tractatus proportionum additi* [Paris, 1528], fol. 92ᵛ. col. 1). In the fifteenth century, Dominicus de Flandria presented the same argument except that he ranked the sun with the three lower planets as essentially a cause for mutations rather than preservation (see his *Questionum super xii libros Methaphisice*, sig. R3ᵛ, col. 2); the Coimbra Jesuits did the same (*De coelo* [see note 12] p. 256) at the end of the sixteenth century.

[16] Buridan, *Metaphysics* [see note 9], fols. 74ᵛ, col. 2 – 75ʳ, col. 1.

moon ought to be moved with fewer motions than other planets."[17]
At the end of the questions, Buridan presents the principle on which
he based his solution. "It is not necessary universally," he declares, "that nobler things should have more or fewer actions; but
[it is necessary] that in striving for the same end [or goal] that
nobler things have fewer actions." Since motions are actions, Buridan had the basis for a solution to the problem posed by Aristotle.
As we have already seen, Buridan had followed a common practice of dividing the celestial region into two parts. The first, and
nobler part, encompassed the sphere of the fixed stars and the three
superior planets. It was their function and goal to control existence, the permanence and duration of things, and the overall well-
being of the universe. To achieve these identical ends, the sphere
of the fixed stars, the outermost and most noble celestial body,
performed only one motion, whereas the other three planets required more than one motion, although whether Saturn required
fewer than Jupiter and Jupiter fewer than Mars is left unstated.
The sun and the three inferior planets, Venus, Mercury, and the
moon, had a different objective, namely to control change in all
inferior things below the moon. Buridan then subdivided them
into two groups with somewhat different ends. The moon's goal
is distinguished from that of the other three. Its cosmic mission is
to prepare sublunar matter to receive the good influences of the
celestial region, and for this "inferior property," as Buridan describes it, the moon requires few motions. Of the other three planets,
the sun, as the most distant from the earth, has fewer motions than
Venus and Mercury; and perhaps Venus requires fewer motions
than Mercury.

But even if astronomers had assigned more motions to Saturn
than Mars, or to Mercury than Venus, Buridan would undoubtedly have preserved the ascending order of perfection by assigning
nobler status to Saturn over Mars and to Venus over Mercury. He
would have invoked Averroes's principle, mentioned earlier, that
excess of absolute nobility did not preclude a less noble body from
exceeding a more noble body in one or more particular properties
or qualities. Thus even if Mars required fewer motions to achieve
the same ends as Saturn, there were numerous other properties in

[17] Buridan, *De caelo* [see note 15], bk. 2, quest. 21, pp. 223–225. Albert of Saxony treated
the same question in his *Questions on De caelo*, bk. 2, quest. 17, fols. 92$^r$, col. 2 – 92$^v$,
col. 2.

which Saturn could be shown to exceed Mars and thus preserve the latter's greater absolute perfection.

Buridan's conclusions reflected a popular conception of celestial perfection that was often repeated.[18] But it was by no means the sole opinion in the Middle Ages. Some opted for the opinion Buridan rejected, namely that the sun was the noblest planet. One emphatic supporter was Nicole Oresme (ca. 1320 – 1382), who insisted that "the perfection of the heavenly spheres does not depend upon the order of their relative position as to whether one is higher than another."[19] Not only is the sun, located in the middle of the planets, "the most noble body in the heavens . . . more perfect than Saturn or Jupiter or Mars, which are all higher up," but "it is probable that Jupiter is more perfect than Saturn, and the moon more so than Mercury." Oresme thus adopted two noteworthy positions: (1) He identified the sun as the most noble planet, and (2) he rejected the ascending order of nobility. Although Oresme's *Le Livre du ciel et du monde,* in which these opinions appeared, was not published until 1968 and was probably not known by early modern Scholastics, one and often both of Oresme's ideas were widely adopted during the late sixteenth and seventeenth centuries.

THE NOBILITY OF THE SUN AND THE ASCENDING
ORDER OF CELESTIAL PERFECTION IN THE LATE
SIXTEENTH AND SEVENTEENTH CENTURIES

Notable support for the first, but not the second, of Oresme's opinions came at the end of the sixteenth century from the Coimbra Jesuits who, in their commentary on *De caelo,* presented a more elaborate justification than did Oresme. The Conimbricenses, as they were called, presented their explanation in terms of celestial spheres.[20] The sun had three orbs to fulfill its objectives. Of these, only the middle orb, in which the sun was actually located, was nobler than all other planets. Indeed with the exception of the sun's

---

[18] As described in note 15.

[19] See *Nicole Oresme: Le Livre du ciel et du monde,* ed. Albert D. Menut and Alexander J. Denomy, C.S.B.; trans. with an introduction by Albert D. Menut (Madison, Wis.: University of Wisconsin Press, 1968), p. 507.

[20] Conimbricenses, *De coelo,* bk. 2, chap. 5, quest. 2 ("Whether as the celestial globes are higher, they are [also] of a nobler nature"), pp. 252–256.17 [see note 12].

middle orb, the other two solar orbs and all other celestial spheres were progressively nobler according to their positions; that is, nobility varied directly as distance from the earth.

But why, if the sun is the noblest planet, was it placed fourth rather than seventh? It was wisely located in the middle so that it could diffuse its power, light and heat, in every direction "as if from the middle of a kingdom. For if it were in the seventh sphere, it would, from such a distance, heat the inferior bodies much more weakly by its light and nearly all things would become hard from the coldness. And if the sun were in the first sphere, all things would burn up from its proximity."[21]

Bartholomew Amicus (1562–1649) retained the two basic ideas of the Coimbra Jesuits – the sun as the noblest planet with the other planets arranged in ascending order of nobility – but presented a more elaborate and interesting rationale for his position. Amicus raised the problem of celestial perfection quite naturally in his commentary on Aristotle's *De caelo,* where he treated the customary question on the order of the planets and their spheres. Upon presenting two rival orders of the planets – the geocentric Ptolemaic (moon, Mercury, Venus, sun, Mars, Jupiter, and Saturn) and heliocentric Aristarchan or Copernican – Amicus asked what preserved the true order of the planets, which for him was

---

[21] Ibid., p. 255. As the source of this statement, the Conimbricenses correctly cite the *Magnum Introductorium* (or *Introductorium Maius*) *in astronomiam,* tract. 3, differentia 3 of the Arabic astronomer and astrologer Albumasar (Abu Ma 'shar), who died around 886. The *Introduction* was twice translated into Latin in the twelfth century, the first time in a literal translation in 1133 by John of Seville, the second time in a somewhat abbreviated version by Hermann of Carinthia in 1140. John of Seville's translation was never published, whereas Hermann's was published three times, twice in Augsburg by Erhard Ratdolt (1489 and 1495) and once in Venice (1506). In the latter edition (*Introduction in astronomiam Albumasaris abalachi octo continens libros partiales* [Venice: per Jacobum pentium Leucensem, 1506]), which is unfoliated, the relevant passage appears on fol. 16ᵛ when the count is begun from the title page. The text is as follows "Ex his itaque patet quod si sol usque ad nonam speram sublimatus esset vel usque ad lunarem orbem humiliatus vel inde frigore vel hinc calore nimio mundum stare non posse. Quamobrem providus auctor omnium dues: solem tanquam universalem corporee vite vomitem in media mundi regione medium posuit."

For information about the translations and editions as well as the considerable influence which Albumasar had in the twelfth century, see Richard Lemay, *Abu Ma 'shar and Latin Aristotelianism in the Twelfth Century: The Recovery of Aristotle's Natural Philosophy Through Arabic Astrology* (Beirut: American University of Beirut, 1962). On p. 113, Lemay summarizes the relevant passage and, in a note, quotes the Latin text from John of Seville's translation. See also David Pingree, "Abu ma'shar al-Balkhi, Ja'far ibn Muhammad," *Dictionary of Scientific Biography,* vol. 1, p. 35, col. 2.

represented by the Ptolemaic system.[22] As was almost always the case, he assumed that the celestial spheres were perfectly ordered and sought only the rationale for that order.

Amicus offers four possible explanations, the last three of which are relevant.[23] The second of the four invoked dignity *(dignitas)* or nobility. The basis of perfection was a hierarchical universe, in which the lowest beings were linked with the highest by a series of graded steps. As the opinion of St. Thomas Aquinas, the arguments in its favor were well known. Such a hierarchical linkage could operate only if inferior beings are subject to superior beings. Astrologers had assigned more perfect and nobler effects to planets and spheres in proportion to their distance from the earth. The eighth sphere of fixed stars is thus noblest of all, "as is obvious from the multitude of stars and especially from the zodiacal circle with its splendid twelve signs under which all planets are moved and from which arise so many generations of things."[24] The container argument, which was one of the mainstays for adherents of celestial nobility, is also mentioned.[25] Each containing sphere is superior to the sphere it contains.

Despite this traditional array of arguments in favor of the most popular theory of perfection, Amicus expressed reservations about it. For "against this opinion stands the perfection of the sun," which is greater than Jupiter and Saturn and yet is placed under them. Indeed it is "placed in the middle, as a king in his kingdom and the heart in the body. Nor indeed is the excellence of the sun over all other planets and the fixed stars to be doubted, since all receive their light, the most perfect of all qualities, from it."

Although he was convinced that the sun was the most perfect

---

[22] Bartholomew Amicus, *In Aristotelis libros De caelo et mundo* . . . Tractatio quinta, quest. 4, dubit. 2 ("De ordine caelorum"), pp. 264, col. 1 – 265, col. 1.

[23] The first explanation assigns the order of perfection to the pure free will of the Creator, thus eliminating any preordained order of perfection. God, it seems, merely placed each sphere and planet where He pleased. If so, "each heaven attained that place which God gave to it, just as is obvious from the location of the whole universe in this imaginary space rather than that [imaginary space]." Ibid., p. 264, col. 1. Amicus rejected this opinion in a single, obscure sentence, the meaning of which is unclear to me. On imaginary space, see my book, *Much Ado About Nothing: Theories of Space and Vacuum from the Middle Ages to the Scientific Revolution* (Cambridge: Cambridge University Press, 1981).

[24] Amicus, *In Aristotelis libros,* p. 264, col. 2.

[25] Reference is also made to Aristotle's *Metaphysics,* bk. 12, text 44, where according to Amicus, "the order of the intelligences follows the order of the motions and mobiles." Ibid.

planet and therefore superior to the planets above it, Amicus did not, as we shall see, abandon the theory of ascending nobility. He used it in conjunction with another theory – the third opinion – which we have not yet seen. In this view, the order of the heavens is not maintained by nobility but by reason of symmetry *(commoditas)* and what is good and proper for the whole *(bonus universi)*. The sun has been placed in the middle of the planets so that it could diffuse its light to the stars and planets above and to the inferior planets and the sublunar region below. Preservation of the whole necessitated a middle position for the sun: if it were too low, "all things would burn up from the heat; and if it were placed too great a distance away, all things would freeze."[26] Moreover, according to the third opinion, the sun ought to be in the middle because it agrees with the three superior planets in its epicyclic motion and with the three inferior planets in the motion of its deferent orb.

The stage was thus set for the fourth opinion to incorporate both of the previous interpretations and win the approval of Amicus. The order of the planets was now primarily determined by symmetry *(commoditas)* and secondarily by nobility *(dignitas)*. Symmetry and the common good of the whole took precedence over nobility whenever necessary, as happened when the sun was placed in the middle of the planets. When preservation of the common good was not in question, however, nobility would always prevail. Thus with the exception of the sun, all the other planets and orbs,[27] from the moon to the sphere of the fixed stars, were presumably arranged in order of nobility.

### DIFFERENCES IN PERFECTION REPUDIATED

As is evident from the ideas of Nicole Oresme, the appearance of the Copernican heliocentric system was not an essential prerequi-

---

[26] Ibid. Amicus cites four authors who held this opinion, including Albumasar in the latter's *Magnum Introductorium,* tract 3, diff. 3 [see note 21]. The others are "Ricc." (perhaps Richard of Middleton), "Saxony" (probably Albert of Saxony), and Christopher Clavius, the famous Jesuit astronomer.

[27] Amicus disagreed with those who would distinguish the planets from their orbs and argued that "the superior orbs are always nobler than the inferior contained [orbs], although the stars [i.e., planets] that are in the superior orbs are not nobler, as is obvious from the sun. . . ." (ibid., p. 265, col. 1). Amicus held that whether or not the planets and their orbs were of the same species, each planet and the orb that carried it were at the same level of nobility or perfection.

site for the abandonment of the theory of the ascending order of celestial perfection, or for the subversion of its credibility. Nevertheless, the rejection of intracelestial perfection probably derived not from the Middle Ages, but from the impact of the Copernican theory and Galileo's telescopic observations on the status of the earth.

From the very nature of its fundamental assumptions, the Copernican theory was bound to have an impact on traditional ideas about celestial nobility and perfection. It made the earth a planet and therefore as perfect or imperfect as the other planets.[28] Consequently, Copernicans had no good reason to perpetuate the traditional medieval belief that the earth was the least perfect body of the universe[29] and in no way comparable to the perfection of the celestial bodies. But it was Galileo's telescopic discoveries that furnished vital empirical support for what had previously been only an implication from the structure of the Copernican theory itself. Observations of celestial bodies led Galileo to conclude that the earth was no less perfect than any of the traditional planets. By insisting that all celestial bodies were as alterable and subject to change as the earth, Galileo had destroyed any basis for differential perfections and nobilities. Indeed Galileo considered "the earth very noble and admirable precisely because of the diverse alterations, changes, generations, etc. that occur in it incessantly. . . . and I say the same of the moon, of Jupiter, and of all other world globes."[30] Because his name rarely occurs in discussions of celestial perfection, the effect of Galileo's radical ideas on Scholastic authors, who retained the geocentric system, is uncertain.

[28] In *De revolutionibus orbium caelestium,* bk. 1, ch. 10, Copernicus declared that "I have no shame in asserting that this whole region engirdled by the moon, and the center of the earth, traverse this grand circle amid the rest of the planets in an annual revolution around the sun." Rosen, trans., *Nicholas Copernicus On the Revolutions* [see note 3], p. 20.

[29] Distinguishing five degrees of perfection, Nicole Oresme declared that "the first and fifth are God and the earth respectively, and both are absolutely motionless, the one because of His great and infinite perfection and the other because of its very small degree of perfection." See *Nicole Oresme: Le Livre du ciel et du monde* [note 19], bk. 2, chap. 22, p. 507.

[30] Galileo Galilei, *Dialogue Concerning the Two Chief World Systems,* trans. Stillman Drake (Berkeley and Los Angeles, 1962), pp. 58–59. In *The Assayer* of 1623, Galileo only hints at a repudiation of celestial perfection, but does not actually discuss it. See, however, his remarks on the noblest shape of the sky on p. 279 of *The Controversy on the Comets of 1618,* trans. Stillman Drake and C. D. O'Malley (Philadelphia: University of Pennsylvania Press, 1960).

But even before Galileo's *Dialogue* of 1632, but after his telescopic discoveries and his statement in *The Assayer,* seventeenth-century Scholastics continued the medieval attacks against the idea that nobility or perfection increases with distance from the earth. Thus Raphael Aversa (1589–1657)[31] adopted an opinion that embraced both aspects of Oresme's attack described above. Like many others, Aversa assumed that the sun was the most noble and excellent of the planets. But unlike most of them, he refused to believe that the nobility of the other planets increased with their distance from the earth. The traditional astrological attributes assigned to the planets made ascending nobility implausible. Saturn, which is the highest planet, is not nobler than Jupiter, nor is Mars, which is higher than Venus, nobler than the latter. Why should we think them nobler when both Saturn and Mars are considered harmful planets, whereas Jupiter and Venus are assumed beneficial? Indeed we could arrange the order of nobility among the planets on the basis of a variety of operations and properties, all of which would be arbitrary. With the exception of the sun, Aversa concluded that "there does not appear any definite rule for [determining] the [nobility of] the other planets."[32]

Like Aversa, Johannes Poncius, O.F.M. (ca. 1599–1661), a Scotistic commentator on the collected works of John Duns Scotus published in 1639, declared that no specific differences could be assigned between the various heavens, even though he considered it very probable that the sun was the most noble of all the planets.[33] In a similar vein, Illuminatus Oddus (d. 1683), another Scotistic commentator, asserted, with little discussion, that although the heavens differ in number – that is, they are distinct orbs and planets – they did not differ in essential perfection.[34] Thus did Poncius, Oddus, and Raphael Aversa challenge the idea of intracelestial differential perfections. They were the Scholastic counterpart to the cosmic leveling process that was inherent in the Copernican system.

---

[31] Raphael Aversa, *Philosophia metaphysicam physicamque complectens quaestionibus contexta,* 2 vols. (Rome: apud Iacobum Mascardum, 1625, 1627), 2, 117, cols. 1–2.

[32] "Aliorum vero astrorum non apparet certa regula." Ibid., 117, col. 1.

[33] John Duns Scotus, *Opera Omnia, 6,* part 2: *Quaestiones in Lib. II Sententiarum,* ed. Luke Wadding (Lyon, 1639; reprinted in facsimile, Hildesheim: Georg Olms Verlag, 1968), 731, col. 1. Johannes Poncius was a supplementary commentator on this volume of Scotus's works. In the comment to which I refer here, his name is given as the commentator on p. 730, col. 1.

[34] Illuminatus Oddus, *Disputationes . . . ad mentem Scoti* [see note 8], p. 36, col. 1.

## IS THE CELESTIAL REGION MORE PERFECT THAN THE TERRESTIAL?

Indeed they did more than that, for they also deviated from, though apparently did not abandon, the traditional Scholastic belief that the celestial region was of greater perfection than the terrestrial, a belief that was mentioned at the outset as the other major aspect of the concept of celestial perfection and to which we must now turn.

Unlike intracelestial perfection, almost all Scholastics who treated the celestial region devoted a question to the relations of the celestial to the terrestrial. Lengthy analyses were commonplace. Most retained the traditional belief that the heavenly region was superior to the terrestrial. But, as we shall see, the conception of that superiority was altered in an important way. Although Copernicans no longer had reason to preserve this cosmic distinction (the transformation of the earth into an orbiting planet not only rendered dubious the assumptions about differences in perfection between the planets, but it also made rather meaningless the traditional distinction between the celestial and terrestrial regions), Scholastic Aristotelians had not yet been compelled to abandon the celestial–terrestrial dichotomy with its assumption of greater celestial perfection. Yet they were not unmindful of the greatly elevated status accorded the earth in the Copernican system. Before long, they too would endow the earth with greater respectability. The changed perception of the earth resulted from two sources: a medieval idea that involved a comparison between living things and celestial bodies, *and* the challenge of the Copernican theory.

## ARE CELESTIAL BODIES MORE PERFECT THAN LIVING THINGS?

For Aristotle, not only was the celestial region more perfect than the terrestrial region, but it was also alive[35] – indeed, it was divine[36]

---

[35] For Aristotle's belief that the celestial bodies are alive, see *De caelo* 2.2.285ª29–30 and 2.12.292ª.20. The triumph of an inanimate celestial region during the Middle Ages to the end of the fourteenth century is described by Richard C. Dales in "The De-Animation of the Heavens in the Middle Ages," *Journal of the History of Ideas*, 41, no. 4 (1980), 531–50. Whether the heavens are animate or inanimate was widely discussed.

[36] *De caelo* 1.3.270ᵇ1–14; 1.9.278ᵇ15–17; *Metaphysics* 1074ᵇ1–14; *On the Parts of Animals* 1.5.644ᵇ23–26 [see also note 37].

– from which he inferred that it was more perfect than all living things, including man.[37] Despite the general acceptance of Aristotle's cosmology by natural philosophers in the Middle Ages, they would eventually challenge all of these claims. Indeed, the divinity of the heavens never formed a part of medieval cosmology because it was incompatible with the Christian faith.

Having denied divinity to the heavens, Scholastics soon inquired whether it was even alive. Although Thomas Aquinas doubted their animation, he did not consider the problem relevant to the faith nor even an important question.[38] During the Middle Ages most natural philosophers denied celestial animation, and by the sixteenth and seventeenth centuries virtually all Scholastic authors repudiated the idea that celestial bodies were alive.[39] Thus when Scholastics compared the heavens with the terrestrial region, they had in mind a comparison between an inanimate heaven and an inanimate sublunar region. In such a comparison, few, if any, would have denied the greater perfection of the celestial over the terrestrial.[40]

At some point in the history of the problem of celestial perfection, Scholastics inquired whethar inanimate celestial bodies were superior to living things on earth. None denied that man, with his rational, spiritual, and intellective soul is more perfect than any inanimate celestial body.[41] Hence the comparison usually focused on things that had a sensitive and/or a vegetative soul. Indeed the

---

[37]"For there are other things much more divine in their nature even than man, e.g., most conspicuously, the bodies of which the heavens are framed." *Nicomachean Ethics* 6.7.1141^b1–2. Trans. W. D. Ross in *The Works of Aristotle* [see note 5].

[38] See Thomas Litt, *Les corps célestes dans l'univers de Saint Thomas d'Aquin* (Louvain: Publications Universitaires; Paris: Béatrice Nauwelaerts, 1963), pp. 108–109; see also Dales, "The De-Animation of the Heavens" [note 35], 543–544.

[39] After denying that celestial bodies are alive, Aversa declared that "this is the common opinion of the theologians and the philosophers" (*Philosophia* [see note 31], 2, 110, col. 1; see also 196, col. 1). Among others who denied life to the heavens, we may mention Conimbricenses, *De caelo*, bk. 2, chap. 1, quest. 2, art. 2 ["it is concluded that the celestial orbs are not animated"], pp. 166–9 [mistakenly paginated 145], where a long list of those who rejected animation is given and where we are also told that almost all Scholastic theologians rejected animation; Franciscus Bonae Spei, *Commentarii tres in universam Aristotelis philosophiam: Commentarius tertius In libros De coelo; De generatione et corruptione; De anima; et Metaphysicam Aristotelis* [Brussels: Apud Franciscum Vivienum, 1652], pp. 11, col. 1 –12, col. 2; and Petrus Hurtado de Mendoza, *Universa philosophia* (Lyon, 1624), p. 366, col. 2.

[40] "Comparando vero coelum cum sublunaribus certissimum est apud omnes corpora coelestia esse omnibus inanimatis perfectiora." Illuminatus Oddus, *Disputationes . . . ad mentem Scoti* [see note 8], p. 36, col. 1.

[41] Ibid.

lowliest of living things – plants, flies, wasps, etc. – were frequently compared to the heavens.

The two major opinions seem to have emanated from Thomas Aquinas (ca. 1225–1274) and John Duns Scotus (ca. 1265–1308). Although Thomas does not appear to have explicitly considered this problem, he seems to have believed that the celestial region was more perfect than animals created by putrefaction.[42] The latter were generated solely by celestial power, especially that of the sun, without the aid of semen. Creatures generated by putrefaction, for example, flies and gnats, were generated by the action of a celestial virtue on the four elements. Since Thomas believed that celestial bodies contained the properties of the elements in a more perfect and excellent manner, it is likely that he considered the heavens more perfect than animals generated by putrefaction.[43] Because Thomas believed that animals generated by semen also required the concurrent action of celestial virtue, it is possible that he also considered the heavens more perfect than all animals except man. Because the heavens played such a powerful causative role in the generation of living things, it is likely that Thomas would have considered the heavens, as a causative agent, more perfect than the terrestrial effects it produces. Thomas's overall interpretation of the influence of the celestial region on the terrestrial agrees rather well with Aristotle's opinions. It was a view that still found occasional supporters in the sixteenth and seventeenth centuries.[44]

John Duns Scotus formulated a radically different approach. Within the context of a discussion on whether any accident in the Eucharist could remain without a subject in which to inhere,[45] Scotus established certain criteria for comparing the absolute perfections of different properties. Thus he compared quantity and quality with respect to substance. On the assumption that substance is the most perfect of beings, and on the further assumption

---

[42] Aquinas's various statements on putrefaction have been collected and analyzed by Litt, *Les corps célestes* [see note 38], pp. 130–143.

[43] Ibid., pp. 135–6.

[44] Aversa (*Philosophia* [see note 31], *2*, 113, col. 1) states that Piccolomini, in his *Liber de caelo*, chap. 23, "affirms that, with the exception of man, the heaven is nobler than animated bodies." Mastrius and Bellutus (*Philosophiae ad mentem Scoti cursus integer* [see note 12], *3: De caelo*, 499, col. 1, par. 109) report that most Thomists believed that the heavens are nobler than sensitive and vegetative life.

[45] What follows is drawn from John Duns Scotus, *Opera Omnia* [see note 33], *8: In Lib. IV Sententiarum*, Distinction 12, quest. 2, pp. 731–2.

that, in absolute terms, quantity is more akin to substance than is quality, Scotus concluded that "quantity is more perfect than quality."

Moreover, he believed it possible to compare two different objects with respect to two different properties if both properties were also possessed by God. Scotus assumes that as a property is more necessary, it is closer to God. Now incorruptibility is necessary and it is close to God, who is the most perfectly incorruptible thing. From this, Scotus inferred that the incorruptible, physical heaven is closer to God than any corruptible thing. But not only is God an absolutely incorruptible entity, He is also an absolutely perfect, intellectual creature. The nearest intellectual creature to God is an angel, after which follows the finite human intellectual nature, followed by all sensitive natures (animals), which approximate more closely to intellectual natures than do nonsensitive things (plants). In the order of things based on intellect, a fly is nearer to God than is the heaven. But in the previous order, based on incorruptibility and corruptibility, the heaven is closer in perfection to God than is the fly.

Using the ideas of Scotus, Illuminatus Oddus insisted that something is said to be more perfect than another thing with respect to some particular quality because it is nearer, and more similar to, the perfect being, God. But when two things are compared to God with respect to two different qualities possessed by God, that is said to be absolutely more perfect which has the greater perfection in the nobler of the two attributes. Thus if we wish to compare the perfection of celestial bodies with a living thing that has a sensitive and vegetative soul, we cannot compare them with respect to degree of life since the celestial bodies are inanimate. However, we might wish to see whether the incorruptibility of a celestial body is absolutely more perfect than the level of life in, say, a fly or wasp. Of these two qualities or attributes, *life* and *incorruptibility,* Oddus argues that "life in God has a more perfect nature than eternity and incorruptibility, for life is of the quiddity of God, [whereas] eternity is [but] an attribute. Therefore animated things, which are nearer to God in degree of life than the heaven, which is nearer [to God] in incorruptibility, will be absolutely more perfect [than the heaven]."[46] For Illuminatus Oddus,

---

[46] Oddus, *Disputationes . . . ad mentem Scoti* [see note 8], p. 36, col. 2. Mastrius and Bellutus (*Philosophiae ad mentem Scoti cursus integer* [see note 12] *3: De coelo,* 500, par. 119) say much the same thing about the comparison between the qualities of life and incorrupti-

then, even a plant, which possess only a vegatative soul, is more perfect than any celestial body.

If we may believe Raphael Aversa, who was not a Scotist, the Scotistic opinions just described formed the common opinion around the time he published his work in 1629.[47] According to Aversa, those who accepted this opinion believed that the incorruptibility of the celestial region

made the heaven more perfect and noble than all inferior bodies, even [those that are] animated. And by reason of its form, [the heaven] is simply and absolutely more perfect than all other inanimate bodies.

But simply and absolutely, the form of the heaven is more ignoble than a soul *[anima]* and the heaven itself is more imperfect than an animated body.[48]

Aversa further argued that incorruptibility favored living things and not the heavens since "our soul is immortal more than the form and substance of the heaven."[49] In his judgment, the heaven did not operate by prudence and wisdom, as a living thing might, but simply as a natural cause.[50] To operate in this manner did not require that the heaven be "alive and essentially more excellent than all inferior substances." Aversa thus accepted the essentials of the Scotistic position.

For some, including Johannes Poncius, it was difficult to explain why "the heavens are more imperfect than any living thing."[51] Poncius found it implausible to believe that plants, which possessed only a vegetative soul, should be more perfect than the heavens. He therefore rejected this claim.[52] But he conceded that

bility. They also concluded that "it must be conceded as absolutely true that a mouse, a flea, and a plant are absolutely nobler than the heaven" (ibid., 501, par. 123).

[47] Aversa, *Philosophia* [see note 31] 2, 112, col. 2.

[48] Among those who allegedly held this opinion, Aversa cites Thomas Aquinas, Ockham, Averroes, and the Coimbra Jesuits. As we saw earlier, it is not likely that Thomas held this opinion but believed rather that the heaven was nobler than sensitive and vegetative life [see also note 44, where Mastrius and Bellutus say that most Thomists also held this opinion].

[49] "Et etiam quoad incorruptibilitatem, anima nostra est immortalis plusquam forma et substantia caeli." Aversa, *Philosopha* [see note 31], 2, 113, col. 1.

[50] "Caelum deinde dicitur regere et gubernare haec inferiora esseque causa superior et universalis, non quasi per sapientiam et prudentiam res disponens, sed influendo et operando tanquam causa naturalis." Ibid., 114, col. 2.

[51] Johannes Poncius O.F.M., *Philosophiae ad mentem Scoti cursus integer* (Lyon, 1672), p. 631, col. 2.

[52] Aversa (*Philosophia* [see note 31], 2, 112, col. 2) also expressed doubts about plants being more perfect than celestial bodies.

sensitive things, which included all animals, were more perfect than celestial bodies because the former could imagine pleasure whereas the latter could not. Poncius also drew the consequence that was implicit, and perhaps even explicit, in Scotus's approach. From the widely accepted principle, drawn perhaps from Duns Scotus, that only a more perfect thing can wholly or partially generate a less perfect thing,[53] he concluded that living things that are more perfect than celestial bodies could not be generated by celestial bodies.

### THE BIG DEPARTURE: THE EARTH IS MORE PERFECT THAN THE SUN

What we have described thus far is a break with an important aspect of the Aristotelian worldview. Not only is man more perfect than the heavens, but so also are animals and, for some, even plants. The celestial region was no longer the most excellent physical entity in the world. But it was still incomparably more perfect

---

[53] "Anima quaecumque est perfectior formis coelestibus, ergo non possunt a corporibus coelestibus produci quia imperfectius non potest producere immediate perfectius." Poncius, *Philosophiae ad mentem Scoti cursus integer* [see note 51], p. 631, col. 2, par. 109. In his *Ordinatio*, bk. 1, dist. 7, question 1, n. 47, in *Ioannis Duns Scoti Opera omnia, 4* (Vatican, 1956), p. 127, Scotus declares that it is necessary that the thing or principle that generates something be more perfect than the form or thing that it produces. Ockham, but apparently not Scotus (at least not in the passage cited in this note), applied this to celestial bodies and their alleged role in generating living things by means of putrefaction. "But as for things generated by putrefaction, it is doubtful whether a celestial body could be a principle of such generated things. John [Duns Scotus] says that this cannot happen because a productive principle ought to be nobler than what it produces; but a non-living thing is not nobler than a living thing, therefore, etc" *(Guilelmi de Ockham Opera philosophica et theologica, 5: Quaestiones in librum secundum Sententiarum (Reportatio),* ed. Gedeon Gal, O.F.M., and Rega Wood (St. Bonaventure, N. Y.: St. Bonaventure University, 1981), 421. Thus from Scotus's principle that a less perfect thing cannot produce a more perfect thing, Poncius drew the same conclusion as did Ockham. But Ockham did not think it absurd that the Heaven could possibly be nobler than an animal. After all, the distance that might separate the perfection of an animal by comparison with the perfection of a celestial body is finite. Therefore, the celestial body could, in principle, have its perfection increased to the point where it equals or exceeds that of an animal. But Ockham also balked at Scotus's conclusion because he was convinced that God acts concurrently with the heavens to produce a terrestrial effect. "And so," Ockham argues (ibid., 423), "although a celestial body could be the cause of things generated by putrefaction, it is nevertheless only a partial cause. But a partial cause can be more ignoble than its effect, although the total cause is always nobler." Thus even if the heaven were less noble or perfect than terrestrial animals produced by putrefaction, the heaven could, nevertheless, be the cause, albeit only a partial cause, of things generated by putrefaction.

than the inanimate earth. It remained for someone to repudiate this traditional belief by uniting into one total being the living things on the earth with the earth itself. Giovanni Baptista Riccioli (1598–1671) took this momentous step in his *New Almagest (Almagestum novum)* of 1651.

In the first of twenty-nine arguments for and against the cosmic centrality of the sun,[54] Riccioli concludes with the electrifying statement that "the earth, with its living, and especially rational, animals is nobler than the sun."[55] To justify his startling claim, Riccioli refers his readers to the first argument of the eighth chapter,[56] where at the outset he enunciates another startling proposition, namely that "the center of the universe is the most noble place in the world, for it is everywhere distant from the extremes and holds the middle position."[57] Riccioli adds, however, that the center of the world is the most noble place only in the natural order of things; in the supernatural order, "the center of the earth is the lowest and most wretched place."[58] But if in the natural order, the center of the world is the most noble place, which body shall occupy it: sun or earth? For Riccioli, of course, the earth occupies the center and does so because

the earth ought not to be taken simply as one pure element of four, or three elements; but [rather it ought to be taken] as one with living plants and animals, but especially with men for whom all the stars were made and [for whom they] are moved, as God attests in *Deuteronomy*. . . . It is

---

[54] Giovanni Baptista Riccioli, *Almagestum novum astronomiam veterem novamque complectens observationibus aliorum, et propriis novisque theorematibus, problematibus ac tabulis promotam; in tres tomos distributam quorum argumentum sequens pagina explicabit. Auctore P. Ioanne Baptista Ricciolo, Societatis Iesu Ferrariensi* (Bologna, 1651), pars posterior, bk. 9, sect. 4, chap. 33, p. 469, col. 1. Only the first of the three volumes was published, and it consisted of two parts each separately paginated. All citations in this article are from the second part (pars posterior). Riccioli titled this section: "29 arguments in favor of the sun's position in the center of the universe and [in favor] of the annual motion of the earth around the center of the universe simultaneously with the daily motion, and their solutions furnished from chapters 8 to 18 inclusively."

[55] "Tellus enim cum viventibus et animalibus praesertim rationalibus est nobilior sole." Ibid.

[56] Ibid., chap. 8, pp. 330, col. 2 – 331, col. 1.

[57] "Non videtur dubitandum quin centrum universi sit locus nobilissimus in mundo; quippe qui aeque undique distat ab extremis et medium obtinet situm." Ibid., p. 330, col. 2.

[58] "Dixi autem loquendo de ordine ac fine naturali, nam si de supernaturali centrum telluris est infimus ac miserrimus locus. . . ." Ibid., p. 331, col. 1. Although Riccioli speaks of the "center of the earth" and not the "center of the world," it is evident that he has equated the latter with the former.

the most excellent body of all the bodies of the world, if we judge by the magnitude of virtue and the dignity of the end [or goal], as is proper, rather than by the magnitude of the mass [of bodies].[59]

In these brief passages, Riccioli has incorporated extraordinary departures from the traditional Scholastic interpretation of Aristotelian cosmology. We notice first the new emphasis on the importance of the center of the universe. Because it coincided with the earth's center, the world center in medieval cosmology had never been accorded much importance except as the alleged point around which the celestial orbs moved with uniform circular motion.[60] With Copernicus, a dramatic change occurred. The center of the universe became the most important part of the cosmos because it was usually assumed to coincide with the center of the sun.[61] As Copernicus explained:

At rest . . . in the middle of everything is the sun. For in this most beautiful temple, who would place this lamp in another or better position than that from which it can light up the whole thing at the same time? For, the sun is not inappropriately called by some people the lantern of the universe, its mind by others, and its ruler by still others. [Hermes] the Thrice Greatest labels it a visible god, and Sophocles' Electra, the all-seeing. Thus indeed, as though seated on a royal throne, the sun governs the family of planets revolving around it.[62]

The lofty praise for the sun as the most noble planet by some medieval and early modern Scholastics was, as we have seen, partially based on its position as the middle, or fourth, planet among the seven. That praise was not unlike that which Copernicus heaped upon the sun as the occupant of the center of a spherical universe,

---

59 "Tellus enim non debet sumi nude pro mero elemento uno ex quatuor, tribusve elementis, sed una cum plantis viventibus et animalibus, se praecipue cum hominibus, in quorum gratiam facta sunt et moventur sidera omnia, Deo id in *Deuteronomio* attestante. . . . Praestantissimum est corpus omnium mundi corporum, si magnitudinem virtutis ac dignitatem finis, ut par est, aestimemus potius quam magnitudinem molis." Ibid. If the comparison between the earth and sun were made on the basis of mass, the sun, whose mass was assumed greater, would have been more excellent than the earth.

60 During the Middle Ages, the role of the geometric center of the world was further compromised by the widespread acceptance of eccentric and epicyclic spheres that moved around other centers. See my article, "Cosmology," in David C. Lindberg, ed., *Science in the Middle Ages* (Chicago: University of Chicago Press, 1978), pp. 280–4.

61 Strictly speaking, Copernicus did not identify the cosmic center with the sun's center, but said only that "near the sun is the center of the universe." Rosen, trans. *Nicholas Copernicus On the Revolutions* [see note 3], bk. 1, chap. 10, p. 20.

62 Ibid., p. 22

around which the other planets moved. Until Riccioli's *New Almagest,* Copernicus's glorification of the world's geometric center seems to have made little or no impact on Scholastic cosmology. With the earth at the center of their world system, they could find little reason to extol the virtues of a center that was judged immeasurably more inferior to anything in the celestial region.

With Riccioli, all this changed. Among the arguments for and against the Copernican theory that involved a comparison of the conditions of the earth as compared to other planets, but which did not involve motion,[63] Riccioli chose to battle the Copernicans on their own ground. To do this, he apparently decided to demonstrate that the earth was indeed superior to the sun. In order to make the comparison plausible, Riccioli accepted the center of the world as the noblest place in the natural order. He may have been genuinely convinced that Copernicus had rightly judged the center of the universe as the most perfect place. If so, it was the best place, only because the center would enable an appropriate body to receive things from other bodies and to communicate things to them in the most effective manner.[64] With the center established as the noblest place in the universe, which of the two contending bodies – earth or sun – was more fit to occupy it? We have seen that at least from the time of Duns Scotus, many Scholastics had argued that living things on the earth were nobler and more per-

[63] Riccioli distinguished four general headings (*Almagestum novum* [see note 54], p. 331, col. 2) under which one might organize the arguments for or against removing the earth from the center of the universe. What has been described above is the first followed by "[2] the motion of the planets themselves; [3] . . . other motions observed in the heavens; or, finally [4] . . . the motions or mutations observed in the sphere of the elements."

[64] The crucial parts of Riccioli's analysis depend on a syllogism that he formulates and parts of which he then refutes. The syllogism reads (*Almagestum novum* [see note 54], p. 331, col. 1): "[Major premise:] To the most excellent of worldly bodies, the most excellent place ought to be assigned. [Minor premise:] But the sun, not the earth, is the most excellent body of the world and the most excellent place of the universe is the center. [Conclusion:] Therefore the center of the universe ought to be assigned to the sun, not to the earth."

Although Riccioli would deny the first half of the minor premise and the conclusion, he concedes the major premise (and the second half of the minor premise) and argues that the most excellent body ought to occupy the most excellent place, which is the center of the world, not because of the "pure geometric excellence" of the center, but "from a physical end and good which such a body ought to receive from others or to communicate to others." Being at the center enables the most noble body to realize these goals and objectives.

fect than the inanimate celestial planets, stars and orbs. All were also agreed that the inanimate heavens were more perfect than the inanimate earth. But no one, it seems, conceived of the earth as a unity consisting of the physical globe *and* the living things in, on, and around it. Riccioli took this momentous step. In so doing, he elevated the earth from an inanimate entity inferior to the celestial region to the most noble body in the physical world. For although the comparison was between earth and sun, Riccioli unambiguously declared the earth to be "the most excellent body of all the bodies of the world." Vain though it was, Riccioli's effort to make the earth worthy of location in the center of the universe by elevating its status over that of the sun and the other celestial bodies is an important event in the history of Scholastic Aristotelian cosmology. In his struggle with the Copernicans, Riccioli departed in two significant ways from traditional medieval cosmology. He made the center of the world the most noble place in the cosmos; and he made the earth more noble and perfect than any celestial body. The first change was apparently the result of Copernicus's emphasis on the center of the world as the only place worthy of serving as the sun's location, the only place from which the sun could properly exercise its dominant role in the universe. The center best served the operations and objectives of the world. Riccioli apparently found these arguments compelling. The second departure, the greater nobility of the earth over the heavens, had, as we saw, a longer history. Ultimately, it depended on the widely held Scholastic belief that the heavens were inanimate, though more perfect than the inanimate earth, and that living things were more perfect than the lifeless heavens. By making living things an integral part of the earth, Riccioli, or someone before him, if he was not the first, made the earth ipso facto more perfect than any celestial body. As the most noble and perfect body in the world, the earth, rather than the sun, was more fit to occupy the center, the noblest of places.

Although Riccioli formulated many arguments against the Copernican theory,[65] he incorporated into one of them two quite significant departures from traditional medieval cosmology. These

---

[65] There are approximately 200 arguments in the ninth book, which range over many categories.

may be added to a growing number of such departures,[66] which, taken together, indicate quite clearly that in the seventeenth century, the last in which medieval cosmology was still a plausible worldview, a number of Scholastic natural philosophers sought to preserve Aristotelian cosmology by abandoning significant bits of the old, while taking on significant bits of the new. Contrary to prevailing opinion, Scholastics were surprisingly flexible as they coped with the new cosmology. Unfortunately, no amount of patchwork could save geocentric cosmology from its inevitable demise. It is, however, worth remembering that to many living in the seventeenth century, that demise was not so evident as hindsight may now reveal. Although Scholastic efforts to adapt Aristotelian cosmology to the new astronomy and cosmology have thus far been grossly neglected, further study of those efforts may not only reveal additional surprises but are sure to provide a more balanced interpretation and understanding of Scholastic cosmological ideas from the Middle Ages to their demise in the late seventeenth century.

[66] For example, the terraqueous sphere (see my monograph, "In Defense of the Earth's Centrality and Immobility: Scholastic Reaction to Copernicanism in the Seventeenth Century," *Transactions of the American Philosophical Society*, 74, part 4 [1984], 22–32); the change from celestial incorruptibility to corruptibility [see references in note 1]; and the change from a solid to a fluid celestial medium (see my forthcoming article, "A New Look at Medieval Cosmology," *Proceedings of the American Philosophical Society*).

# 7

## Baptizing Epicurean atomism: Pierre Gassendi on the immortality of the soul

MARGARET J. OSLER

During the early decades of the seventeenth century, European intellectuals interested in the new science were actively concerned with formulating a philosophy of nature to replace the traditional Aristotelianism, which had suffered major setbacks in the wake of the Copernican revolution, the skeptical crisis following the Protestant Reformation, and the revival by the humanists of various alternative philosophies of nature.[1] The main contenders in this search for foundations for the new science were the mechanical philosophy of René Descartes and Pierre Gassendi, the animistic philosophies associated with the hermetic tradition, and the chemical philosophy espoused by the followers of Paracelsus. Although the mechanical philosophy ultimately emerged triumphant, the outcome was not certain during the 1630s and 1640s. Many of the points of contention dividing these philosophies of nature were theological. To a seventeenth-century Christian of reasonably orthodox belief, each of the views was theologically problematic. The mechanical philosophy was perceived by some as leading to materialism and atheism. The chemical and Hermetic view that

I am grateful to Lisa Sarasohn who read earlier drafts of this paper and suggested important ways to improve it. J. J. MacIntosh has been a constant source of critical suggestions, informed with wit and useful knowledge. William A. Wallace also read an earlier draft and made helpful comments. Allison Chapman, in an unlikely setting, helped me to organize the mass of material that went into the writing of this paper.
[1] E. A. Burtt, *The Metaphysical Foundations of Modern Science* (first published, 1924, New York: Doubleday and Company, 1954); Richard H. Popkin, *The History of Scepticism from Erasmus to Spinoza* (Berkeley: University of California Press, 1979); Frances Yates, *Giordano Bruno and the Hermetic Tradition* (New York: Random House, 1964); Allen G. Debus, *The Chemical Philosophy*, 2 vols. (New York: Science History Publications, 1977).

matter possesses its own internal sources of activity also seemed to threaten God's role in nature.[2] Theology was a central concern to seventeenth-century thinkers, and any proposal to provide new metaphysical foundations for science had to be shown to be theologically acceptable. It is in this context that we must understand Gassendi's profound concern with the immortality of the human soul.

The entire corpus of Gassendi's writing, particularly his major work, the *Syntagma Philosophicum,* can best be understood as an attempt to restore and Christianize Epicurean atomism.[3] The Epicurean philosophy was traditionally considered to be materialistic and atheistic. Gassendi sought to expunge the atheistic components of Epicureanism and to retain the atomism as a philosophy of nature underpinning the new science. One of the Epicurean

[2] Allen G. Debus, *Science and Education in the Seventeenth Century: The Webster–Ward Debate* (London: MacDonald, 1970); Thomas Harmon Jobe, "The Devil in Restoration Science: The Glanvill–Webster Witchcraft Debate," *Isis, 72* (1981), 343–56; Charles Webster, *From Paracelsus to Newton: Magic and the Making of Modern Science* (Cambridge: Cambridge University Press, 1982).

[3] The most recent, thorough discussion and analysis of Gassendi's views is René Olivier Bloch, *La philosophie de Gassendi: nominalisme, matérialisme, et métaphysique* (La Haye: Martinus Nijhoff, 1971). Bloch claims that Gassendi was a materialist who disguised his genuine views in a mask of theological language. He perceives "a rupture between the man and his sentiments, on the one hand, and the philosopher and his thought, on the other. Gassendi constantly searched for a more or less precarious accord between a scientifically inspired materialism and the demands that drew him to his faith. The more or less systematic character of this research confirms the sincerity of these demands, but also what were the avatars in the time of this research and what were the philosophical limits of these systematizations" (p. 299). Bloch seeks to show that Gassendi was less than honest in his frequent discussions of theological issues. "Gassendi's lack of candor brings to mind the methods of Bayle or the tactics of the Encyclopedists . . ." (p. 288). Bloch sees Gassendi's problem as the modern one of a *conflict* between science and religion: "How is Gassendi going to make his 'Epicurean philosophy' coexist with the affirmations which presuppose religion, those of monotheism, creationism, and Providence, those of the finitude in space and time of a single created world, and of the elements which constitute it, that of the finality of the cosmic, biological, and human universe, that of the existence of immaterial and immortal souls?" (p. 301). My translation.

I maintain, on the contrary, that an interpretation of Gassendi truer to his real aims and to the intellectual context of the early seventeenth century is one which sees his theological concerns as genuine and having a formative influence on his philosophical and scientific thought: Gassendi wanted to revive Epicurean atomism; but to do so, he found it necessary to modify the ancient philosophy in ways that rendered it compatible with Christian orthodoxy. This approach to Gassendi's thought requires fewer mental gymnastics in interpreting his writings. It is also shared by the major, twentieth-century Gassendi scholar. See Bernard Rochot, *Les travaux de Gassendi sur Epicure et sur l'atomisme* (Paris: J. Vrin, 1944).

doctrines most offensive to orthodox Christians was his claim that the human soul is corporeal and mortal. In order for Gassendi to propose Epicurean atomism as a viable philosophy of nature, he had to modify the Epicurean philosophy to incorporate the Christian position on the immortality of the soul.

The aim of this chapter is to elucidate Gassendi's views on the immortality of the soul and to attempt, as far as possible, to trace his views to their sources in the writings of Epicurus, Lucretius, Aristotle, and Thomas Aquinas. I claim, as a working hypothesis, that Gassendi took as much as he could from the atomists without violating the tenets of Christian belief, and supplemented their views with ideas drawn from Aristotle and the Scholastics in order to preserve the Christian doctrine of the immortality of the soul.[4] The result is eclectic and not always consistent, but it gives us insight into the theological and ideological struggles the mechanical philosophy faced at its inception.

Pierre Gassendi (1592–1655) is most frequently remembered for reintroducing the philosophy of Epicurus into the mainstream of European thought.[5] Gassendi's version of Epicurean atomism and his adaptation of Epicurean hedonism exerted a major influence on seventeenth-century developments in science and political philosophy.[6] Before European intellectuals could embrace the philosophy of Epicurus, however, his views had to be purged of the accusations of atheism that had followed them since antiquity.[7] Gassendi, a Catholic priest, assumed the task of baptizing Epicurus by identifying the objectionable elements in his philosophy of nature and modifying them accordingly. For example, he denied

---

[4] The Aristotelian philosophy of the late Middle Ages and Renaissance was hardly a monolithic view. For an account of its diversity and eclecticism, see Charles B. Schmitt, *Aristotle and the Renaissance* (Cambridge, Mass.: Harvard University Press, 1983).

[5] Bloch, *La philosophie de Gassendi* [see note 3]. See also Richard H. Popkin, "Gassendi, Pierre," in Paul Edwards, ed., *The Encyclopedia of Philosophy*, 8 vols. (New York: Macmillan and Free Press, 1967), III, 269–73.

[6] L. T. Sarasohn, "The Influence of Epicurean Philosophy on Seventeenth-Century Ethical and Political Thought: The Moral Philosophy of Pierre Gassendi," (unpublished dissertation, UCLA, 1979); L. T. Sarasohn, "The Ethical and Political Philosophy of Pierre Gassendi," *Journal of the History of Ideas, 44* (1983), 549–60; Howard Jones, *Pierre Gassendi (1592–1655): An Intellectual Biography* (Nieukoop: B. de Graaf, 1981); Richard W. F. Kroll, "The Question of Locke's Relation to Gassendi," *Journal of the History of Ideas, 45* (1984), 339–60.

[7] For a thorough discussion of Epicurus on the gods and religion, see J. M. Rist, *Epicurus: An Introduction* (Cambridge: Cambridge University Press, 1972), pp. 140–63, 172–5.

the Epicurean doctrines of the eternity of the world, the infinitude of atoms, and the existence of the *clinamen,* or swerve, which Epicurus had introduced in order to account for the impact among atoms in an infinite universe;[8] he also took great pains to emphasize God's providential relationship with the creation.[9]

One of the most troublesome aspects of Epicureanism was his denial of the immortality of the soul. Epicurus had considered the soul, like everything else in the cosmos, to be composed of atoms and the void.[10] Epicurus's Roman disciple, Lucretius, had argued for the corporeality of the soul on two grounds: The soul has the power to move the body, which is material; and "our mind suffers along with the body, and is distressed by the blow of bodily weapons."[11] For Lucretius, these facts proved that the nature of the mind is corporeal and that the soul is mortal. ". . . when the body has perished, you must needs confess that the soul too has passed away, rent asunder in the whole body."[12]

The Epicurean theory of the soul is not compatible with orthodox Christian belief, which includes as articles of faith the survival of the soul after death, divine punishment and reward in the afterlife, the resurrection of Christ, and the possibility of human resurrection at the time of the second coming.[13] Gassendi understood that the nature of the soul was a serious problem for a Christian

---

[8] Pierre Gassendi, *Syntagma Philosophicum,* part II, sec. III, books III, IV, V. In *Opera Omnia,* 6 vols. (Lyons, 1658; reprinted Stuttgart-Bad Constatt: Friedrich Frommann Verlag, 1964), II.

[9] Osler, "Providence and Divine Will" [see note 6]. For background on the question of divine providence and the laws of nature, see Francis Oakley, *Omnipotence, Covenant, and Order* (Ithaca: Cornell University Press, 1984).

[10] Epicurus, "Letter to Herodotus," *The Stoic and Epicurean Philosophers,* ed. Whitney J. Oates (New York: Modern Library, 1940), p. 10. See also Rist, *Epicurus* [note 7], chap. 5.

[11] Lucretius, *De Rerum Natura,* trans. Cyril Bailey, 3 vols. (Oxford: Oxford University Press, 1947), *I*, 311.

[12] Ibid., 343.

[13] Some Christian theologians, most notably Tertullian, had argued that the soul has a corporeal nature. See Frederick Copelston, *A History of Philosophy,* 9 vols. (Garden City, New York: Doubleday, 1962), 2, part I, 37–9; see also the article entitled "Soul" in *The Catholic Encyclopedia* (New York: Encyclopedia Press, 1912). The heterodoxical nature of Epicurean philosophy is underscored by the fact that Dante placed Epicurus in the Sixth Circle of Hell: "In this part Epicurus with all his followers, who make the soul die with the body, have their burial place." (*Inferno, X,* 14) Dante Alighieri, *The Divine Comedy,* trans. Charles Singleton, 3 vols. (Princeton, N.J.: Princeton University Press, 1970), *I,* 99. Interestingly, Dante placed "Democritus, who puts the world on chance," only in limbo along with the virtuous pagans (*Inferno, IV,* 136).

rendition of atomism, and he devoted many pages of his monumental *Syntagma Philosophicum* to a consideration of this issue.[14] Significantly, the final book of the "Physics" (Part II of the tripartite *Syntagma Philosophicum*) is entitled "On the Immortality of Souls" and deals comprehensively with this issue.

Gassendi discussed the soul systematically. He took as his starting point the nature of the soul in general. Then he considered the souls of animals, their nature and function, and the human soul, and he concluded by attempting to demonstrate the immortality of the human soul. His arguments and analysis are eclectic, drawing on the views of Epicurus, Aristotle, and Thomas Aquinas, among others. He culled what he could use from each of these thinkers. His concern with refuting the heterodoxical portion of the Epicurean position is clear: He devoted an entire chapter to a point-by-point refutation of Epicurus's objections to the immortality of the soul.[15]

The bulk of Gassendi's discussion of the soul occurs in the section of the *Syntagma Philosophicum* entitled "On Living Earthly Things, or On Animals." Most of this section of 440 double-column pages of dense Latin prose deals with topics in physiology and the theory of sensation. After a general discussion "On the Variety of Animals" and "On the Parts of Animals," Gassendi addressed the question of the nature of the soul. The soul is what distinguishes living things from inanimate things.[16] At the outset, Gassendi adopted the distinction between *anima* and *animus,* which he drew directly from Lucretius.[17] "The *anima* is that by which we are nourished and by which we feel; and the *animus* is that by which we reason."[18] Gassendi agreed with Lucretius that the *anima,* or sentient soul, is present throughout the body; he disagreed with Lucretius on the locus of the *animus,* or rational soul, placing it in the head rather than the chest.[19] Gassendi also disa-

---

[14] Pierre Gassendi, *Syntagma Philosophicum,* part II, sec. III, posterior part, books III, IX, XIV, XV. In *Opera Omnia, II.*

[15] Ibid., *II,* 633–50. The major commentators largely ignore this issue. Bloch, for example, devotes only three pages to Gassendi's extensive discussion. Bloch, *La philosophie de Gassendi* (note 3), 397–400.

[16] Gassendi, *Opera Omnia II,* 237.

[17] Lucretius claimed that the soul consists of two parts, which he called the *anima,* or irrational part responsible for vitality and sensation, and the *animus,* or rational part. *De Rerum Natura* [see note 11], 309.

[18] Gassendi, *Opera Omnia ii,* 237. My translation.  [19] Ibid., 445–6.

greed with the ancient atomists, who had maintained the corporeality of the soul. Epicurus had said that the soul is corporeal because there is nothing incorporeal except the void. Thus, if the soul were not corporeal, it could neither do anything nor suffer anything bodily. Gassendi's rebuttal of the Epicurean argument draws on the voluntarist theology that provides the conceptual background for his approach to natural philosophy.[20] He stated that "the actions of God are not necessary"; even if nothing incorporeal can be imagined except the void, it does not follow that God's creative act is restricted by the limitations of human imagination.[21]

The human soul consists of two parts, the rational soul or *animus,* and the irrational, sentient, vegetative soul, or *anima.*[22] Gassendi began his discussion of the soul with a long account of the souls of animals. Because humans differ from animals by uniquely possessing rational souls, the souls of animals correspond to the irrational parts of the human soul.[23] The soul of animals is "something which can be said to live in the body of the animal while it is alive, leaving it at death. Clearly, life is its presence in the body of the animal, and death is its absence."[24] Because of the fact that it is in the various parts of the body while the animal is alive, "the soul seems to be something very fine."[25] Denying that it is a form or merely a symmetrical disposition of matter, as various Peripatetics had maintained, Gassendi claimed that

---

[20] For a fuller account of Gassendi's voluntarism and its role in his natural philosophy, see Osler, "Providence and Divine Will" [see note 6].

[21] Gassendi, *Opera Omnia, ii,* 246. This is the sort of argument that Bloch regards as evidence of the fact that Gassendi allowed faith to dictate the content of philosophy, even when the conclusions of that philosophy taken on their own stood opposed to the conclusions known by faith: "it is clear that the immateriality of the rational soul is simply postulated in philosophy in the name of theological exigency." Bloch, *La philosophie de Gassendi* [see note 3], pp. 369, 374.

I do not agree. Just because arguments based on the nature of divine providence do not seem adequate in light of modern philosophical analysis, it does not follow that Gassendi did not take such arguments seriously nor that certain theological claims did not form basic assumptions within which his philosophical views were formed.

[22] Gassendi, *Opera Omnia, ii,* 256.

[23] Ibid., 250. At this point Gassendi incorporated the Aristotelian notion of the soul, consisting of vegetative, sensitive, and rational parts. Man alone possesses all three; animals possess both sensitive and vegetative souls; plants possess only the vegetative soul. W. D. Ross, *Aristotle* (New York: Meridian Books, 1959), pp. 128–9; Aristotle, *De Anima,* $414^a2-4$, $415^a3-6$, $435^a12$.

[24] Gassendi, *Opera Omnia, ii,* 250.     [25] Ibid.

the soul seems to be a very tenuous substance, just like the flower of matter *[florem materiae]* with a special disposition, condition, and symmetry holding among the crasser mass of the parts of the body. . . .[26]

. . . such a substance seems to be made of a most subtle texture, extremely mobile or active corpuscles, not unlike those of fire or heat; indeed, whether they are spherical, as the authors of atoms wish, or pyramidical as Plato thought, or some other figure, they seem from their own motion and penetration through bodies to create the heat which is in the animal. Since for this reason they are said to be excited or to feel hot and the heat of animals manifestly depends . . . on the motion or actions of the soul, it follows that cold and death are the cessation of such action. Finally, the soul seems to be like a little flame or a most attenuated kind of fire, which . . . thrives or remains kindled while the animal lives, since, if it no longer thrives or is extinguished, the animal dies.[27]

Besides the fact that vital heat is a sign of life, evidence for the claim that the soul of animals is like a little flame can be found in the fact that "just as a snuffed out candle is repeatedly rekindled . . . so a suffocated and strangled animal, not yet dead, having been led from the water or heavy smoke or released from the halter, can repeatedly be brought to inhale the air"[28] and can thus be revived. Like a flame, the soul is in constant motion, not only when the animal is awake, but also in sleep, a fact confirmed by the existence of dreams.[29] It is "the principle of vegetation, sensation, and every other vital action."[30]

The soul of animals and the animal part of the human soul presented no problem for Gassendi, the mechanical philosopher. Like every other version of the mechanical philosophy, he held as a

[26] Ibid. Note the striking similarity between Gassendi's account of the animal soul as "the flower of matter" and his description of the principle of motion in individual objects such as boys or atoms: "For when a boy runs to an apple offered to him, what is needed to account for the apple's attraction to the boy is not just a metaphorical motion, but also most of all there must be a physical, or natural, power inside the boy by which he is directed and impelled toward the apple. Hence it may apparently be said most plainly that since the principle of action and motion in each object is the most mobile and active of its parts, a sort of bloom of every material thing *[quasi flos totius materiae]* and which is the same thing that used to be called form, and may be thought of as a kind of most rarefied tissue of the most subtile and mobile atoms – it may therefore be said that the prime cause of motion in natural things is the atoms, for they provide motion for all things when they move themselves through their own agency and in accord with the power they received from their author in the beginning; and they are consequently the origin, and principle, and cause of all the motion that exists in nature." Ibid., *I*, 337. Craig B. Brush (ed. and trans.) *The Selected Works of Pierre Gassendi* (New York: Johnson Reprint, 1972), pp. 421–2.

[27] Gassendi, *Opera Omnia, II*, 250–1. [28] Ibid., 252. [29] Ibid. [30] Ibid.

fundamental tenet that all the phenomena of nature must be explained in terms of matter and motion. For Gassendi the facts of biology and perception presented no special problem, as they could be readily explained (or so he thought) in terms of the motion of atoms. His insistence on the special tenuousness, subtlety, and mobility of the atoms comprising the animal soul signals the difficulty that these phenomena present to any attempt at materialistic reduction. Gassendi, however, was content to state simply that they were so reducible. His claim that the rational soul, in contrast to the animal soul, is incorporeal establishes the boundaries of his mechanization of the world. In this respect, Gassendi's philosophy resembles that of Descartes and differs from that of Hobbes. Descartes's radical distinction between *res extensa* and *res cogitans* created a boundary that falls along the same lines as Gassendi's distinction between the animal and rational souls. Hobbes, however, sought to explain even the human soul and social processes in terms of matter and motion alone. In this respect, the difference between Hobbes, on the one hand, and Gassendi and Descartes, on the other, represents a difference in the domain to which mechanization applied.[31]

For Gassendi, the animal soul is transmitted from one generation to the next from the moment of conception. The animated portion of the semen kindles the soul in the embryo, in the same way that a burning torch can kindle the flame of a new torch.[32]

> . . . thus from the creation of animals at the beginning of the world . . . the soul was propagated from that time; and when the first were extinguished, there were always others being made, and so it remains even now in those that exist . . . .[33]

Thus, for the souls of animals, only one act of creation was necessary. Since the beginning, the souls of animals have been transmitted from one generation to the next by the biological process of reproduction.

The soul of animals, then, is material. It is the principle of vitality shared by all animals, including human beings. Correspond-

---

[31] For a fascinating comparison between Gassendi and Hobbes in this respect, see Lisa Sarasohn, "Motion and Morality: Pierre Gassendi, Thomas Hobbes, and the Mechanical World View," *Journal of the History of Ideas*, forthcoming July, 1985. I am grateful to the author for allowing me to see the manuscript of this paper.

[32] Gassendi, *Opera Omnia*, II, 252.    [33] Ibid., 253

ing to the vegetative and sensitive souls of Aristotle, it has the corporeality ascribed to the soul by Epicurus. But the soul of animals is only part of the human soul. There is another part of the human soul, which separates humans from other animals and which is the subject of theological as well as philosophical concern.[34]

After rehearsing the various theories about the nature of the human soul put forward by ancient and medieval philosophers and theologians, Gassendi declared that

the human soul is composed of two parts: . . . the irrational, embracing the vegetative and sensitive, is corporeal, originates from the parents, and is like a medium or fastening *[nexus]* joining reason to the body; and . . . reason, or the mind, which is incorporeal, was created by God, and is infused and unified as the true form of the body. . . .[35]

Gassendi argued for the double composition of the human soul on several grounds. In an interesting aside, he noted that if one were to hold the position that the soul develops along with the body, as the Epicureans maintained,[36] then one would not "commit homicide according to either civil or canon law [by procuring] an abortion in the first days after conception."[37] To the seventeenth-century Catholic priest, this view was patently absurd.

One of Gassendi's chief arguments for the fact that the soul consists of two parts, "one rational, the other sensitive," is that

it seems to be congruent with what the theologians distinguish in the two parts of the soul, one higher, the other lower; they prove this distinction especially from that place in the Apostle, "I see another law in my limbs repugnant to the law of my mind." Evidently, since one simple thing cannot be opposed to itself, it is argued from the battle that exists between sense and mind that sense and mind, or the rational and sensitive souls, are different things.[38]

If man did not have a rational soul, Gassendi argued, there would be no accounting for the fact that he is "a little less than the angels and remains the same after death"; otherwise, he would not differ from the brutes in either life or death. How could it be said "that some live the intellectual or angelic life," and some "the animal

---

[34] Ibid., 255.   [35] Ibid., 256.
[36] "Moreover, we feel that the understanding is begotten along with the body, and grows together with it, and along with it comes to old age." Lucretius, *De Rerum Natura* [see note 11], I, 325.
[37] Gassendi, *Opera Omnia*, II, 256.   [38] Ibid., 257.

and beastly?"[39] It is with regard to the rational soul that man can be said to have been made in the image of God; the material, sensitive soul is communicated to the fetus from the father's semen.[40]

It is not incongruous to speak of man as having two souls, "since beyond the functions of the sensitive soul which man shares with the animals . . . there are also special and eminent functions."[41] This situation is not peculiar to man. After all, animals can be said, in Aristotelian terms, to possess the vegetative soul along with the sensitive soul. In common parlance; we often speak of a single, unitary human soul; when doing so, we are designating only the rational soul.

> This is certainly what must be understood when Christ the Lord said, "Refuse to fear them who strike the body and cannot strike the soul"; indeed he understood only the rational soul, for otherwise the sensitive [soul], as not separate from the body, would be dead . . . from the striking, while the man would be destroyed and the body would cease to be animate.[42]

Throughout his discussion, Gassendi's terminology is anything but consistent. Although he spoke in terms of the three souls described by Aristotle in *De Anima*,[43] he also used the Epicurean distinction between *anima* and *animus*. I think he can be understood as amalgamating the two conceptualizations: The functions of the *anima* correspond to those of Aristotle's vegetative and sensitive souls, whereas the functions of the *animus* correspond to those of Aristotle's rational soul. Gassendi differed from Epicurus in denying the materiality and mortality of the *animus* or rational soul.[44]

Having argued that man, like the animals, possesses a sensitive soul, Gassendi proceeded to discuss the nature of the rational soul:

> . . . in agreement with the holy faith, we say that the mind, or that superior part of the soul (which is appropriately rational and unique to man) is an incorporeal substance, which is created by God, and infused into the body; . . . it is like an informing form . . .[45]

His argument is directed at establishing that the rational soul is incorporeal. He began by considering the faculties of the mind or rational soul, namely, intellect and will. The intellect is the pri-

[39] Ibid.    [40] Ibid.    [41] Ibid.    [42] Ibid., 258.    [43] Aristotle, *De anima*, 413ᵇ4–13.
[44] See Gassendi's lengthy discussion of the Aristotelian account of the soul, *Opera Omnia*, II, 259.
[45] Ibid., 440.

mary faculty, according to Gassendi, since the will is rooted in the intellect. His discussion, therefore, focuses on the actions and objects of the intellect.[46] At this point, Gassendi's discussion shifts from assertion of faith to philosophical argument.

First, he wanted to establish that "the intellect is distinct from the imagination."[47] The imagination, or phantasy, for Gassendi and other seventeenth-century philosophers, is the faculty by means of which we have images of the objects of our thought.[48] Now, Gassendi claimed that not all human thought involves the use of imagination:

> . . . there is in us a species of intellection by which we carry on reasoning so that we can understand something which is imagined or of which we cannot have observed the image . . . I use an example . . . from the magnitude of the sun. Having been led by reasoning, we understand that the sun is 160 times greater than the earth; yet the imagination is thwarted, and however much we try . . . , our imagination cannot follow such vastness. . . .[49]

Many similar examples establish the proposition that there are things which "we understand but which cannot be imagined, and thus the intellect is distinct from the phantasy."[50] The phantasy, however, imagines with material species. Therefore, the intellect must be incorporeal.[51]

Gassendi's second argument for the incorporeality of the intellect begins with the observation that the intellect can know itself. This ability goes "beyond all corporeal faculties, since something corporeal is in a certain place . . . so that it cannot proceed towards itself but only towards something different." For this reason, it makes no sense, for example, to speak of sight seeing itself or knowing its own vision. This ability to reflect on itself is unique to the human intellect and a sure sign of its immateriality.[52] This

---

[46] Ibid.    [47] Ibid.

[48] Ibid., 398–424. According to the *OED:* "Fantasy, Phantasy. 1. In scholastic philosophy: a. mental apprehension of an object of perception; the faculty by which this is performed . . . b. The image impressed on the mind by an object of sense. . . . 4. Imagination: the process and faculty of forming mental representations of things not actually present."

[49] Gassendi, *Opera Omnia, II,* 440.    [50] Ibid.    [51] Ibid., 441.

[52] Ibid. Compare Gassendi's argument to that of Thomas Aquinas: "The principle of the act of understanding which is called the soul of man, must of necessity be some kind of incorporeal and subsistent principle. For it is obvious that man's understanding enables him to know the natures of all bodily things. But what can in this way take in things

argument is not unlike Descartes's use of the *cognito* to draw an absolute distinction between mind and matter. Gassendi's third argument for the incorporeality of the soul is that not only do we form concepts of universals, but we also perceive the reason for their universality.[53] Even if animals on occasion seem to form universal notions – for example, when they recognize a stranger as a human being – nevertheless, they never apprehend the universal purely abstractly, but always with some degree of individuality and concreteness. Furthermore, unlike humans, they do not understand the nature of universality per se. Possessing only a corporeal faculty, the phantasy, animals are restricted to fairly concrete aspects of universals.

Truly, I add that there is no corporeal faculty which is not limited to a certain kind of thing; and however much phantasy seems to extend to several kinds, nevertheless they all are contained under the rubric of the sensible. That is why the rest of the animals which possess only the phantasy are limited to sensible things.[54]

The corporeality of the sense organs provided Gassendi with yet another argument for the incorporeality of the soul. The intellect is the organ of reason:

. . . inasmuch as . . . the organ is a medium between the faculty and the object . . . , the organ cannot act in itself . . . , so the intellect cannot be engaged in the knowing organ . . .[55]

If the intellect were corporeal, like the sense organs it could perceive corporeal things. Nevertheless, this line of argument does not imply that an incorporeal intellect would be unable to know corporeal things. For just as the souls of animals contain all the functions of the inferior souls of plants, so the human soul possesses all the abilities and functions of the souls of animals as well

must have nothing of their own nature in its own, for the form that was in it by nature would obstruct the knowledge of anything else. For example, we observe how the tongue of a sick man with a fever and bitter infection cannot perceive anything sweet, for everything tastes sour. Accordingly, if the intellectual principle had in it the physical nature of any bodily thing, it would be unable to know all bodies. [It is] impossible, therefore, that the principle of understanding be something bodily." *Summa Theologiae, 11: Man (Ia, 75–83)*. Trans. Timothy Suttor (London: 1970), 11.

[53] Gassendi, *Opera Omnia, II,* 441.     [54] Ibid., 442.     [55] Ibid.

as the higher faculty of intellection.[56] Despite the fact that the soul dwells in the body, it is nonetheless incorporeal.[57]

Having established to his satisfaction that the "rational soul or human mind" is incorporeal, Gassendi proceeded to argue that "it has God as its author by whom it was brought into being or who created it from nothing."[58] The passage from nothing into something is infinite; execution of this transformation requires more than the finite force which belongs to natural things, "but an infinite [force], which is God alone, so that the rational soul, which is an incorporeal substance, can only know God as author."[59] One might object, argued Gassendi, that to assert that there is a cause surpassing all of nature which can create something from nothing is to transgress physical dogma. But,

is that not what Thales, Pythagoras, Plato, the Stoics, and many great philosophers have asserted? Therefore, it is not abhorrent that there is a power in the author of the world which surpasses all of nature; and which, forming and ordering particular things, produces the individual forces by which they operate from the beginning. Why therefore is it not a physical dogma that there is a power in the maker of nature by which something can be made from nothing?[60]

Indeed to argue that it is impossible for something to be created from nothing on the grounds that we never observe such an event would be as absurd as to assert that "Daedalus had no power within himself except such as he observed in his automata."[61] In this respect, Gassendi departed from Epicurus and Lucretius whose physics had begun from the principle *"ex nihilo nihil fit."*[62]

Truly there is nothing which does not proclaim that there "was an acting infinite force in the great maker of all, who alone depends on nothing else, who is not limited by the action of any force. . . ."[63] It is no reproach to physics to resort to the author of nature, especially in the case of the rational soul which, "since it is immaterial and unless [it is created from nothing], cannot be created by any other cause than God."[64] Since the soul was created by God, like all other things in the world, "it is within the order of things which God constituted in nature and protects by his

[56] Ibid.   [57] Ibid., 442 and 629.   [58] Ibid., 442.   [59] Ibid., 443.   [60] Ibid.   [61] Ibid.
[62] Lucretius, *De Rerum Natura* [see note 11], I, 156–73.
[63] Gassendi, *Opera Omnia*, II, 443.   [64] Ibid.

providence. . . ." [65] In this sense, the soul is nothing extraordinary or beyond the natural order, "since wherever and whenever a man is born [God] creates a rational soul which he infuses into his body." [66] Unlike the material components of mechanical nature, which are all composed of atoms and the void and are produced by second causes, the rational soul of each person is individually and directly created by God.

A further question arises, "How can something incorporeal be joined with the body . . . and [how can it] be the informing form?" [67] Even if it is argued that the sensitive soul serves as an intermediary between the body and the rational soul, the problem remains; for the sensitive soul "is nevertheless corporeal and infinitely distant from the incorporeal." [68] Thus the difficulty of explaining the connection between the incorporeal rational soul and the corporeal body remains, however fine and attenuated the body may be. Unlike angels or pure intelligences, which subsist independently as pure acts of intelligence, the sensitive soul is such that "its nature has a destination and inclination to the body and the rational soul." [69] If the sensitive soul can be regarded as the form of the body, then the rational soul, even though it is a substance, can be regarded as the form of the individual person. [70] And although the sensitive soul is dispersed throughout the body, the rational soul resides in the brain.

. . . because the influence of . . . the nerves from all the senses and parts of the body is in the brain, the seat of the Phantasy is established in the brain. . . . it seems congruous that also the mind or rational soul . . . is united not without another corporeal or sensitive soul so that it can think

---

[65] Ibid.

[66] Ibid. Gassendi's views on the different origins of the sensitive and rational souls are similar to those of Thomas Aquinas: ". . . the souls of brutes are produced by a certain material force, whereas the human soul is produced by God." Thomas Aquinas, *Summa Theologiae*, 11: *Man*, 31. [See note 52.]

[67] Gassendi, *II*, 443–4.    [68] Ibid., 444.    [69] Ibid.

[70] Ibid. This point has a long history in Scholastic discussions of the nature of the soul. Christian philosophers were concerned to preserve both the substantiality of the soul, its immortality, and the unity of man, doctrines which are not prima facie compatible. Gassendi's position closely resembles that of Thomas Aquinas. For an extended discussion of the complex metaphysical problems surrounding the question of the nature of the soul, see Anton Charles Pegis, *St. Thomas and the Problem of the Soul in the Thirteenth Century* (Toronto: Pontifical Institute of Mediaeval Studies, 1934).

and understand by the intervention of the Phantasy; it is congruous, I say, that it has the same seat as the Phantasy, the brain.[71]

At this point, Gassendi's discussion raises many of the issues involved in the perplexing mind–body problem that was much more precisely articulated by Descartes. To the extent that Gassendi retained Aristotelian terminology and conceptualizations, such as considering the mind as form, he did not raise the question in the explicit way that Descartes's radical distinction between mind and body necessitated.[72] Nonetheless, the problem existed for both philosophers, and, likewise, its solution eluded them both. Descartes's invocation of the pineal gland as link between mind and body is no less arbitrary or question-begging than Gassendi's use of the sensitive soul as intermediary between the body and the rational soul.[73]

Having discussed the nature of the rational soul and having argued for its incorporeality, Gassendi then addressed the question of the immortality of the soul. He regarded this issue as the "crown of the treatise" and "the last touch of universal physics."[74] As a statement of faith, Gassendi declared,

that the rational soul is singular and in one and the same singular man and it is incorporeal and was created by God and infused in the body, so that it is in itself as informing and not simply assisting; truly after death it survives or remains immortal; and as it bore itself in the body, either it will be admitted to future happiness in heaven, or it will be thrust down unhappy in hell, and it will regain its own body in the general resurrection, just as it was in itself and will receive its good or evil.[75]

Although "the divine light shines for us from this sacred faith," nevertheless, theologians have been accustomed to discuss argu-

---

[71] Gassendi, *Opera Omnia, II,* 446.

[72] "It was Plato who was the first to make a sharp distinction between the mind and the body, holding that the mind could exist both before and after its residence in the body and could rule the body during that residence. St. Augustine further developed this distinction and theorized in more detail about the two. But it was Descartes who first developed a systematic theory of the natures and interrelationship of mind and body." Jerome Shaffer, "Mind–Body Problem," *The Encyclopedia of Philosophy,* ed. Paul Edwards (New York: Macmillan and Free Press, 1967), 8 vols, *V,* 336–7.

[73] René Descartes, *Treatise of Man,* trans. Thomas Steele Hall (Cambridge, Mass.: Harvard University Press, 1972), 79 ff.

[74] Gassendi, *Opera Omnia, II,* 620.      [75] Ibid., 627.

ments for and against the immortality of the soul.[76] Consequently, despite his acceptance of the faith, Gassendi addressed the arguments concerning the immortality of the soul. His strategy was as follows: First, he supported the immortality of the soul on grounds drawn from faith, from physics, and from morals; then he refuted the Epicurean arguments against the immortality of the soul as presented by Lucretius.

Having begun with a statement of faith, Gassendi immediately undertook to argue for the immortality of the soul on what he called physical grounds. He stated his argument succinctly: "the rational soul is immaterial; therefore it is immortal." The reason why, according to Gassendi, an immaterial thing is also immortal or incorruptible is that, "lacking matter, it also lacks mass and parts into which it can be divided and analysed. Indeed, what is of this kind neither has in itself nor fears from another . . . dissolution."[77]

Gassendi anticipated several objections to this apparently simple argument and dealt with them in turn. One possible objection could derive from "the Ethnics" (the Greeks), who believed that there are demons and other incorporeal spirits that both come into being and pass away. In support of this position, one might go on to maintain that "since only the author of nature is creative, for that reason he alone is said to have immortality. . . . [I]t must be allowed that the whole world, even the incorporeal things which he produces from nothing . . . can absolutely be returned to nothing if he wishes it."[78] But, Gassendi objected, in spite of God's absolute power to annihilate all things that he created, "from the supposition that nothing works against the order of nature and this state of things which he instituted most wisely, likewise he preserves it constantly; and it is evident that incorporeal things are preserved eternally."[79] Divine providence, in this case, explains the coherence of the natural order. It should be noted that this emphasis on God's providence as preserver of the natural order departs from Gassendi's more usual voluntarism, which emphasizes God's absolute power. The language of the present passage

---

[76] Ibid.    [77] Ibid., 628.    [78] Ibid.

[79] Ibid. For a full discussion of God's relationship to nature in Gassendi's work, see Osler, "Providence and Divine Will" [note 6].

closely resembles that of Descartes's discussion of the status of eternal truths.[80]

Another possible objection to the incorporeality of the soul is "that there are Fathers who linked immortality with corporeality," and there are "philosophers who claim that the heavenly bodies are incorruptible."[81] Gassendi rejoined by saying that "those Fathers who have immortal corporeal things have them only by Divine Grace and not really from their own nature . . . ; and whatever the meaning of those philosophers is, . . . our consequent is not contradicted, . . . because there can be some incorruptible bodies no less than there are incorruptible, incorporeal things."[82] In other words, the possible existence of corporeal, immortal things does not weaken the argument that incorporeal things must be immortal.

To support his physical argument for the immortality of the soul, Gassendi asserted that despite differences about the state of the soul after death, there is universal agreement about the fact that the soul is immortal, as evidenced by the frequency of superstitions about death. In a moment of astonishing lack of skepticism, Gassendi quoted Cicero: " 'everything about which all people agree must be considered a law of nature': it is indeed fitting that the sense of immortality is endowed by nature . . ."[83] And anyone who denies the soul's immortality is violating the principles of nature. Purported examples of "certain . . . wild people . . . in the new world, in whom no opinion about immortality is inbred . . ." do not stand close examination. For they all turn out to fear demons or nocturnal spirits and the like, an indication, according to Gassendi, that they really do believe in immortal souls.[84] Even if one could find a few counterexamples, they would not weaken the generality of the belief in immortality, any more than the fact that some people "are born with one leg or desire their own death" allows us to argue that "no men are bipeds or have a natural appetite for propagating life."[85] Not only is belief in the immortality of the soul universal, but so too is the natural

[80] See Margaret J. Osler, "Eternal Truths and the Laws of Nature: The Theological Foundations of Descartes' Philosophy of Nature," *Journal of the History of Ideas,* forthcoming, July, 1985.

[81] Gassendi, *Opera Omnia, II,* 628. [See note 14.]     [82] Ibid., 629.     [83] Ibid.     [84] Ibid.

[85] Ibid.

appetite to survive death. Since nature does nothing in vain, people would not be endowed with such an appetite if it represented a vain desire. Consequently, the soul must be immortal.[86]

Gassendi called his third line of argument for the immortality of the soul moral. It rests on an assumption that might well be called the principle of the conservation of justice. Gassendi argued that "to the extent that it is certain that God exists, so it is certain that he is just. It is appropriate to the justice of God that good happens to the good and evil to the wicked." But, in this life, anyway, rewards and punishments are not justly distributed. Consequently, "it is required that there be another life in which rewards for the good and punishment for the evil are distributed."[87] One might possibly object, he argued, that this reasoning would also entail immortality for the souls of animals; for there are many examples of injustice in the animal kingdom, as when "a peaceful herd of sheep is mangled by a wolf or a simple dove without evil is attacked by a hawk."[88] But we need not conclude from these examples that there is an afterlife for animals; the argument follows only for the human realm because of God's special providence for mankind. Further evidence for this conclusion is derived from the fact that "men alone, among animate beings, implore, know, . . ., venerate, praise, and love their creator."[89] Men also differ from the animals in having knowledge of their future state, desiring pleasure, and fleeing pain. "For which reason it is not remarkable if providential justice distributes rewards and punishments to men which are not the same for other animals."[90]

Gassendi continued that Stoics might criticize his argument by maintaining that the afterlife is not necessary to ensure a proper distribution of justice in the world, because virtue is its own reward and vice its own punishment. Gassendi objected, claiming that it is the incentive of future rewards and fear of future punishments that cause men to seek virtue and shun vice. If this were not the case, it would not be necessary to pay soldiers before a difficult battle or a day laborer before he completes the job. "No one, while he acts depraved, fears only that vice is its own punishment; rather, he fears infamy, prison, torture, and the gallows."[91]

Having thus argued for the immortality of the soul on the basis of faith, physics, and morality, Gassendi faced one major task, to

[86] Ibid., 631.    [87] Ibid., 632.    [88] Ibid.    [89] Ibid.    [90] Ibid.    [91] Ibid.

refute Epicurus, who had maintained that the human soul is material and mortal. His refutation of Epicurus's arguments was crucial to Gassendi's primary enterprise, the restoration of Epicurean atomism. He clearly wanted to separate himself from the atheistic connotations of Epicureanism and to present atomism as a doctrine thoroughly compatible with the major tenets of Christianity. Gassendi devoted the penultimate chapter of the "Physics" to this task. Entitled "Epicurus' Objections to the Immortality of Souls Refuted" and quoting extensively from Lucretius, this chapter enumerates and refutes some thirty arguments against the immortality of the soul.[92]

Gassendi concluded his painstaking discussion of the immortality of the soul by stating that the reasons presented for establishing it are not known with the certainty of mathematical evidence. Although they cannot replace "the Sacred Faith, as if it needs the light of reason," nevertheless, they can support the truths of faith by overcoming some of the obstructions in its path.[93]

Lack of mathematical certainty did not deprive Gassendi's arguments for the immortality of the soul of any sound foundation at all. On the contrary, it gave them the same epistemological status as all the rest of empirical knowledge: high probability.[94] Arguing that the lack of certainty, which followed from the skeptical arguments, does not entail lack of any knowledge at all, Gassendi advocated a probabilistic account of scientific knowledge, a "mitigated scepticism" to use Popkin's phrase, which was sufficient for the needs of life in this world.[95] The fact that Gassendi thought that articles of faith, such as the immortality of the soul, do not require rational proof, reveals the influence of fideism on his thought. In this respect, Gassendi's philosophical and theological views can be understood in the context of the debates about skepticism, faith, and reason that followed the Reformation and

[92] Ibid., 633–50.   [93] Ibid., 650.

[94] For a full discussion of the probability–certainty controversy in the seventeenth century, see Richard H. Popkin, *The History of Scepticism* [note 1]; Henry van Leeuwen, *The Problem of Certainty in England, 1630–1690* (The Hague: 1963); Barbara J. Shapiro, *Probability and Certainty in Seventeenth-Century England* (Princeton, N.J.: Princeton University Press, 1983); and Margaret J. Osler, "Certainty, Scepticism, and Scientific Optimism: The Roots of Eighteenth-Century Attitudes Towards Scientific Knowledge," *Probability, Time, and Space in Eighteenth-Century Literature*, ed. Paula Backscheider (New York: AMS Press, 1979).

[95] Osler, "Providence and Divine Will," [see note 6], 555–60.

the revival of the ancient skeptical philosophers in the sixteenth and seventeenth centuries.[96]

Gassendi's concern with the immortality of the soul must be understood in the context of his desire to restore Epicurean atomism as a viable philosophy of nature. To do so, it was necessary for him to rid it of the stigma of atheism. To the seventeenth-century Christian natural philosopher, one of the most objectionable aspects of Epicureanism was its claim that the human soul is material and mortal. Eclectic as he was, Gassendi utilized as much of the Epicurean philosophy as he could, in this case retaining the distinction between *animus* and *anima*. Rejecting the portions he found objectionable, he superimposed bits of Aristotelianism and Scholasticism onto Epicurean foundations in order to render his atomism theologically acceptable. Gassendi's arguments were not always consistent and clear, and his writing is tediously prolix. But his ultimate goal – to restore a theologically sound atomistic philosophy of nature – was always in clear view.

This discussion of Gassendi's views on the immortality of the soul illuminates certain general characteristics of his thought as well as tying his work into the broader context of early attempts to formulate a theologically acceptable version of the mechanical philosophy. His discussion of the immortality of the soul reveals the eclectic and transitional nature of his position. On the one hand, he was clearly and quite self-consciously trying to articulate a philosophy of nature to replace the Aristotelianism he despised and to provide metaphysical foundations for the new science which he promoted so actively.[97] On the other hand, his thought is deeply marked by the Humanist adulation of the classics. In contrast to Descartes, the other seminal mechanical philosopher, who presented his ideas to the world as the product of his own reflections, Gassendi felt obliged to base his views on the work of an ancient model. Having chosen Epicurus, whose atomism seemed compatible with the spirit of the new science, Gassendi then had to modify the ancient system so that it met the demands of Christian theology. Deeply immersed in the traditions of Scholasticism,

---

[96] Popkin, *A History of Scepticism* [see note 1], 1–42, 129–51.

[97] For Gassendi's hostility to Aristotelianism, see his earliest published work, *Exercitationes Paradoxicae Adversos Aristoteleos* (1624), in *Opera Omnia, III*. See also, Bernard Rochot, trans. and ed., *Dissertations en forme des paradoxes contre les aristotéliciens* (Paris: J. Vrin, 1959).

Gassendi derived the inspiration for his theological repairs of the pagan Epicureanism from the thought of Thomas Aquinas and other traditional Christian theologians. Neither ancient nor modern, Gassendi contributed to the new worldview by relying on traditional methods of argument.

Much like twentieth-century discussions about artificial intelligence, seventeenth-century discussions about the immortality of the soul were concerned, at the deepest level, with defining human nature and demonstrating the characteristics that render it unique. The traditional arguments, going back through the Scholastics to Aristotle, dealt with the issue of how the human soul differs from the souls of animals. Gassendi's arguments are cast in this mold. Following Descartes's more radical proposal that only man has a soul and that animals are simply machines or automata, later discussions took place clearly within the context of the mechanical philosophy.[98] Gassendi can be understood as a traditional figure in this development: By appealing to traditional arguments, he sought to make Epicureanism safe for Christianity. Once the mechanical philosophy was widely accepted in the following generations, its wider implications for the traditional debate about human nature were more thoroughly explored.

[98] For a full account of these discussions, see Leonora Cohen Rosenfeld, *From Beast-Machine to Man-Machine: Animal Soul in French Letters from Descartes to La Mettrie* (first published, 1940; New York: Octagon Books, 1968).

# 8

## The manifestation of occult qualities in the scientific revolution

RON MILLEN

### THE PROBLEM OF OCCULT QUALITIES

"Occult" has become something of a pejorative term among historians of the scientific revolution. According to the dominant view of seventeenth-century science, occult qualities were banished from nature by the new philosophy of mechanism. Even historians of science who have emphasized the hermetic component of the scientific revolution have not really come to grips with the issue of occult qualities in the seventeenth century. The persistence of occult modes of natural action in the writings of prominent virtuosi is viewed as an anomaly, an instance that simply proves that the new science was not altogether new. But the real importance of hidden qualities for the mechanical and experimental philosophers remains unclear.

A recent article "What Happened to Occult Qualities in the Scientific Revolution?" has broken new ground.[1] It presents the thesis that far from banishing these qualities, the new philosophy actually incorporated them into modern science. If the question posed in the title has been too long in the asking, the answer at least does much to clarify contemporary attitudes toward occult qualities. However, the contrast drawn by the author between the views of mechanists and Aristotelians is somewhat misleading; it is indicative of a common tendency to view the scientific revolution in terms of a clash between ancients and moderns. To clarify the issue still further, one must consider the changing attitudes toward

[1] Keith Hutchison, "What Happened to Occult Qualities in the Scientific Revolution?" *Isis, 73* (1982), 233–53.

occult qualities in the tradition of Renaissance Scholasticism. As the present essay will show, Scholastics in the sixteenth and early seventeenth centuries confronted the problem of how to deal scientifically with the occult, and the methods they developed were not essentially different from those utilized by mechanical philosophers. The philosophical method of explicating causes and the experimental method of investigating effects were the two roads to knowledge of occult virtues, and the approaches of Scholastics as well as mechanists tended in one direction or the other, depending upon which of the two methods had the upper hand. By taking into account the status of occult qualities in the Scholastic tradition, one can get a clearer insight into the attitudes of leading virtuosi and how they incorporated these qualities into the framework of the new philosophy.

To begin, it will be helpful to say something about the meanings associated with the term "occult" in Scholastic philosophy. In the most common sense, the word signified that which was unintelligible. Occult qualities were properties and powers for which one could offer no rational explanation. In the Middle Ages, philosophers attempted to account for corporeal properties in terms of the four elements and their primary qualities: heat, cold, wetness, and dryness. The class of occult or "specific" virtues, as they were called, comprised the powers of bodies that could not be reduced to a certain temperament of the elementary qualities.[2] The source of specific virtues was frequently assigned to the body's immaterial substantial form. For example, some physicians did not regard the power of scammony to purge bile as occult since they believed that the effect could be explained by the drug's particular heat and dryness. Almost everyone, however, admitted that the magnet's capacity to attract iron was occult because magnetic virtue did not result from the specific mixture of the four elements. Obviously, the notion of unintelligibility was somewhat subjective. A property like the purgative virtues of scammony might be treated as a manifest quality by some physicians, as an occult quality by others. Much depended upon one's faith in the ability of reason to comprehend the workings of nature.

---

[2] Cf. Thomas Aquinas, "On the Occult Works of Nature" in J. B. McAllister, *The Letter of Saint Thomas Aquinas* (Washington, D.C.: Catholic University Press, 1939), p. 21. See also Lynn Thorndike, *A History of Magic and Experimental Science* (New York: Columbia University Press, 1959), *V*, 550–62, and passim.

A second usage of "occult" emphasized the notion of *insensibility*. A quality was called "occult" or "hidden" if it was not directly perceptible. Unlike the preceding case, there was nothing subjective about this division of corporeal properties into manifest and occult; the demarcation was purely technical. Typical manifest qualities included colors, odors, and tastes, whereas the class of occult qualities contained such entities as the magnetic power of the loadstone, influences emanating from the planets, and the specific virtues of medicines. One finds this technical sense persisting in the seventeenth century even among philosophers who thought that almost all qualities were epistemologically occult. Sennert, for example, wrote:

> Now these Qualities are called Occult, Hidden, Abstruse to distinguish them from the manifest qualities discernable by the external Senses, especially the Feeling; whereas on the contrary these are not perceivable, although their operations are: So we see the attraction made by the Loadstone; but we do not perceive the qualities causing that motion of the Iron.[3]

There was an essential link in medieval philosophy between the connotations of insensibility and unintelligibility. According to Aristotelian epistemology, true scientific knowledge – *scientia* – was derived from sensible images. During sensation, the sensory organs received the forms of manifest qualities abstracted from their subject matter. After the imagination had further abstracted the sensible form, the intellect sifted them, separating the accidental from the essential features until at last it obtained the universal form or essence. According to this sensationalist epistemology, insensible entities were ipso facto unintelligible and therefore outside the domain of true scientific knowledge. One may object that none of the active powers of bodies could be perceived directly by the senses, and yet Scholastics believed that they could give a scientific account for many of them. However, the powers that could be explained were all derived from the primary qualities – heat, cold, wetness, and dryness. In Aristotle's philosophy, these were the principles of all perceptible properties. Thus, Scholastics brought at least some active powers within the scope of natural philosophy by explicating them in terms of the immediately intelligible qualities of nature. Only nonelementary virtues, operating through

[3] Daniel Sennert, *Opera* (Paris: Apud Societatem, 1641), *I*, 142.

insensible means, could not be accommodated. As we shall see, the incorporation of occult properties into natural philosophy during the sixteenth and seventeenth centuries depended upon breaking the link between the connotations of insensibility and unintelligibility.

To comprehend more clearly the change in attitude toward occult qualities during this period, it will be helpful to consider briefly their status in medieval natural philosophy and medicine. In general, Peripatetics and Galenists were divided over the reality and importance of occult qualities. Whereas Christian natural philosophers tended to deny that they existed, physicians admitted that certain medicines possessed specific virtues beyond the powers of the elementary qualities. Moreover, an experiential knowledge of these properties was important. In the absence of any detailed study of occult qualities in the Middle Ages, it is not clear to what extent one can pursue the division between philosophers and physicians. However, there are general grounds for believing that the differences were significant, reflecting the combined influence of several factors. Not the least of these was that Aristotle never discussed occult virtues. Whatever his silence implied, it discouraged medieval commentators from taking up the subject. Galen, in contrast, alluded to the powers of specific remedies operating not by contrary qualities but by the similitude of their "total substance." Exactly how he interpreted occult properties is not clear, but the sanction given to them by his authority is evident in the frequent citations of Galen by defenders of occult qualities in the Renaissance.

Peripatetic philosophers had surprising difficulty admitting the existence of insensible entities. Thomas Aquinas denied that there could be animals, or even parts of animals, insensible because of their smallness.[4] The point was of some importance, given the passage in Genesis, where Adam named all the animals at the Creation. Aquinas conceded, as Augustine had earlier, that man's senses may have suffered some loss of acuity as a punishment for the Fall, but the tendency to deny the possibility of insensible material entities seems to have been a common feature of medieval natural

---

[4] Thomas Aquinas, *Commentary on Aristotle's Physics*, trans. R. J. Blackwell et al. (New Haven: Yale University Press, 1963), pp. 33–4.

philosophy. Aquinas did admit certain insensible natural actions such as magnetic attraction, which he claimed was due to an inexplicable occult virtue. However, he insisted that many similar phenomena that seemed to be natural actually resulted from the supernatural activity of angels and demons.[5] He did not claim that man had absolutely no access to insensible causes since God was the prime example of an occult power, and Aquinas wished to show that man could obtain some natural knowledge of Him. However, he never equated the defective knowledge based on sensed effects with true scientific knowledge of causes derived through the powers of sensation and reason. Thomism, which provided the basis for Christian natural philosophy in the Middle Ages, could not resolve the problem presented by occult qualities, namely, how could a science based upon sensation deal with entities by definition insensible? Before Scholastic philosophers began to discuss occult qualities in the sixteenth century, they had to be prodded by thinkers outside the tradition of Christian Aristotelianism.

Galenists did not manage to solve the problem either. Nevertheless, they included specific virtues within the branch of medicine dealing with materia medica. Traditionally, physicians were less apt to regard their ignorance of the ultimate causes of natural actions as a defect of their knowledge. Sennert noted this when he remarked:

And therefore those who traduce these Qualities by calling them the Sanctuary of Ignorance are long since worthily refuted by Avicenna . . . when he thus writes: As he that knows fire warms by reason of the heat therein truly knows and is not ignorant, so he that knows the Load-stone draws Iron because it hath a virtue whose nature is to draw Iron without doubt is knowing and not ignorant.[6]

The difference between Avicenna's and Aquinas's attitude toward occult qualities seems to be representative of the greater acceptability of specific virtues within the medical tradition. However, even physicians did little more than note the existence and uses of these powers; there does not seem to have been any attempt to study them methodically.

---

[5] Aquinas, "Occult Works," pp. 20, 22.  [6] Sennert, *Opera, I,* 142.

## OCCULT QUALITIES IN RENAISSANCE
## ARISTOTELIANISM

In the sixteenth and seventeenth centuries, a serious effort was made first by Scholastics and later by proponents of the new science to bring occult qualities within the scope of natural philosophy. Two distinct approaches were taken. Philosophers attacked the thesis that the insensible is unintelligible by attempting to devise causal *explanations* for occult actions. It may seem that by offering rational accounts for marvelous phenomena, they in effect denied the existence of the occult. However, a closer look reveals that philosophers did not challenge the reality of the occult in nature; they challenged the belief that man could not reduce occult properties to intelligible causes. The experimental approach, in contrast, accepted that certain causes could not be explained; nevertheless, they could be brought within the domain of science by investigating their *effects*. Instead of trying to account for the properties of bodies, proponents of the experimental method tended to treat every quality as hidden. With the elimination of the epistemological distinction between manifest and occult qualities, the way was open for the incorporation of the latter into an experimental philosophy.

The principal influence prompting Peripatetics in the Renaissance to admit the existence of occult qualities was the rise of natural magic. Proponents of natural magic portrayed it as an occult philosophy of nature capable of explaining in terms of natural causes many marvelous phenomena commonly attributed to angels and demons.[7] Their explanations utilized the activity of insensible – though not necessarily incorporeal – agents. To magicians, insensibility did not carry the connotation of unintelligibility because they emphasized that through special powers of imagination, the adept could grasp the workings of hidden causes. Only common philosophers relying upon the limited faculties of sensation and reason regarded the insensible realm as beyond man's comprehension. Such bold, dogmatic claims presented a challenge to Peripatetic philosophers to come to terms with entities beneath the level

---

[7] D. P. Walker suggests a logical scheme for Renaissance theories of natural magic in *Spiritual and Demonic Magic* (Notre Dame, Ind.: University of Notre Dame Press, 1975), pp. 75–84.

of perception. Incorporeal spirits and demons appeared more and more as a refuge of the ignorant.

Christian Aristotelians tended to view natural magic with suspicion. Even though adepts stressed the natural basis of occult actions, they found it difficult to sustain the distinction between natural and supernatural magic. Many opponents followed Giovanni Pico in rejecting natural magic not on the grounds of any inherent implausibility but because it threatened Christianity. However, among Peripatetics of the naturalistic school, conditions were favorable for the incorporation of magical doctrines, including occult qualities, into Aristotelian philosophy. At Padua, the medical rather than the theological faculty exerted the principal influence on the course of studies, and hence the emphasis fell upon Aristotle's scientific writings. The challenge to the Peripatetic view of nature represented by natural magic could not be ignored or put off with superficial arguments; it had to be met head on. In contrast to the confidence with which magicians dealt with occult phenomena, the silence of Scholastic philosophers suddenly seemed embarrassing.

The immediate response of Peripatetics can be seen in one of the influential works of Aristotelian naturalism: Pietro Pomponazzi's *De Naturalium effectuum admirandorum causis et de Incantationibus* (1556).[8] The treatise opened with Pomponazzi's reply to a physician from Mantua who had written concerning the case of two infants, one suffering from erysipelas, the other from burns. It seems that both had been cured by a man who used only words and no other apparent remedies. How, the Mantuan asked, could the Peripatetic philosophy account for such phenomena? Theologians might explain the effect as the result of demons, but Aristotle apparently did not believe in demons. Some might claim that the words used by the man in healing were the instrument of higher powers such as the stars, but it was difficult to see how celestial bodies could have produced such marvels. The physician concluded that Peripatetics must be very embarrassed because they lacked an explanation for such phenomena.

Pomponazzi conceded that it was difficult to indicate what Ar-

---

[8] Although the work was first published at Basle in 1556, it was probably written between 1515 and 1520. See the introduction by Henri Busson in his translation, *Les causes des merveilles de la nature ou les enchantements* (Paris: Rieder, 1930).

istotle actually thought about marvels because he never said anything about them. Perhaps, however, it would be possible to find
a satisfactory account of the phenomena in keeping with the *spirit*
of Aristotle's philosophy. Pomponazzi never stopped to question
the reality of the marvelous cures reported by his correspondent,
and throughout the treatise, he continued to display the same un-
critical attitude. He tended to take his facts where he could find
them, reserving his ingenuity for the task of devising explanations.

Before tackling the case of the two infants, Pomponazzi prepared the ground by setting forth several "Peripatetic hypotheses." The first postulated the existence of invisible, occult
powers in natural bodies. The second claimed that the number of
occult properties was nearly infinite. In contrast with the reluc-
tance of medieval philosophers to recognize hidden properties,
Pomponazzi virtually steeped natural bodies in the occult. His hypotheses represented essentially a magician's view of nature, expressed within the framework of the Peripatetic philosophy.

Lost in the translation was the sense of occult qualities as incorporeal virtues. Traditionally, these powers were associated with
substantial forms, not matter. Agrippa described occult qualities
as nonelementary virtues originating in the ideas of God and residing in the celestial intelligences.[9] In the process of generation, matter was disposed to receive divine powers, which emanated from
the stars to the various stones, herbs, metals, and other bodies in
the sunlunar world. Peripatetics divested many occult qualities of
their Platonist and Neoplatonist trappings. According to the naturalistic doctrine of the origin of forms, the essential properties of
a body resulted from a mixture of the four elementary qualities.
Hence, specific virtues could be explained by the body's temperament. Such a theory, in effect, put all corporeal properties on the
same plane. One could denominate certain properties as occult,
but the term did not mean "unintelligible," simply "not understood." The category of occult qualities grew or shrank depending
upon one's faith in the accounts offered for individual effects. In
Pomponazzi's writings, the emphasis was upon man's ability to
have true scientific knowledge of the occult.

---

[9] Cornelius Agrippa, *Three Books of Occult Philosophy or Magic,* ed. and rev. W. Whitehead
from 1651 English trans. (London: Aquarian Press, 1975), p. 62.

Pomponazzi's conception of the philosopher in *De Incantanta-tionibus* strongly resembled the popular notion of the hermetic adeptus. In one of his hypotheses, he described man as an inter-mediary being, situated between the realms of corruptible and eternal things.[10] Man did not simply participate in the corruptible realm; he contained it within his human nature and therefore rep-resented a microcosm of the greater world. Through his natural faculties of sensation and reason, man could acquire knowledge of hidden causes. Exactly how one was to obtain sensible images of insensible causes Pomponazzi did not say. Perhaps he believed that man could have only probable knowledge of the occult. In re-sponse to the problem posed by the Mantuan physician, he sub-mitted three alternative explanations, as if to say that any one might be true and the philosopher should not seek after certainty. How-ever, skepticism was hardly the theme of the treatise. The portrait of the philosopher drawn by Pomponazzi revealed a man whose powers of cognition extended far above the level of ordinary men, almost to the level of divinity:

And it follows that there are certain men who have accomplished many prodigies thanks to their science of nature and the stars, which is some-times attributed to their sanctity or to necromancy, when they are nei-ther saints nor necromancers. It follows also, if it is true as many ap-proved authors have stated, that there are herbs, stones, or other means of this sort which repel hail, rain, winds, and that one is able to find others which have naturally the property of attracting them. Assuming that men are able to know them naturally, it follows that they are able, in applying the active to the passive, to induce hail and rain and to drive them away; as for me, I do not see any impossibility.[11]

How was man to acquire this science? Pomponazzi recommended study and experience, but the real key to his method – if one may call it that – seems to lie in the powers of introspective reason. Occult phenomena in nature corresponded to occult processes in the microcosm of man's psyche, and therefore the hidden causes of things could be discovered by looking within. Despite Pom-ponazzi's assertions that knowledge required a sensible image, the tendency to elevate philosophers to the level of semidivinities sug-gested that the true adept could dispense with phantasms and sim-ply intuit the causes of occult actions. Such bold claims were heard

[10] Pietro Pomponazzi, *Les causes des merveilles* [see note 8], p. 124.　[11] Ibid., p. 133.

again in the seventeenth century from a modern spokesman for the power of introspective reason, René Descartes.

A more solid philosophical approach to the problem of explicating occult qualities can be found in Girolamo Fracastoro's small treatise *De Sympathia et Antipathia Rerum* (1546). Although Fracastoro was Pomponazzi's student, he showed a keener awareness than his teacher that natural philosophy must be based on experience. He also realized that to keep experience from degenerating into fantasy and magic, it must be collected and unified by stable philosophical concepts. Fracastoro found a unifying concept in the Aristotelian notion of cause. According to his analysis, there were three kinds of cause, distinguished according to their proximity to things.[12] He criticized contemporary philosophers and physicians for being preoccupied with universal and remote principles such as matter and form, the four elementary virtues, and the four humors. Although Fracastoro accepted the validity of these ancient concepts, he recognized that they could not be used effectively to order experience. The concern for ultimate principles had left many important topics of physics and medicine unexplored and others not clearly explained.

Fracastoro recommended that philosophers concentrate on the genus of more immediate and particular causes capable of unifying all the pehnomena associated with a given subject. The two topics he chose to explore were the nature of contagion and the sympathies and antipathies of things. The subject of contagious diseases was on the minds of many physicians in the early sixteenth century after the pandemic outbreak of syphilis. Galenic remedies had proved to be ineffective in treating the new disease, and the field was left open to empirics. In self-defense, perhaps, Fracastoro encouraged Galenists to pursue a more empirical approach. The selective pattern of the disease's transmission suggested that some occult sympathy was at work, and so the investigation into the causes of contagion required a preliminary inquiry into sympathies and antipathies. By finding causal principles capable of explaining apparent actions at a distance, one would have a foundation upon which to construct the theory of contagious diseases.

Fracastoro warned against pushing the inquiry into the realm of

---

[12] Girolamo Fracastoro, *De Sympathia et Antipathia Rerum Liber Unus* (Venetiis, 1546), Introduction.

immediate causes since knowledge of these was proper to God and divine intelligences. Man should strive after perfect cognition, but he should remember that anyone who presumed to possess it would only reveal himself as inept and arrogant. Fracastoro's words seemed to be aimed in part at philosophers who attempted to explain occult qualities in terms of a specific proportion of the elements. Since man could not apprehend the internal composition of mixture in bodies, he could not have the kind of perfect scientific knowledge characteristic of higher intelligences. Reason must not go beyond the limits of experience.

In the fifth chapter, Fracastoro took up the question of the principle of motion responsible for attraction between similar things. Included in this class of phenomena were all those that Galenists had described as actions from similitude. Fracastoro rejected the concept of similitude, claiming that if bodies attracted each other in virtue of their similar natures, they should do so no matter what distance separated them. In many cases, however, motion occurred only when the distance was quite small. There was no action, he said, except through the medium of contact. Since similar things attracted toward each other were originally separated, each body had to emit something that touched the other and caused motion. Fracastoro rejected the atomistic theory of an efflux of particles because occasionally the action occurred over distances corpuscles could not reasonably be expected to travel. Since the mechanism of effluvia was not universal, he turned to the theory of spiritual species. He described the species as "simulacra" – tenuous, superficial entities emitted from the crass matter of the body and having the same form. Although the concept of spiritual species was common in Scholastic philosophy, particularly in the theory of optics, their nature was still surrounded with problems. Fracastoro acknowledged that there was some question how a spiritual substance could cause motion in material bodies. He did not pretend to have the answer but suggested a theory in which natural heat operating through spiritual species provided the moving power.

On the whole, Fracastoro's treatise appears as a progressive attempt to explicate the causes of occult qualities. There was a recognition that sympathies and antipathies existed and that the basic principles of Aristotelian matter theory were inadequate to account for them. Although the concept of spiritual species gave way to atomistic theories in the seventeenth century, the two theories were very similar. It is not surprising that a confirmed atomist like Walter Charleton praised Fracastoro for his handling of sympathies.[13]

The influence of Fracastoro's ideas can be detected even among his immediate contemporaries. Jerome Cardan took up the subject of the occult in a brief work, *De secretis*.[14] In the beginning, he wrote that although Fracastoro, and even he himself, had written about arcane matters, the "hidden rules of the dialectic" for treating the occult had not yet been found and reduced to order; his book aimed to make clear the method (ratio) for investigating hidden causes.

Cardan's ideas about method reflected his interests in mathematics. On several occasions, he compared the secrets of nature to those of mathematics. In both cases, cognition was based upon causes. However, man had a clearer understanding of the principles of mathematics and hence was able to have greater certainty about mathematical secrets than about occult phenomena. Nevertheless, Cardan believed that one could utilize general principles to explicate certain marvelous natural effects. It was impossible to have perfect cognition; for that depended upon a grasp of the ultimate principles. Only the gods had such knowledge since only they knew the recesses of things. Man's perception was limited to externals. He could know but a few secrets, and those through analogy with mechanical things. Cardan mentioned the example of certain machines and vases in which apparently marvelous effects were produced by the flow of water and steam through the internal parts. He recommended that one follow the example of Hero and Archimedes by utilizing both natural and mathematical principles to give an account of such occult phenomena.

Cardan's suggestions seem reasonable enough, but when we get down to cases, it is clear how little explanatory value his dialectical method actually possessed. He illustrated its application with reference to the paradigm of occult actions: magnetic effects. To explain the phenomena, he set forth certain premises the first two of which were grounded in experience: (1) When the magnet is ap-

[13] Walter Charleton, *Physiologia Epicuro-Gassendo-Charltoniana* (London: Thomas Heath, 1654), p. 348.
[14] Jerome Cardan, *Opera Omnia*, facsimile of 1636 Lyons ed. (Stuttgart: Fromann, 1966), II, 537–51.

plied to the end of a bar of iron, it draws the iron toward itself, and (2) when the magnet approaches the bar from the side, it orients the iron so that one end is directed toward the north, the other toward the south. In addition to these empirical premises, Cardan also supposed three general axioms applicable to all natural things: (1) Similar things attract, and contraries repel; (2) if something is not able to draw a like object toward itself, it is drawn toward that object; and (3) anything attached to the attracted object is necessarily drawn along with it. On the basis of this set of five premises, Cardan said, one can explain the causes of all magnetic phenomena. Exactly how was never made clear.

The three figures examined in this section all shared the belief that the Peripatetic philosophy could not ignore the existence of the occult. Some method had to be found to explain marvelous phenomena in terms of natural causes. Although the explanations devised by Scholastic philosophers were swept away by the new philosophy, the attitudes toward the intelligibility of the occult persisted. As long as the effects were produced by natural causes, they fell within the province of the natural philosopher. The mechanical philosophy was marked by a greater skepticism toward the reality of occult effects and by a greater self-assurance in dealing with the occult. It was only in the seventeenth century that occult qualities were fully incorporated into modern science.

## OCCULT QUALITIES AND THE MECHANICAL PHILOSOPHY

Almost all historians of the scientific revolution regard occult qualities as a casualty of the mechanical philosophy. It is argued that the modern concept of nature based on the principles of matter and motion banished every kind of occult power. Although there are reasonable grounds for maintaining this position, it tends to misrepresent the attitudes of leading members of the new science. In this section, we shall take a closer look at what might be called the Cartesian method for dealing with occult phenomena. The emphasis was upon explicating these effects by means of particular mechanisms.

Descartes faced a challenge somewhat like the one that had confronted Pomponazzi nearly a century and a half earlier: Given the reality of certain marvelous phenomena in nature, how could one explain them? In the mid-seventeenth century, the challenge came primarily from occult philosophers who boasted that they had knowledge of the underlying causes of occult effects. For Descartes, to deny the existence of occult qualities altogether would have been to concede that the new philosophy was incapable of handling them. Descartes, of course, made no concessions. On the contrary, he regarded the mechanical philosophy's capacity to account for any and all occult qualities as one of the strong points of the new system. This does not mean that he believed that every marvelous effect described by his opponents was indeed genuine, but for occult properties well documented, he held forth the possibility of a scientific explanation. Aristotelianism had not promised as much. There were many properties of bodies whose derivation from the mixture of the elements Peripatetics did not profess to know. Where Aristotelians hesitated, Descartes plunged ahead. Just after his lengthy mechanical explication of magnetic virtues in the *Principia,* he announced:

there are no powers in stones or plants so occult, no sympathies or antipathies so miraculous and stupendous, in short, nothing in nature (provided it proceeds from material causes destitute of mind and cognition) that its reason cannot be deduced from these [i.e., mechanical] principles.[15]

Far from denying the existence of occult qualities, Descartes boasted that he had brought them within the province of natural philosophy.

Historians of the scientific revolution commonly point to such passages as this one to support the claim that Descartes rejected the notion of occult qualities. But a closer consideration of the point of Descartes's remarks suggests a different conclusion. The lesson to be learned from the discussion of the loadstone was that one could find explanations for documented cases of occult actions such as magnetic attraction; explanations that did not rely on any other principles besides matter and motion. Having shown how to proceed with the paradigm case of an occult virtue, Descartes pointed to other such properties and told natural philosophers to go and do likewise. To say that he rejected occult qualities is to miss the positive methodological prescription. In the example of

---

[15] René Descartes, *Oeuvres,* ed. Charles Adam and Paul Tannery (Paris: Vrin, 1969–74), 8, pt. 2, 314–15.

the loadstone, he was really setting out a program for the manifestation of the occult.

The celebrated dispute later in the century between Cartesians
and Newtonians over universal gravitation illustrates the point.
When Cartesians charged that gravity was an occult quality, they
were, at bottom, objecting to the way Newtonians treated the
property. Newtonians failed to develop an official account of the
particular mechanism underlying the observed effects of gravitation; instead, they insisted that it was possible to mathematize the
motions of gravity without explaining the causes. Cartesians re-
sponded with accusations that their opponents wished to reintro-
duce occult qualities into physics. Beneath the rhetoric, it is clear
that what perturbed the Cartesians was the absence of an explanatory mechanism for the property "gravitas." The Newtonian belief that one could deal scientifically with properties whose causes
were unknown ran counter to the whole program of Cartesian
physics. What better way to vent one's frustrations than to hurl
the epithet "occult" at the enemy? Unfortunately, the use of the
term in this way created the impression that Cartesians rejected
the existence of these qualities. Neither the Newtonians nor the
Cartesians really banished the occult from nature. What the Cartesians prohibited was the notion of an occult quality whose causes
were not made manifest. This was the real sin of the Newtonians:
They had introduced an occult quality without first explaining it.

Against the interpretation presented here one might object that
the Cartesian program undermined the very essence of occult
qualities by attempting to make them intelligible. Perhaps if all
mechanical philosophers shared Descartes's faith in the powers of
human reason, one could make a stronger case for this claim.
However, there were many who adopted the Cartesian method
yet took a more skeptical attitude toward the possibility of discovering nature's innermost secrets. Walter Charleton spoke for this
group in a chapter of his *Physiologia* entitled "Occult Qualities Made
Manifest":

Not that we dare be guilty of such unpardonable Vanity and Arrogance
as not most willingly to confess that to *Ourselves all the Operations of
Nature are meer Secrets;* that in her ample catalogue of Qualities, we have
not met with so much as one which is not really Immanifest and Abstruse when we convert our thoughts either upon its Genuine and Prox-
ime Causes, or upon the Reason and Manner of its perception by that

Sense, whose proper Object it is and consequently that as the *Sensibility* of a thing doth no way presuppose its *Intelligibility,* but that many things, which are most obvious and open to the *Sense,* as to their *Effects,* may yet be remote and in the dark to the *Understanding* as to their *Causes.*[16]

Charleton denied that the criterion of sensibility could be used to demarcate qualities into manifest and occult; there was no epistemological ground for the distinction. From a skeptic's vantage point, all qualities in nature, including sensible ones, were really occult.

If so, why did Charleton dislike the word "occult"? He called it an unhappy and discouraging epithet by which a certain class of qualities was set apart from those whose causes were presumed to be known. He described occult qualities – in particular, sympathies and antipathies – as "windy terms" and a "refuge for the idle and ignorant." Again, historians of science have commonly taken such references as evidence of a rejection of occult qualities. However, a closer look at the sense of Charleton's words reveals a different attitude:

For no sooner do we betake ourselves to Either [i.e., occult qualities or sympathies and antipathies] but we openly confess, that all our Learning is at a stand, and our Reason wholly vanquisht, and beaten out of the field by the Difficulty posed. We deny not that most if not All of those Admired Effects of Nature, which even the Gravest Heads have too long thought sufficient Excuses of their Despair of Cognition, do arise from some Sympathy or Antipathy betwixt the Agent and Patient, but for all that have we no reason to concede that Nature doth institute or Cause the Sympathy or Antipathy, or the Effect resulting from either, by any other Lawes, or Means, but what she hath ordained and constantly useth in the production of all other Common and familiar Effects. We acknowledge also that *Sympathy* is a certain *Consent* and *Antipathy* a certain *Dissent* betwixt Two Natures from one or both of which there ariseth some such Effect as may seem to deserve our limited Admiration; but is it therefore reasonable for us to infer that those Natures are not subject unto, not regulated by the General and Ordinary Rules of Action and Passion, whereto Nature hath firmly obliged Herself in the rest of Her Operations?[17]

The passage clearly expresses Charleton's belief that the problem with occult qualities lay not in the qualities themselves but in the attitude of certain "grave philosophers." Instead of taking mar-

---

[16] Charleton, *Physiologia* [see note 13], pp. 341–2.    [17] Ibid., p. 343.

velous effects as a special challenge to man's reason, they gave up the hunt for possible explanations. The class of occult qualities offered a sanctuary to lazy and ignorant minds who lacked the sporting instinct aroused in the mechanist by an occult quality.

Whether one believed that he could actually capture the quality or only surround it with plausible mechanisms, the important thing was that the Cartesian program offered a method for dealing with the marvels of nature scientifically. With the advantage of hindsight, we can see that the value of the method was extremely limited. Devising explanatory models required ingenuity, but when it was done, what did one really have? Hypothetical mechanisms were as fragile as Don Quixote's helmet. The real value of the various fairy-tale-like hypotheses for those who believed in them lay in their capacity to satisfy a fundamental need: the need to know the cause of things. It was the same motive that had prompted Pomponazzi to search for an Aristotelian explanation of the marvels described by the physician of Mantua; and that had prompted Fracastoro to try to explain the causes of sympathies and antipathies. The mechanists' program for dealing with occult qualities was not essentially different from the Aristotelian in the goal of understanding the insensible realm of nature. It was different because mechanical models satisfied the need for clarity better than anything Aristotelians had to offer.

### THE PATH OF EXPERIENCE

As the Cartesian program for manifesting occult qualities captured the fancy of intellectuals, it became fashionable to laugh at Scholastics who still talked in terms of powers and faculties. However, Scholastics would be the ones wearing the smiles today; for we no longer require that science provide the kind of ultimate explanations demanded by Descartes. Take, for example, the properties of a familiar drug such as aspirin. It is white; it dissolves in water; but most important, it relieves pain. If we press the investigation into its pain-relieving properties, we find they reside in a so-called "active ingredient." Having isolated this chemical substance, we do not press for a further account of how the molecular structure of the substance produces its effects; it is sufficient to regard the pain-relieving property as something like an occult quality. This does not mean that we have no true scientific knowledge of it; the effects of aspirin on the brain and other parts of the body can be studied experimentally. A great deal can be learned not by attempting to give an account of aspirin's properties, but by studying its effects.

Occult qualities have become part of modern science. We take for granted that we can have true scientific knowledge of them based on an experimental method. However, it was not always so. In the medieval period, the pain-relieving powers of a drug might have been ascribed to an occult virtue known through experience alone. Although physicians valued the knowledge gained by experience, it did not have the status of *scientia*, which demonstrated effects from true causes. In the remainder of the essay, we shall examine a different program for manifesting occult qualities from that developed by Cartesians. The immediate roots of the experimental program lay in the sixteenth-century medical tradition, where one finds the beginnings of a more practical concept of *scientia*.

Jean Fernel, perhaps the greatest physician of the sixteenth century, incorporated occult qualities into the theory of physiology and pathology as part of a program to reform the discipline of medicine. Fernel believed that the medical theory inherited from previous generations was not complete:

> Many tell us that the art of healing, discovered by the labour of our forefathers, and brought to completion by dint of Reason, has now attained its goal. They would have us, who come after, tread in the same footsteps as did the Past. It were a crime, they tell us, to swerve a hair's breadth from the well-established way. But what if our elders, and those who preceded them, had followed simply the path as did those before them? . . . Nay, on the contrary, it seems good for philosophers to move to fresh ways and systems; good for them to allow neither the voice of the detractor, nor the weight of ancient culture, nor the fullness of authority to deter those who would declare their own views. In that way each age produces its own crop of new authors and new arts.[18]

Fernel proceeded to describe the current revival of arts and sciences after twelve centuries of decline. He pointed to many modern achievements but especially to the voyages of discovery, which had given man "a new globe." Were Plato, Aristotle, Ptolemy,

---

[18] Preface to Book I of *De Abditis Causis*, cited from Charles Sherrington's *The Endeavor of Jean Fernel* (Cambridge: Cambridge University Press, 1946), pp. 16–17.

and other ancient philosophers who had a knowledge of geography to return today, they would not recognize the world, so much had been discovered. In all of this, we hear the voice of a Humanist scholar embued with the spirit of the Renaissance. Fernel would have set out to reform whatever discipline he took up; as it happened, he chose medicine.

The current theory of medicine was dominated by the ancient doctrine of the elementary qualities and their temperaments. In a word, the doctrine stated that the health of every body consisted in a certain balance of the contrary qualities. Disease was a disturbance of the temperament resulting from the excess or defect of humors. Fernel proposed a new category, diseases of the total substance, in which the idea of an imbalance played no part. In the preface to his influential dialogue *De Abditis Causis* (1548), he related that he had been impressed twenty years earlier with the question in Hippocrates: Is there not in disease something preternatural? Fernel answered that there were certain diseases of an occult nature, namely plague and pestilence. Whereas they were spread by invisible means and were known by various names, they all had one feature in common: Their causes could not be explained by the traditional Galenic theory. Such diseases were not natural in the sense that they resulted from a disturbance in the corporeal humors; instead, they resulted from the weakening of the total substance – the divine substantial form – of the organism. The pathological agents were occult qualities. When Hippocrates asked the question about preternatural diseases, he supposedly had in mind the class of occult diseases.

In the first book of the dialogue, Eudoxus (Fernel's spokesman) established that all natural things consisted of three components: elements, the temperament of elementary qualities, and a divine substance derived from the heavens. The nature of spiritual substance was hidden from man; its powers were called occult because they could not be perceived by the senses and therefore could not be explained by reason. In the second book, Eudoxus applied the general theory of the composition of bodies to human physiology. The result was a radical transformation of the Galenic doctrine of natural faculties. According to orthodox medical theory, the biological functions of the body were performed by various natural faculties utilizing powers produced through the temperament of elementary qualities. In Fernel's theory, the living body was governed by a spiritual soul whose subordinate faculties consisted of spiritual substances or forms. For some operations, the substantial forms utilized the temperament, and it was possible to account for the effects in terms of elementary qualities. However, the more important operations were above the powers of the elements. These functions required forms acting directly on the body by means of specific virtues or occult qualities. Eudoxus tried to convince his listeners that the doctrine of divine forms was really a Galenist theory. Not only was it not a Galenist theory; it also undermined the foundations of Galenic medicine by introducing a dualism of body and spirit into the theory of biological operations. To Fernel, life was not the activity *of* an organic body – as it was for both Aristotle and Galen – but the activity of a vital principle *within* an organized body. The source of life, the soul, was hidden, and hence, knowledge of biological phenomena could be gained only by observation of living organisms.

Fernel took the reform of medicine seriously. After finishing the *Dialogue,* he began writing the treatises in the *Universa Medicina.* His *Physiology* and *Pathology* broke new ground and established for him a reputation lasting until the eighteenth century. In the preface to the *Therapeutics,* written toward the end of his life, we find a statement of the philosophy underlying the new medicine: "Nothing whatever is discoverable in man which does not obey Nature and Nature's Laws, save and except only man's understanding and man's free-will. Nature throughout is one eternal law, and Medicine is a book written within that law."[19] Belief that the body had its own laws led Fernel to exclude magic from the domain of physiology and medicine. Nothing could be above the necessity governing the operations of the human body except the unique powers belonging to man's immortal soul. It was a bold vision of the living organism. It incorporated the occult at the same time that it subordinated the magical features of occult qualities to the laws of nature.

For all that, Fernel remained a systematizer and not a discoverer of new truths in the discipline of physiology he helped to create. The future of the science lay with those who aimed to explain the processes of life in physical and chemical terms. In the seventeenth

---

[19] Cited from Sherrington, *Endeavour,* p. 95; cf. Fernel, *Universalis Therapeutices* in *Universa Medicina* (Genevae: Apud Petrum Chouet, 1638), pp. 1–2.

century, William Harvey used experimental methods to discover the circulation of the blood, thereby overturning the foundations of Galenic physiology. Although modern physiologists like Harvey turned away from the idea of spiritual faculties, they nevertheless owed an important debt to proponents of these forms. The experimental method of modern science rests upon the assumption that it is possible to gain scientific knowledge of hidden causes by observing their effects. The aim, of course, is to go beyond the level of phenomena and discover causal explanations capable of unifying a variety of effects. Although Fernel did not believe that one could explain the important processes of life, he did believe that one could acquire legitimate scientific knowledge of hidden causes whose effects were observable. More important, he embodied his belief in a systematic treatise presenting physiology for the first time as an integral subject. The work represented a new and more practical idea of *scientiae* than the one found in the Scholastic philosophical tradition. Experience rather than reason provided the basis for a physician's knowledge. However, Fernel realized that mere experience could be deceiving. Eudoxus spoke of the need to observe the effects of medicines under different circumstances and at different times with a view toward obtaining "the perfection, faith, and constancy of experience." Moreover, reason played a part in the process. Eudoxus remarked that the perfect knowledge of experience could not be obtained except by reason.[20] We are not yet at the stage where such prescriptions had coalesced into a working scientific method. But Fernel's words indicate a newly found faith in man's power to chart the terrain of the unexplored world of hidden causes.

The leading Scholastic proponent of hidden qualities in the seventeenth century was Sennert. His ideas were much influenced by Fernel, although Sennert rejected the Neoplatonist theory that forms emanated from the heavens. In the section on occult qualities in the *Hypomnemata physica* (1636), he followed a Scholastic mode of inquiry, asking first what occult qualities are and whether they exist in nature.[21] His definition of occult qualities emphasized the notion of insensibility: "Now these qualities are called occult, hidden, or abstruse to differentiate them from manifest qualities discernable by the external senses, especially touch." The category of

---

[20] Fernel, *Universa Medicina,* p. 242.   [21] Sennert, *Opera* [see note 3], *I,* 141–9.

manifest properties included primarily the qualities of the four elements and the tangible properties resulting from the particles of bodies such as rarity, density, hardness, softness, roughness, smoothness, and so forth. Other sensible qualities included colors, odors, and tastes, although Sennert seemed hesitant to treat them as manifest. They were sensible, to be sure, but they were derived from nonelementary forms and hence were closely related to the active powers in the class of occult qualities.

In an epistemolgical sense, all qualities were occult. One of the principal themes of Sennert's writings was that forms exceeded man's limited powers of comprehension; not even the simple manifest quality, heat, could be explained. Although such a claim might have been used to undermine the possibility of scientific knowledge, Sennert drew a more positive conclusion: Man could obtain the same degree of certainty about occult qualities as about manifest ones:

Meanwhile all the more learned philosophers and physicians have undertaken to defend the truth and taught that the causes of many things in natural philosophy and medicine depend upon hidden qualities; and that we are frequently glad to fly to the saving sanctuary (as Scaliger calls it in his *Exercit.* 218, sect. 8) of an occult propriety. Although it is called ignorance by some, they rather accuse the weakness of our understanding to dive into the secrets of nature than blame these hidden qualities. For if the true origin of these qualities be sought into (which few have taken care to do) the knowledge thereof will produce as certain science as that of the first qualities. For the natural philosopher knows no more of heat but that it heats and that it flows from and depends upon the form of fire; and this form is as unknown to man as those forms from which the hidden qualities arise.[22]

Sennert chose the example of heat to make a point. If magnetic virtue was the paradigm of an occult quality, heat was the Aristotelian paradigm of a manifest quality. Scholastic philosophers were confident they knew its nature; indeed, if this quality were not known perfectly, none of the other corporeal properties could be known since they were explained in terms of the primary qualities. In the passage above, Sennert undermined Aristotelian ef-

22 Ibid., 142.

forts to explicate specific virtues by claiming that the very qualities used to manifest the occult were occult qualities themselves.

That occult qualities existed could be proved from two considerations: (1) Many actions in nature were totally different from those of the elements; and (2) the manner in which occult agents acted differed from the elements' mode of operation. Examples of nonelementary actions included magnetic attraction, the dormative effect of opium, the convulsions caused by poisons, and the quieting of convulsions through antidotes. According to Sennert, there were a thousand such actions different from the elementary operations of heating, cooling, humidifying, and drying common to the elements. Regarding the manner of acting, Sennert noted that a sufficient bulk of the elements was required before they could exercise their powers. In addition, it required a period of time for contrary qualities to act upon their opposites. However, extremely small amounts of certain poisons could kill a man almost instantaneously. It did not seem correct to attribute such effects to elementary virtues. The proofs Sennert offered for occult qualities were based on familiar phenomena – at least to physicians – rather than rare and marvelous events. There was certainly something marvelous about the actions described, which is why Sennert chose them, but he might have chosen the action of fire since it was equally occult.

What did it mean, then, to say that one could have scientific knowledge of occult qualities? Sennert maintained that if one investigated their origin, the knowledge acquired was no less certain than the knowledge of sensible properties. He also stated that few had taken the care to make the investigation. Peripatetics had approached the problem with philosophical methods; explaining the origin of a quality meant giving an account based on the primary qualities. Sennert, in contrast, had in mind an empirical investigation. Instead of devising explanations, one should attempt to isolate the material substances in which qualities inhered. The analytical methods of chemistry offered the best means for unlocking natural bodies; indeed, without a knowledge of chemistry, Sennert said, one could not hope to excel in investigating the occult.[23] Sennert's program for the use of empirical methods to dis-

---

[23] Ibid., 916.

cover forms was based on a standard of scientific knowledge closely linked to experience. Experience rather than reason led to the underlying causes of corporeal properties, at least to the extent that man could know them.

Reason had to order experience if one was to obtain genuine knowledge of nature. Since occult qualities could not be explicated, the role of reason was limited to devising ways of classifying the observable effects. The latter half of the chapter on the origin of occult qualities described six classes in which all occult qualities could be arranged. For our purposes, the classes themselves are less important than the general method. Sennert, in effect, approached the subject of occult qualities in the manner of an encyclopedist. The totality of experience was fitted into a tidy systematic schema. It was hoped that by arranging and ordering occult qualities, one could transmit the knowledge of them to future generations. In this way, the investigation into nature's secrets would become a continuing project, drawing upon the work of many naturalists.

## HIDDEN QUALITIES IN BACONIAN SCIENCE

Sennert's ideas reflect the contemporary movement to base natural philosophy on the disciplines of natural history. The most well known proponent of the empirical restoration of the sciences was, of course, Francis Bacon. Bacon criticized Scholastics for giving up on the discovery of forms; nevertheless, his ideas owed much to those he criticized. For Bacon, as for Sennert, nature was a realm of occult qualities, and the naturalist's first task was to study their effects. In this section, we shall look more closely at the Baconian approach to the secrets of nature.

References to occult qualities are scattered throughout the *New Organon*. Without exception, they have an unfriendly tone. For example: "Operations by consents and aversions . . . often lie deeply hid. For what are called occult and specific properties, or sympathies and antipathies, are in great part corruptions of philosophy."[24] In another aphorism, Bacon stated:

Now the human understanding is infected by the sight of what takes place in the mechanical arts, in which the alteration of bodies proceeds

---

[24] Francis Bacon, *The New Organon* (Indianapolis: Bobbs-Merrill, 1960), p. 261.

chiefly by composition or separation, and so imagines that something similar goes on in the universal nature of things. From this source has flowed the fiction of elements, and of their concourse for the formation of natural bodies. Again, when man contemplates nature working freely, he meets with different species of things, of animals, of plants, of minerals, whence he readily passes into the opinion that there are in nature certain primary forms which nature intends to educe, and that the remaining variety proceeds from hindrances and aberrations of nature in the fulfillment of her work, or from the collision of different species and the transplanting of one into another. To the first of these speculations we owe our primary qualities of the elements; to the other our occult properties and specific virtues; and both of them belong to those empty compendia of thought wherein the mind rests, and whereby it is diverted from more solid pursuits.[25]

Such passages give a fairly clear idea of Bacon's complaint against occult qualities. He did not challenge their existence, although he did question many reported instances of magical effects; he was "almost weary of the words sympathy and antipathy on account of the superstitions and vanities associated with them."[26] When one had weeded out the false ascriptions and fables, there remained a small store of genuine sympathies approved by experience, such as those between the magnet and iron, or gold and quicksilver. Chemical experiments involving metals revealed some. But the greatest number were derived from the specific virtues of medicines. For Bacon, the medical tradition constituted the best source of faithful, honest descriptions of occult qualities necessary for a solid, empirical foundation of scientific knowledge.

Scholastic physicians, however, were content merely to describe and catalogue occult virtues. Bacon wanted to do more. His real complaint against Scholastics was that they broke off the investigation into occult qualities too soon; their program resulted in compendia or encyclopedias. Bacon had a grander vision for the reform of science in which compiling natural histories was just the starting point. He hoped to discover the universal laws, the forms, behind occult phenomena. That required *dissecting* both the specific virtues and the configuration of bodies into their simple components. One could not have "much hope of discovering the consents of things before the discovery of forms and simple con-

---

[25] Ibid., p. 263.   [26] Ibid.

figurations. For consent is nothing else than the adaptation of forms and configurations to each other."[27]

The key to the discovery of occult virtues was inductive logic. Bacon devoted the entire second book of the *New Organon* to explaining his method of induction. At the conclusion, he boasted that "my logic aims to teach and instruct the understanding, not that it may with the slender tendrils of the mind snatch at and lay hold of abstract notions (as the common logic does), but that it may in very truth dissect nature and discover the virtues and actions of bodies, with their laws as determined in matter."[28] Without logic and true induction, the analysis of bodies was fruitless. Bacon criticized those who relied exclusively on the power of fire to resolve compounds; one had to pass from Vulcan to Minerva to discover the true textures and configurations upon which properties depended.

Bacon illustrated the method of investigating forms with the example of heat. As with Sennert, the use of heat contained an implicit lesson. Bacon aimed to show that the form of heat, the prototype of a manifest quality, was more elusive than Scholastics imagined. He drew up three tables of observational data containing instances where the nature of heat was present and absent and where it exhibited variations of degree. Such a collection had to be made in the manner of a natural history "without premature speculation, or any great amount of subtlety." In the next stage, inductive logic entered the picture. Bacon stated that it was necessary to proceed first in a negative way by excluding natures not belonging to the form of heat:

Then indeed after the rejection and exclusion has been made, there will remain at the bottom, all light opinions vanishing into smoke, a form affirmative, solid, and true and well-defined. This is quickly said; but the way to come at it is winding and intricate.[29]

Before the mind negotiated the maze, it was permitted to take a "first vintage" or "commencement of interpretation." This consisted in a preliminary definition on the basis of the evidence considered. Bacon described the genus of heat as the motion of a body's particles, although he rejected the mechanical doctrine that motion is the universal cause of heat. Rather, the "essence and quiddity"

[27] Ibid., p. 261.    [28] Ibid., p. 265–6.    [29] Ibid., p. 152.

of heat was motion limited by certain specific differences. After discussing four such differences, Bacon finally arrived at the form of heat:

Heat is a motion, expansive, restrained, and acting in its strife upon the smaller particles of bodies. But the expansion is thus modified; while it expands all ways, it has at the same time an inclination upwards. And the struggle in the particles is modified also; it is not sluggish but hurried and with violence.[30]

All this, one should remember, was just the first vintage. Whatever heat was, after Bacon's subtle mind had finished with it, clearly it was no longer a manifest quality. Its form resided in the midst of a labyrinth, a place traditionally reserved for occult powers.

If discovering the simple nature of heat was so complicated, it is discouraging to imagine the complexity of discovering specific virtues. Nevertheless, Bacon saw his method of induction as a means for acquiring solid scientific knowledge of occult qualities. His method did not explicate them in terms of primary qualities as Peripatetics and mechanists did. Neither did it rest with describing and cataloging. Instead, Bacon aimed to acquire a deeper knowledge by dissecting specific virtues into their simple forms and corporeal structures. The program came to naught because none of his followers knew exactly what to make of the fantastic method outlined in the *New Organon*. However, the idea that experimental philosophy could deal scientifically with occult qualities prevailed. In the later seventeenth century, it was taken up by Robert Boyle, who incorporated the investigation of the forms underlying occult qualities into his own program for experimental science.

Boyle illustrated the experimental philosopher's approach to occult qualities in his *Experiments, Notes, &c. About the Mechanical Origin or Production of Divers Particular Qualities* (1675).[31] He divided all corporeal properties into four classes: (1) "primary qualities" such as heat and cold, (2) sensible qualities such as tastes and odors, (3) secondary or chemical qualities, and (4) occult qualities. The purpose of the treatise was to show that each type could be explained in terms of the mechanical principles, matter and motion. Rather than demonstrate this for a great number of qualities,

---

[30] Ibid., p. 162.
[31] Robert Boyle, *Works*, ed. Thomas Birch (London: Andrew Millar, 1744), *III*, 565–652.

Boyle chose certain specimens typical of an entire class. For example, occult qualities were represented by magnetism and electricity. Thus, although the treatise dealt with only a few properties, it was universal in scope, "there being scarce one sort of qualities of which there is not an instance given in this small book." Boyle did not regard the class of occult powers as a special case; he placed them on the same plane with sensible qualities. There was no ground for separating the two since both were produced by mechanical means.

Previous mechanists had said the same thing, but Boyle did not advocate the earlier program for explicating occult qualities by devising hypotheses. In the preface, he clearly expressed his attitude toward the business of inventing mechanisms:

I do not undertake, that all of the following accounts of particular qualities would prove to be the very true ones, nor every explication the best that can be devised. For besides the difficulty of the subject and incompleteness of the history we yet have of qualities, may well deter a man, less diffident of his own abilities than I justly am, from assuming so much to himself, it is not absolutely necessary to my present design.[32]

Boyle's design was to show that the mechanical philosophy could be extended to a wider range of phenomena than many critics of mechanism allowed. He noted that there were many who preferred mechanical explanations but claimed that such accounts applied only to machines and a limited number of natural phenomena. To remove this prejudice, it was not necessary that the explications proposed be the best possible; it was sufficient to treat them as types or models without claiming that they represented the actual modes in which nature operated. Above all, Boyle stressed that he was free to change his mind as the empirical study of qualities progressed:

I intend not therefore by proposing the theories and conjectures ventured at in the following papers to debar myself of the liberty either of altering them or of substituting others in their place, in case a further progress in the history of qualities shall suggest better hypotheses or explications. . . . my purpose in these notes was rather to shew it was not necessary to betake ourselves to the scholastick or chemical doctrine of qualities, than to act the umpire between the differing hypotheses of the Corpuscularians; and provided I kept myself within the bounds of the mechan-

[32] Ibid., 569.

ical philosophy, my design allowed me a great latitude in making explications of the phaenomena I had occasion to notice.[33]

Given Boyle's attitude toward hypotheses, it is not surprising that the mechanism invented to explain electrical attraction, for example, was rather clumsy. He imagined that viscous effluvia emitted by the electrical body fastened upon the attracted object; when the particles of the electric body were no longer agitated, attraction ceased, and the strings contracted back into the body. Although the hypothesis was crude, it may be said in Boyle's defense that he was not really interested in devising elegant mechanisms; he was primarily concerned with devising experiments to reveal the actions of electrical and magnetic bodies under a variety of conditions. As he noted, the title of the work did not promise discourses, but rather experiments and notes. The two treatises on electricity and magnetism were filled with a numbered series of experiments and observations from which one could infer, in general terms, a mechanical explanation. The form of the treatises contained an implicit statement: Instead of writing theoretical discourses on the origin of qualities, whether occult or manifest, naturalists should try to discover their causes through *experiments*. It was Bacon's program for the discovery of forms incorporated into the general framework of the mechanical philosophy.

Boyle recognized that one could not always, or even generally, obtain clear mechanical accounts for corporeal properties. His only advice was to continue the search. In the following passage, we hear Boyle the natural philosopher speaking:

The chymists are wont . . . to content themselves to tell us, in what ingredient of a mixt body the quality inquired after does reside, instead of explicating the nature of it, which . . . is as much as if in an inquiry after the cause of salivation, they should think it enough to tell us, that the several kinds of precipitates of gold and mercury, as likewise of quicksilver and silver . . . do salivate upon the account of the mercury, which though disguised abounds in them; whereas the difficulty is as much to know upon what account mercury it self, rather than other bodies, has that power of working by salivation. Which I say not, as though it were not something (and too often the most we can arrive at) to discover in which of the ingredients of a compounded body the quality, whose nature is sought, resides; but because, though this discovery

33 Ibid.

it self may pass for something, and is oftentimes more than what is taught us about the same subjects in the schools, yet we ought not to think it enough, when more clear and particular accounts are to be had.[34]

Clearly, Boyle did not disdain the effort to locate the proper material subjects of qualities; the kind of program for analyzing bodies recommended by Sennert yielded valuable knowledge. But like Bacon, Boyle felt an urge to go further and discover the principles from which qualities were derived. He was not satisfied to leave them occult, although he was a good enough naturalist and chemist to realize that in most cases, one could not do better. He was fascinated by the siren song promising clear and particular accounts of occult qualities; only a greater desire to observe nature closely prevented him from yielding to the temptation.

In the seventeenth century, John Locke excluded the Cartesian dream of perfect knowledge from the sphere of experimental philosophy. Locke's views, expressed in the *Essay Concerning Human Understanding* (1690), reflect his close association with Boyle and other experimental philosophers as well as his own training as a physician. He accepted the corpuscular doctrine that corporeal properties emanate from the internal constitutions of bodies or, to use Locke's term, their *real essences*. However, forms were hidden from man, and therefore, the particular means by which they produced empirical properties remained occult. If, Locke said, one knew the mechanical affections of the particles in rhubarb, hemlock, and opium in the same way that a watchmaker knows the parts of a watch, he could obtain absolutely certain demonstrations that rhubarb will purge, hemlock will kill, and opium will induce sleep. But man lacked senses acute enough to discover the motions of atoms; hence he had to be content to remain ignorant of the particular causes behind corporeal properties. One could not obtain a degree of certainty about qualities beyond what was warranted by a few experiments. Locke concluded:

And therefore I am apt to doubt that, how far soever human industry may advance useful and *experimental* philosophy in *physical* things, *scientifical* will still be out of our reach: because we want perfect and adequate *ideas* of those very bodies which are nearest to us and most under our command.[35]

---

[34] Boyle, *Works, II*, 41.
[35] John Locke, *An Essay Concerning Human Understanding* (London: Dent, 1961), *II*, 161.

To Locke, the quest for absolute certainty was a delusion. He said, "as to a perfect science of natural bodies . . . we are, I think, so far from being capable of any such thing that I conclude it lost labour to seek after it."[36] Clearly, he was not recommending that one abandon the enterprise of science; he was merely stating the limits inherent in the nature of scientific knowledge. Experimental science could gain knowledge of causes only through the study of their sensible effects. Any attempt to discover the forms behind the qualities would seduce man from the rigorous path of experience.

In Locke's epistemology, the distinction between manifest and occult qualities was largely superseded by the doctrine of primary and secondary qualities. All the properties of a body except the size, shape, texture, and motion of the particles belonged in the class of secondary qualities. Only vestiges of the manifest–occult distinction remained in Locke's division of secondary qualities into the sensible powers and the active and passive powers. There were also shades of Galenic faculties in the term "power," by which he described the causes of all secondary qualities, sensible as well as active. He remarked:

For the colors and taste of *opium* are as well as its soporific or anodyne virtues, mere powers, depending upon its primary qualities, whereby it is fitted to produce different operations upon different parts of our bodies.[37]

In ascribing the soporific virtue of opium to a power, Locke was not intimidated by the Cartesians' ridicule; his epistemology was his reply.

By the end of the seventeenth century, occult qualities not only had been incorporated into modern science; they had become the foundation. Since it is not possible to give an ultimate explanation for corporeal properties, all attempts to account for them must terminate with entities whose properties are, in a sense, occult. This does not mean that they can never be explained; it simply means that the properties of the more fundamental entities are unaccounted for. Today, we take such occult qualities for granted. But prior to the scientific revolution, the existence of unexplained properties constituted an anomaly for natural science, a place where the system broke down. To have scientific knowledge of proper-

[36] Ibid., 164.    [37] Locke, *Essay, I,* 249.

ties, one had to explain them in terms of perfectly manifest qualities. The program for manifesting occult qualities began with the Aristotelians in the Renaissance and flourished in the writings of mechanical philosophers. Only gradually was an experimental method capable of dealing with occult qualities developed. The first step came with the recognition of their importance – a process in which Scholastic physicians and naturalists played an essential part. As the experimental philosophy came into its own, largely through the work of Boyle, occult qualities were incorporated into a program of research. The difference between them and other types of so-called manifest qualities lost its significance. In the end, occult qualities took their place at the foundations of modern science. That we take them for granted represents one of the triumphs of the scientific revolution.

# 9

## Piety and the defense of natural order: Swammerdam on generation

EDWARD G. RUESTOW

Bizarre as it now seems, the belief that all living things had existed invisibly small since the Biblical Creation exercised a powerful appeal in the late seventeenth and eighteenth centuries. It has been argued prrsuasively of late that this appeal rested on a perception of nature as barren mechanism, a perception that agreed nicely with a pious insistence on God's monopoly of all creative power.[1] Indeed, such a union of mechanism and piety would have constituted a compelling synthesis that renders the broad and enduring grip of the doctrine of preexistence more comprehensible.

But a doctrine of such wide appeal doubtless offered several facets capable of speaking to a variety of needs, and the concerns that engendered such a concept in the minds of its earliest proponents need not have been those that underlay its subsequent and more general espousal. The first to have proposed the idea of preexistence in print was Jan Swammerdam, and, responding to the achievements and promise of his own research, he appears to have invested preexistence with a distinctive significance.

In 1669, in his *Historia insectorum generalis,* Swammerdam announced that he had at last penetrated the darkness that had until then surrounded the generation of animals.[2] He now believed, he

---

[1] See Jacques Roger, *Les sciences de la vie dans la pensée française du XVIII<sup>e</sup> siècle,* 2nd ed. (Paris: Armand Colin, 1971), pp. 325–84; Peter J. Bowler, "Preformation and Pre-existence in the Seventeenth Century: A Brief Analysis," *Journal of the History of Biology,* 4 (1971), 221–44; Shirley A. Roe, *Matter, Life, and Generation: Eighteenth-Century Embryology and the Haller–Wolff Debate* (Cambridge: Cambridge University Press, 1981), pp. 2, 5–9, 89–102, 151.

[2] Jan Swammerdam, *Historia insectorum generalis, ofte algemeene verhandeling van de bloedeloose dierkens* (Utrecht: Meinardus van Dreunen, 1669), pt. 1, pp. 51–2. It might be noted

wrote, that no generation took place at all, and what one saw in nature was no more than a propagation and growth of parts. How a parent lacking all its limbs could produce a perfect fetus was now easily explained, for all the parts, Swammerdam added three years later, were contained within the egg.[3] The meaning of the Biblical assertion that Levi, in the loins of his father, had paid a tithe long before his birth was now also clear, and a certain savant with whom Swammerdam had shared his discoveries had further suggested, as Swammerdam related it, that the basis of original sin had perhaps been found as well, because all men would have been similarly enclosed in the loins of their initial ancestors.

These remarks on preexistence are brief and problematic, however, and, although Swammerdam implied that he would have more to say after further research, he added little throughout his life to those few explicit lines in the *Historia*. Three years later, in the *Miraculum naturae sive uteri muliebris fabrica,* he specified the female egg as the site of preexistence, but otherwise simply restated in Latin what was earlier said in Dutch.[4] The statements from the *Historia* stand virtually unaltered in the "great work" he was completing in the last years of his life, published only posthumously in 1737–8 as the *Bybel der Natuure.*[5]

Nonetheless, other passages scattered throughout his work seem also to reflect his adherence to preexistence. Just as a plant grew from a seed already containing leaves and a tiny sprout, he wrote in the *Historia,* so an insect emerged from an egg that likewise contained all its parts.[6] Repeatedly insisting that the egg was the animal itself wrapped in a skin, Swammerdam subsequently described the parts of the insect still within the unlaid egg as growing

---

that Swammerdam's close friend, Melchisédech Thévenot, understood Swammerdam to mean what he said, that there was no generation but only the emergence of what already existed; see Harcourt Brown, *Scientific Organizations in Seventeenth Century France (1620–1680)* (Baltimore: Williams & Wilkins, 1934), p. 281.

[3] Swammerdam, *Miraculum naturae sive uteri muliebris fabrica* (Leiden: apud Severinum Matthaei, 1672), p. 21.

[4] Ibid., pp. 21–2.

[5] Swammerdam, *Bybel der Natuure* (Leiden: Isaak Severinus, Boudewyn Vander Aa, and Pieter Vander Aa, 1737–8), I, 34. He often anticipated this work in his letters as the "great work"; see for instance G. A. Lindeboom, ed., *The Letters of Jan Swammerdam to Melchisédec Thévenot* (Amsterdam: Swets & Zeitlinger, 1975), p. 75.

[6] *Historia,* pt. 1, p. 64. Swammerdam spoke here of the insect "seed," *saat,* as well, but the context and other passages equating "seed" and insect "egg" make his meaning clear; see ibid., pt. 2, pp. 2, 18. *Bybel,* I, 273.

from invisible but nonetheless "real," "substantial," or perhaps "essential" *(wesentlyke [wezenlijk])* parts that preceded them.[7]

He conceived of the egg in these terms even before it was fertilized, a necessary assumption with respect to preexistence. He reported in the *Historia* that he had discovered the beginnings of the eggs of the moth *Orgyia antiqua* already within the caterpillar and had observed their development through the chrysalis and into the imago.[8] Concluding again that the egg was the animal itself, Swammerdam took these observations as particularly enlightening evidence that no essential change took place in the insect's parts.[9] The logic is obscure, but not the fact that the egg he was referring to – growing visibly with its parent, as he also noted[10] – included the egg before its fertilization. In the *Miraculum,* he declared not only that the black dot in the frog's egg was the frog complete with all its parts, but that the animal itself was hiding in that dot when the egg was still in the ovary, necessarily before fertilization as well, as Swammerdam knew.[11] The *Bybel* reaffirmed that the "entire animal" was fully perfected in the frog's ovary.[12]

Apparently influenced by William Harvey's "incontrovertible demonstration" that no semen entered the uterus,[13] Swammerdam in effect minimized the significance of fertilization for the physical formation of the animal. He ascribed to fertilization the communication of some enhanced quality of motion and perhaps "life" itself, though he was ambiguous on this latter point. In the *Historia,* he emphasized that the eggs of the female insect were not only living before fertilization but endowed with feeling and movement as well, which feeling, movement, and life the male semen simply enabled to persist.[14] The human egg in the ovary is also described in the *Miraculum* as already alive, receiving from

---

[7] *Bybel,* I, 41; II, 568, 603–4, 728. *Historia,* pt. 2, p. 30.

[8] *Historia,* pt. 2, p. 30. *Bybel,* II, 565. A. Schierbeek, *Jan Swammerdam (12 Februari 1637 – 17 Februari 1680), zijn leven en zijn werken* (Lochem: "de Tijdstroom," [1947]), p. 267 (regarding plate 33). I am grateful to L. Lavenseau of the Laboratory of Neuroendocrinology, the University of Bordeaux, for confirming in a letter dated 26 September 1984, that Swammerdam, despite the primitive optical instruments at his disposal, might indeed have observed the beginnings of the future oocytes as spots in the ovarioles of the female *Orgyia antiqua* at the end of its last larval stage.

[9] *Bybel,* II, 568.  [10] *Historia,* pt. 2, p. 30.  [11] *Miraculum,* p. 21. See *Bybel,* II, 809–10.

[12] *Bybel,* II, 470.

[13] *Bybel,* II, 515–16. See William Harvey, *Exercitationes de generatione animalium* (London: typis Du-Gardianis, 1651), pp. 226–9.

[14] *Historia,* pt. 1, pp. 64–5. Also *Bybel,* I, 44. Swammerdam wrote here of insect "seeds" *(saadkens),* not "eggs," but see note 6.

fertilization only a "more perfect motion" by which it could proceed to live independently of the ovary, but in a later study of a species of viviparous snails, he now spoke of the semen's providing life as well as movement.[15] It was clear, nonetheless, that, whatever fertilization offered the egg, the physical form of the animal was already there, for Swammerdam imagined the snail egg as a watch that, though subsequently given life and movement by the semen, was brought forth with all its parts by the female alone.

Needless to say, Swammerdam's understanding of the egg could not be demonstrated by observations, and when he did presume to do so, his logic was shaky and obscure. Ultimately, to the extent that he continued to adhere to the idea of preexistence, he did so for other reasons than the dictates of what he had observed.

Like the moth's eggs within the caterpillar, certain observations were surely very suggestive to a mind suitably inclined, and the microscope in particular provided what must have been exciting moments. Swammerdam discovered the ovaries in the future queen bee still within its cell, ovaries that were already closely packed with what he estimated to be ten to twelve thousand eggs.[16] Behind the eggs ready to be laid by the mature queen bee, the microscope also revealed the beginnings of further eggs, progressively diminishing in size until they escaped his finest lens.[17] He assumed that they continued on in invisible smallness, however, to be seen and counted, he wrote, only by their Creator.

But the microscope had little else to offer. Opening beetle eggs on the verge of being laid, Swammerdam could apparently see nothing beyond a whitish fluid, and, although he was thrilled by the sight of tiny embryos developing in the eggs of the viviparous snail, the embryos disappeared from sight in the smaller eggs, even though those eggs had already left the ovary.[18] He attributed the invisibility of the embryos in the smaller eggs to their transparent delicacy, but it took a greater act of faith to disregard what the frog egg ultimately revealed to the microscope.

In studies published only in the *Bybel*, Swammerdam systematically observed the beginnings of the frog's embryonic development. He described the newly laid egg as a small, black globule whose most pronounced feature by the second day was a deep

---

[15] *Miraculum*, p. 24–5. *Bybel*, I, 178.    [16] *Bybel*, II, 477–8.
[17] Ibid., 471–2. See also ibid., 604.    [18] Ibid., I, 175, 303.

furrow that ran around it and nearly divided it in two.[19] He could at this point see no entrails with his microscope and no other contents than small, yellowish granules that made up the substance of the egg. Vague entrails subsequently began to appear, and by the fifteenth day a distinct intestine had taken shape, but, like the rest of the emerging animal, loosely composed, it seemed to Swammerdam, of the globular granules of the original egg.

The critic of preexistence might have anticipated arguments against preformation put forward a century later, and insisted that the visibility of the granules before the larger parts they subsequently composed testified that those larger parts were not originally present even in rudimentary form.[20] Swammerdam gives no hint of sensing such implications, however. He still asserted in the *Bybel* that the black spot in the frog's egg was the little animal, the tadpole in its initial vestment *(rok)*, the frog itself, and, although he no longer described it as complete with all its parts, he still maintained that the entire animal in the form of the egg reached its full perfection in the ovary.[21]

Such an instance, in which Swammerdam clung to the concept of preexistence in virtual defiance of his observations, underscores a predisposition to preexistence that had roots deeper, indeed, than empirical experience. But Swammerdam's thinking was *not* deeply colored by the mechanist commitment associated with the new sense of nature's barrenness, her inability to produce the complex, purposeful structure of living things. He had, it is true, grasped and exploited the fruitfulness of a mechanistic outlook in physiological experimentation and explanation. The disputation with which he completed his medical training at Leiden argued through instrumental demonstrations and mechanical analogies that, in respiration, the air was inhaled as well as exhaled by being pushed, not by attraction or an imaginary avoidance of a vacuum.[22] He embraced Descartes's doctrine of rarefaction and condensation and often referred in that disputation to a "subtle matter," a hallmark of the Cartesian influence in medicine and physics.[23]

---

[19] Ibid., *II*, 789, 813–19.

[20] Howard B. Adelmann, *Marcello Malpighi and the Evolution of Embryology* (Ithaca, N.Y.: Cornell University Press, 1966), *II*, 921–4.

[21] *Bybel*, *I*, 42–3; *II*, 470, 789, 812–15.

[22] Swammerdam, *Tractatus physico-anatomico-medicus de respiratione usuque pulmonum* (Leiden: apud Danielem, Abraham. et Adrian. à Gaasbeeck, 1667).

[23] Ibid., pp. 46, 49, 71, 72, 119.

Nonetheless, the mechanistic emphasis of the disputation is atypical of his other writings, just as the characterization of the snail egg as a watch with all its parts was uncharacteristic imagery for Swammerdam.[24] Even in his disputation, moreover, Swammerdam's mechanism pertained essentially to the macrostructure of the body rather than to the kinds of processes beyond the reach of the senses with which the mechanical philosophy was preoccupied.

Perhaps nothing better illustrates Swammerdam's indifference to the comprehensibility that was the goal of mechanism than the way he expressed himself with respect to preexistence, embraced by many of its proponents precisely in order to render generation comprehensible. Unlike their predecessors, whose mechanism had been more dynamic, late seventeenth-century mechanists increasingly found generation to be inconceivable in mechanistic terms. Granting that all things in the world were produced only by the laws of motion, Malebranche, often linked with Swammerdam in the authorship of preexistence, would not allow that those laws could form so complex and purposeful a machine as a plant or animal body.[25] But those laws *were* adequate, he maintained, for a simple growth and unfurling (*pour développer*) of parts that were already organized in a body; hence, he concluded, to explain generation, the preexistence of the plant or animal in miniature had to be assumed. In his own remarks on preexistence, however, Swammerdam shows no such concern with the explanation or comprehensibility of the processes involved.

It is not clear what Swammerdam conceived to be happening as the parts of the preexistent animal developed. His image of the snail embryo as a watch with all its parts conforms nicely to the more developed explanation of preexistence provided by Malebranche, as does Swammerdam's assertion that the parts within the egg come from invisible but *wezenlijk* antecedents. Deriving in large part from the verb *groeien,* "to grow," however, his choice of words to characterize the process by which those parts emerged is not without its problems. In the passage on the growth of parts

---

[24] Besides the analogy between the snail embryo and a watch, the only reference that I can cite in Swammerdam's entomological works to the animal or, in this instance, the human body as a machine is *Bybel, II,* 859.

[25] Nicolas Malebranche, *Entretiens sur la métaphysique et sur la religion,* in *Oeuvres complètes,* dir. André Robinet (Paris: J. Vrin, 1958–70), *XII,* 228–9, 252–3.

from *wezenlijk* beginnings, that process was expressed in terms of the verbs *aanwassen* and *algroeien;*[26] the former also means simply "to grow or increase," but *algroeien* is uncommon and perhaps idiosyncratic and its innuendos concerning the nature of the process at issue are unclear. The growth of the parts of the eggs within the caterpillar is rendered by a derivative of *bijgroeien,*[27] also an unusual verb at the time. It is presently translated as signifying "to grow again" or "heal," as in the case of a wound, a meaning that makes little sense in the context in which Swammerdam uses it. In the early eighteenth century, however, Hieronymus David Gaubius of the medical faculty at Leiden translated the word in question as "epigenesis."[28] In his explicit statement of preexistence, moreover, when he wrote that no more than a propagation or growth of parts was observed in nature, Swammerdam used a derivation of the verb *aangroeien,* for what I have translated as "growth," and although the verb *aangroeien* is also straightforward in its meaning "to grow or increase," Swammerdam uses its derivatives – characteristically with *toesetting,* to be sure – when explicitly translating the word "epigenesis" himself.[29]

"Epigenesis" was a term apparently coined by Harvey to represent a process of generation that, as explained by him, was quite alien to the mechanistic mind. For Harvey, it signified the successive emergence of differentiated parts from a homogeneous substance, and, rather than attempting to explain how this happened in terms of comprehensible natural activity, Harvey likened the whole process to the enactment of successive commands of God.[30] The concept of epigenesis aroused little support at the time,[31] but Swammerdam appears to have had no reservations about it and explicitly embraced it in another context, as we shall see. His own conception of preexistence, consequently, may have entailed the acceptance of this obscure and nonmechanistic process of organic development. At a minimum, Swammerdam was being none too precise in portraying the processes he had in mind.

Indeed, although Swammerdan denied the generation of the an-

---

[26] *Bybel, I,* 41.    [27] Ibid., *II,* 568.    [28] Ibid.
[29] *Historia,* pt. 1, pp. 43, 51, and compare the end of the quotation from Harvey and Swammerdam's translation of the same on pp. 37, 40.
[30] Harvey, *De generatione* [see note 13], pp. 121–2, 124–5.
[31] See Walter Pagel, *William Harvey's Biological Ideas, Selected Aspects and Historical Background* (New York: Hafner, 1967), pp. 233, 237, 239; F. J. Cole, *Early Theories of Sexual Generation* (London: Oxford University Press, 1930), p. 144.

imal itself, he retained more than a vestige of the idea of generation with respect to the development of the animal's parts. In his initial statement of preexistence in the *Historia*, Swammerdam wrote in Dutch that he believed no *teeling* [*teling*] took place but only a growth or *voortelling* [*voortteling*] of parts.[32] He himself rendered *teling* as the Latin *generatio* three years later and *voortteling*, which I translated above as "propagation," as *propagatio*.[33] In fact, Swammerdam otherwise typically used *voortteling* to denote animal reproduction and both *teling* and *voortteling* interchangeably to translate the title of Harvey's book on generation.[34] If Swammerdam, like the mechanists, was troubled by the incomprehensibility of the generation of complex and purposeful structures, it had not prevented him from immediately suggesting it again at another level.

Swammerdam's epistemological priorities were indeed fundamentally opposed to the essential thrust of the mechanical philosophy and to the elevation of reason over sense experience, to which Malebranche and other subsequent adherents of preexistence frequently resorted.[35] Swammerdam emphasized his own distrust of reason divorced from observation and, twisting a Cartesian catchword, asserted that clear and distinct knowledge must rest on experience and the senses.[36] Contrary to the mechanist thrust (as well as to the scientific orientation that long lay behind the attraction of preformationary doctrines) was also his insistence that only effects, not causes, could be known.[37] To Swammerdam, even the causes of the growth of animals and their parts were impenetrable,[38] and his empirical researches constantly reinforced a sense of nature's ultimate inscrutability, an inscrutability that he actually relished for its religious suggestiveness.[39]

If a commitment to the mechanical philosophy was not a source of Swammerdam's attraction to preexistence, neither can his piety alone provide a sufficient explanation. Confronted with the every-

---

[32] *Historia*, pt. 1, p. 51.    [33] *Miraculum*, p. 21.    [34]*Bybel*, I, 23–4, 34.

[35] See the Malebranche, *La Recherche de la vérité*, in *Oeuvres complètes*, I, 80–3; Nicolas Andry, *De la génération des vers dans le corps de l'homme* (Paris: Laurent d'Houry, 1700), 296–8; Henry Baker, *The Microscope Made Easy* (London: R. Dodsley, 1742), p. 252.

[36] *Historia*, pt. 1, pp. 147–8, 153–4; pt. 2, pp. 5–10. *Bybel*, II, 625, 868–9.

[37] *Historia*, pt. 1, p. 148; pt. 2, pp. 5, 8. *Bybel*, II, 625, 867–8, 870–1.

[38] *Bybel*, II, 784–5.

[39] Ibid., I, 169, 230; II, 498–9, 594, 597–8, 664. Swammerdam, *Ephemeri vita* (Amsterdam: Abraham Wolfgang [1675], pp. 78–9, 92.

day experience of the continuing emergence of living things, Augustine had resorted to a doctrine of preexistent *rationes seminales* or *rationes causales* to uphold the Biblical account of a Creation that lasted only six days, and Malebranche also wove the brevity of the Biblical Creation into his treatment of preexistence.[40] But the temporal conclusion of Creation does not appear as an important consideration for Swammerdam, who did not even allude to it in his impassioned attacks on spontaneous generation; insisting on the abeyance of God's creativity may have clashed too sharply with his sense of divine omnipotence. Even the creation of the preexistent plants and animals themselves, a point developed more emphatically in later preexistence doctrine, is alluded to only indirectly in Swammerdam's reference to original sin.

Preserving the immediate derivation of all living things from God's creating hand was nonetheless an important point to Swammerdam, rendered more acute, perhaps, by an urgent personal need. His religious crisis of the mid-1670s revolved in part around his revulsion for the personal pleasure and thirst for recognition that had increasingly become the motivation, he then confessed, for the major preoccupation of his life, his scientific research.[41] For some time, consequently, he had perhaps been struggling to justify his scientific passion all the more insistently as the pious contemplation of the Creator's handiwork.

If that religious need was the primary source of his recourse to preexistence, however, why did he fail to consider the possibility that generation might indeed be a new creative act of God? Protesting in pious indignation against Scholastic teachings that sharply distinguished between generation as natural and creation as not, Johannes Baptista van Helmont had insisted earlier in the century that generation was, after all, the continuing creation of new forms from nothing by God, and Harvey's elaborate treatment of epigenesis was designed to demonstrate the ultimate guidance of divine intelligence.[42] In a poetic vein, Swammerdam himself wrote of God's pulling and tugging at the early human embryo, and he

[40] R. A. Markus, "Marius Victorinus and Augustine," in A. H. Armstrong, ed., *The Cambridge History of Later Greek and Early Medieval Philosophy* (Cambridge: Cambridge University Press, 1970), pp. 397–9. Malebranche, *Entretiens* [see note 25], XII, 229.

[41] *Ephemeri* [see note 39], pp. 180, 239, 245–6.

[42] Johannes Baptista van Helmont, *Oriatrike or, Physick Refined*, trans. J. C. (London: Lodowick Loyd, 1662), pp. 130–3. Harvey, *De generatione* [see note 13], pp. 90, 113, 125, 127–8, 141, 144–7, 169–170, 249.

cited the Biblical account that spoke of the miraculous engendering of lice throughout Egypt as an irrefutable sign of the finger of God.[43] Swammerdam's religious preoccupations were indeed intense and deeply interwoven with his view of nature, but his insistence on a preexisting animal that forestalled any new creative act suggests that their relevance to the doctrine of preexistence is problematic.[44]

Since Swammerdam does not conform to the usual explanation of that doctrine's appeal, and since, moreover, his own remarks about preexistence are too brief to provide the basis for an alternative approach, let us temporarily set the question of preexistence aside and turn to other aspects of Swammerdam's thought that appear to be related to preexistence but are, at the same time, more fully developed. In contrast to the sparseness of his passages on preexistence, his commentaries on metamorphosis are recurrent, elaborate, and often ardent; moreover, they frequently echo his formulations regarding preexistence. By exploring his treatment of metamorphosis, perhaps we can discover decisive concerns that may help us to understand his attraction to preexistence as well.

Throughout his entomological researches, Swammerdam maintained a continuing campaign against the common understanding of metamorphosis as the transformation of one animal into another, and one of his most exhilarating moments as a naturalist was when he found the butterfly and its parts "hidden in the caterpillar and all rolled up and locked within its skin"; it was the discovery Swammerdam prized above all others, wrote his close friend and patron Melchisédech Thévenot in 1669, the year the *Historia* appeared.[45] As Swammerdam explained in that work, the brandy and vinegar in which he soaked a caterpillar on the verge of pupation not only killed the insect but hardened its parts, so that, when the skin was removed a few days later, the parts of the butterfly were revealed folded beneath.[46] The transformation

[43] *Ephemeri*, pp. 421–2. *Bybel, I*, 67.

[44] As we have seen, Swammerdam did associate preexistence with the exegesis of the Biblical passage on Levi's being in his father's loins and with the doctrine of original sin, but in both cases these religious associations seem to be afterthoughts rather than expressions of a formative religious concern.

[45] *Historia*, pt. 1, p. 27; pt. 2, pp. 41–2. *Bybel, II*, 603. Brown, *Scientific Organizations* [see note 2], p. 281.

[46] *Historia*, pt. 2, p. 47; *Bybel, II*, 612. Swammerdam also offered an alternative technique whereby the caterpillar, suspended by a thread, was dipped in boiling water (*Historia*, pt. 2, pp. 43, 44; *Bybel, II*, 605).

of the caterpillar into a chrysalis was no more, therefore, than the growth or filling out of the parts of the caterpillar that provided the shape of the future butterfly, and the process of metamorphosis was only the gradual shedding of coverings that had concelaed those parts.[47] Those who had assumed metamorphosis to be a change on the inside as well as the outside of the animal were wrong, and the moth or butterfly was not a new, transformed animal but only an old animal that had shed its skin.[48]

What was true of the metamorphosis of the moth or butterfly was true of all insect change as well. Although much depended on the hand and eye of the observer, wrote Swammerdam, the form of the future imago could be revealed in every pupa, as he had shown in a variety of insects.[49] Hence, that the "whole stuff" of the worm was transformed into a pupa and the same again into a creeping, flying, or swimming animal was wholly false.[50] The pupa was "the animal itself" and contained all the parts of the future imago; this was the most important truth in the theory of insects, he wrote, and on it rested the doctrine of all their changes.[51]

The significance of Swammerdam's conception of metamorphosis extended even beyond the realm of insects, however. It characterized the development of plants and red-blooded animals as well, so that Swammerdam identified not only the development of the tadpole into a frog with insect metamorphosis but that of the chick into a hen as well, there being no other change than the growth of its limbs, he asserted.[52] In the *Bybel,* he would liken the development of the human embryo in the womb to metamorphosis.[53]

Swammerdam also likened the swelling of hidden parts and their final revelation in metamorphosis to the burgeoning blossom within the bud.[54] Hence, the true nature of metamorphosis that Swammerdam believed he had discovered became a prototype for all organic change; it revealed that all the "natural changes" of living things were fundamentally alike, he wrote, so that the same rules appeared to underlie all God's work.[55]

[47] *Historia*, pt. 2, pp. 29–30. *Bybel, I*, 6, 37.
[48] *Historia*, pt. 1, p. 19. *Bybel, I*, 14, 37; *II*, 567.
[49] *Historia*, pt. 1, pp. 13–17, 20–2, 26–7. *Bybel, II*, 715.
[50] *Bybel, I*, 28.  [51] *Historia*, pt. 1, pp. 7, 14, 21, 30, 33, 42, 44, 145. *Bybel, I*, 6, 9, 19.
[52] *Historia*, pt. 1, pp. 8–9, 62–3; pt. 2, pp. 35–7, 40. *Bybel, I*, 5–6.  [53] *Bybel, II*, 792–3.
[54] *Historia*, pt. 1, pp. 62, 64, 87, 95, 100, 101–2; pt. 2, pp. 38–40. *Bybel, I*, 5, 19, 21–22, 44, 210–11, 221, 271–2; *II*, 416–17.
[55] *Historia*, pt. 1, p. 30; pt. 2, p. 40. *Bybel, I*, 19, 21–2.

Swammerdam was clearly unwilling to consider generation any less of a "natural change" than metamorphosis, and the similarities between his characterization of metamorphosis, on the one hand, and his few comments pertinent to preexistence, on the other, are difficult to overlook. He often likened the insect egg to the pupa and noted early in the *Historia* that a clear understanding of the pupa was the foundation of the understanding of the egg as well.[56] In both cases, what occurred was only an unveiling of parts that had grown beneath a concealing cover and then dried out from a state of fluidity to firmness,[57] and it was in similar terms and with similar emphasis that he insisted that both egg and pupa were the "animal itself."

As in the case of preexistence, Swammerdam's representation of the processes at work in metamorphosis was not without its difficulties for the mechanistic imagination. He characteristically rendered the process of growth in metamorphosis with derivative forms of *aangroeien,* but the growth or addition of parts (*een toevoeging ofte toedoening*) that preceded chrysalid metamorphosis was now explicitly identified in the *Historia* with epigenesis, and in the *Bybel* the causes of the growth of a larva into a pupa and that again into a fly were declared to be as inscrutable as the Divinity that made it happen.[58]

Swammerdam had initially touted his new doctrine of metamorphosis as replacing obscurity with understanding. For centuries, he wrote, writers had searched in vain for the reasons governing metamorphosis, and in the common view, he added, nothing was more amazing than the transformation of a caterpillar into a winged beast.[59] Metamorphosis as conceived specifically by the "immortal Harvey" was not only contrary to experience, according to Swammerdam, but incomprehensible as well.[60] But once the true nature of this change was recognized, he declared, it was no more amazing than the growth of other animals or the burgeoning of a flower bud, so that all its incomprehensibility lay only "in our imaginings."[61]

---

[56] *Historia,* pt. 1, pp. 7–8, 148–9, 86, 103. *Bybel,* I, 273; II, 603–4, 728.

[57] *Historia,* pt. 1, pp. 15, 42–3, 55–6, 58, 59, 60–1, 72, 86, 103, 143, 144, 148; pt. 2, p. 21. *Bybel,* I, 28, 37, 41–2; II, 716.

[58] *Historia,* pt. 1, 38, 40, 101. *Bybel,* II, 680. Swammerdam dropped the explicit identification with epigenesis in the *Bybel* (see I, 272), however, although he did commit himself, with respect to metamorphosis, to precisely what Harvey meant by epigenesis, a successive growth of parts (*Bybel,* I, 19, 28).

[59] *Bybel,* I, 5; II, 580.    [60] *Bybel,* I, 14.    [61] *Historia,* pt. 1, pp. 27, 29. *Bybel,* I, 5.

His continuing study of insect metamorphosis changed his outlook dramatically, however. Having cultivated new skills in insect
dissection, Swammerdam declared in the *Bybel* that the growth of
the grub of the soldier fly *Stratiomyia furcata* into the fly itself sur-
passed all human understanding.[62] In addition, the changes he had
now observed taking place in the internal anatomy of the soldier
fly compelled him to defy observations for the sake of his concept
of metamorphosis as he had done for the sake of preexistence.
Those internal changes struck him as miraculous, as if there had
been a new creation – "or, rather, generation" – or a resurrection
in another body.[63] He acknowledged that what he now had seen
might well have provided good grounds for earlier conclusions
that metamorphosis was a radical transformation. Nonetheless, he
himself still maintained that the same animal persisted throughout
and that even the members of its body remained the same, though
some were lost and others grew in their place.[64] Despite his vig-
orously asserted empirical epistemology, Swammerdam appears
to have also had compelling nonempirical reasons for his insistence on the enduring animal in metamorphosis.

Central to Swammerdam's concern with metamorphosis was
its association with spontaneous generation. It was widely assumed that the caterpillar died in metamorphosis and that the butterfly was engendered from its putrefaction,[65] and the general cur-
rency of this understanding of metamorphosis is reflected in some
strange reports of ostensible observations. Thomas Moffett related that in the chrysalis the butterfly's tail appeared where the
caterpillar's head had been, and Johannes Goedaert testified that
what had been the belly of the one became the back of the other.[66]

But other reports that seemed even more bizarre rested on substantial fact. At the turn of the century, accounts had appeared of
caterpillars that on occasion produced not butterflies but flies, and,
midcentury, Goedaert cited a number of instances of caterpillars

[62] *Bybel*, *II*, 666. Schierbeek, *Jan Swammerdam* [see note 8], p. 267 (regarding plates 39–
42). See also *Bybel*, *II*, 680.

[63] *Bybel*, *II*, 666, 680–1.

[64] Ibid., 666. Anticipating some twentieth-century interpretations, Swammerdam remarked that it was as if a grown animal had reverted to its infancy to grow up again in
a more glorious body; ibid., 681.

[65] See for instance Teresa of Avila, *The Interior Castle*, trans. Kieran Kavanaugh and Otilio
Rodriguez (New York: Paulist Press, c.1979), 91–4.

[66] Thomas Moffett, *Insectorum sive minimorum animalium theatrum* (London: ex Officina ty-
pographica Thom. Cotes, 1634), pp. 182, 319. Johannes Goedaert, *Metamorphosis naturalis* (Middelburg: Jaques Fierens, [1662?–9?]), *I*, 9.

or chrysalides that produced large numbers of flies or worms that pupated and then engendered flies.[67] How strange and amazing it was, mused Goedaert's editor Johannes de Mey, that the same kind of caterpillar generated in two different ways.[68] It only strengthened his belief – who would doubt it, he asked – that the flying imagoes were a different kind of insect than the worm or caterpillar from which they sprang.[69] The Amsterdam physician Steven Blanckaert wrote of opening chrysalides crammed with up to 500 worms, and Swammerdam himself remarked that it seemed as if the life and movement of a few animals had divided into a multitude of others.[70]

But it was Harvey who framed the terms of the issue of metamorphosis for Swammerdam. In 1651, Harvey had not only ascribed the emergence of the butterfly to the putrefaction of the caterpillar but had virtually identified a broadened conception of metamorphosis with spontaneous generation.[71] He insisted on the absolute nature of the transformation in metamorphosis and on the complete absence of internal differentiation and structure within the chrysalis from which the butterfly was formed. Moreover, he specified that the organs of the butterfly emerged within the chrysalis in a transformation of the whole mass of the homogeneous "putrescent matter" of the former caterpillar.

Swammerdam was among those in the late seventeenth century who now found the concept of spontaneous generation profoundly irreligious. He repeatedly associated it with atheism and denounced it as a godless opinion sustained by a heathen philosophy.[72] Consequently, he could perceive the idea of one animal becoming another through metamorphosis as pointing to godlessness as well.

The heart of the issue was chance; for Swammerdam in particular, it was a choice between the Creator and "all the atheists who say the generation of these small animals is by accident."[73] Always conceiving spontaneous generation as generation by accident, he

---

[67] Moffett, *Insectorum theatrum*, p. 192. Ulisse Aldrovandi, *De animalibus insectis libri septem* (Bologna: apud Ioan. Bapt. Bellagambam, 1602) p. 253. Goedaert, *Metamorphosis* [see note 67], I, 4, 42–5, 146–7; II, 239–241.

[68] Goedaert, *Metamorphosis* [see note 67], I, 42n.    [69] Ibid., 201.

[70] Steven Blanckaert, *Schou-burg der rupsen, wormen, ma'den, en vliegende dierkens daar uit voortkomende* (Amsterdam: Jan ten Hoorn, 1688), pp. 20, 57, 76, 88. *Bybel*, I, 280.

[71] Harvey, *De generatione*, pp. 121–2, 188–9.

[72] *Bybel*, I, 171; II, 394, 432, 669, 708, 712, 713.    [73] *Bybel*, II, 669.

gave himself no less to the destruction of "accidental generation" than to the destruction of the concept of radical metamorphosis.[74] They were virtually the same thing, after all, and Harvey identified his own doctrine of metamorphosis not only with spontaneous generation but explicitly with chance as well.[75] According to Swammerdam, consequently, the discovery that metamorphosis was no more than an unveiling of existing parts also overturned the accidental generation of animals.[76] Underlying the urgency of Swammerdam's assault on the traditional concept of metamorphosis was in part, then, a broader concern to deny chance any place in the formation of living things.

If the specter of chance was the goal behind Swammerdam's opposition to older views of metamorphosis, we might well ask what chance meant to Swammerdam and why it was so abhorrent. On occasion, by associating spontaneous generation with the "accidental confluence" of particles,[77] Swammerdam seemed to echo the association of chance with the random motion of classical atomism, in which case the ominous implication in many key passages would seem to have been the denial of purposeful design.

Not only was it inconceivable and absurd to Swammerdam that the intricacy of insect anatomy could be the product of chance, it was impious as well,[78] for the purposeful design that was denied was God's. The denial of God's guiding hand in the formation and metamorphosis of insects – as opposed to the epigenesis of higher animals – could easily be read into Harvey's account,[79] but Swammerdam protested that a parentage of chance and putrefaction belied the wisdom and providence in the lives and design of insects and cast into doubt his insistence that they testified to the Deity.[80] For Swammerdam, it was indeed the prospect that one animal could become another through accident (and putrefaction) that

[74] Ibid., 1, 274.   [75] Harvey, *De generatione* [see note 13], p. 122.

[76] *Historia*, pt. 1, p. 44.   [77] Ibid., pt. 2, pp. 394, 432.

[78] *Ephemeri*, p. [11] of "Ernstige Aanspraak." *Bybel*, I, 85, 106, 171, 346; II, 394, 432, 707–8, 713.

[79] Harvey's whole approach to the study of epigenesis as exemplified in the developing chick egg had been to mount thereby to "the first and supreme efficient cause" (*De generatione*, [see note 13], 141), which turned out, of course, to be God [see note 40], but the first cause of metamorphosis – or spontaneous generation, that is – was not an external efficient cause but matter itself, hence, he continued, the diminished perfection of the animals so produced (*De generatione*, p. 122). Elsewhere, however, Harvey did affirm that the engendering of *all* animals was to be ascribed to God (ibid., pp. 146–7).

[80] In addition to the passages cited in note 77, see *Historia*, pt. 1, 23–4.

rendered Harvey's doctrine of metamorphosis a straight road to atheism.[81]

Swammerdam was deeply committed, it is clear, to preserving God's direct role in the creation of all creatures, insects no less than man. It was undeniably a compelling motivation in his assault on spontaneous generation and, hence, on the common understanding of metamorphosis. As we asked in the case of preexistence, however, if Swammerdam's essential concern was piety, why was he unwilling simply to leave the transformation within the chrysalis in the hands of God, as others, he knew, had done?[82] Swammerdam himself declared that the transformation of the tadpole revealed in splendor the almighty arm of God, and he similarly saw God moving and visibly revealing His divinity in the metamorphosis of the soldier fly.[83] Why, then, despite the dramatic transformation he observed, did he insist on the persistence of the animal rather than allowing that almighty arm to frame another?

The mechanists and other savants of the late seventeenth century were no more inclined than the Scholastics had been to allow the direct intervention of divinity in generation.[84] The reasons adduced for this scruple, when they were adduced at all, were generally solicitous of the image and dignity of God, to safeguard Him from responsibility for nature's deformed "monsters" or to avoid demeaning His omnipotence with the slow processes of nature.[85] But Swammerdam resorted to none of these arguments to keep God at a distance.[86]

Swammerman simulataneously extolled and stayed the hand of God in metamorphosis, I suggest, because he was concerned to preserve not only purposeful design in every creature but a stable order in nature as well. Moreover, I shall argue that the defense of order no less than design lay behind his campaign against chance

[81] Bybel, II, 669.    [82] See Historia, pt. I, p. 32.    [83] Bybel, II, 666, 828–9.

[84] Regarding the early seventeenth-century Scholastics, see Franco Burgersdijk (who taught at Leiden), Collegium physicum, 2nd ed. (Leiden: ex Officinâ Elziviriorum, 1642), p. 255.

[85] John Ray, The Wisdom of God Manifested in the Works of the Creation (1691; reprint, New York: Garland, 1979), 34–5. Nicolaas Hartsoeker, "Extrait Critique des lettres de feu M. Leeuwenhoek," Cours de physique (The Hague: Jean Swart, 1730), p. 20. See also Ralph Cudworth, The True Intellectual System of the Universe (1678; reprint, Stuttgart-Bad Cannstatt: Friedrich Frommann, 1964), p. 150.

[86] Nor, as John Ray had done, did Swammerdam resort to the Biblical testimony that God had finished all his labors, after all, on the sixth day of Creation; John Ray, Synopsis methodica animalium quadrupedum et serpentini generis (London, 1693), p. 15.

as well – for he did protest that to ascribe the birth of insects to accident and putrefaction also rendered the firm order of nature completely accidental[87] – and, though order and design might often seem nearly indistinguishable, order acquired a distinctive significance for Swammerdam as a research naturalist.

Swammerdam also vested the word *order* with diverse connotations. He observed it in the intricate construction of small creatures and in both the arrangement and the ingenious precision of the operation of their parts.[88] In such usages, order was close in meaning to that purposeful complexity that bespoke the hand of God. But of no less importance to Swammerdam was the order he perceived in the regularity and specificity of insect life cycles.

Each animal, he wrote, had its ordained time, place, manner of life, and particular food, and generation repeatedly recurred according to an unalterable rule and order.[89] Though it was reflected even in the consistency of a moth laying its eggs, Swammerdam found the most vivid manifestation of that "very wonderful order of nature" in his observations of parasites.[90] Several different but consistent patterns in the life cycles and behavior of the flies and wasps that parasitized caterpillars and chrysalides conformed every year to an inflexible order, he noted, and the flies responsible for the galls on willow leaves deposited their eggs in the buds according to a constant order that was likewise repeated every year.[91] That firm and inviolate order was revealed in the very identification of specific parasites with specific hosts.[92]

Revealing his sense of the fundamental opposition between chance and ordered life cycles, he repeatedly cited that order as belying the assumption that chance or accident played a role in the generation of these parasites.[93] But I would urge that, although order is cited more often than not as an argument against chance, his denial of chance, conversely, was in no small part a defense of order, preventing the firm order of nature, as he had written, from becoming accidental. That even environmental circumstances (dampness and weather) might shape the regular course of metamorphosis was rejected by Swammerdam as chance, and assum-

[87] *Historia,* pt. 1, p. 3.    [88] Ibid. *Bybel, II,* 708, 713, 827, 909.    [89] *Bybel, II,* 710.
[90] *Historia,* pt. 1, pp. 152, 154; pt. 2, p. 30.    [91] *Bybel, II,* 711–2, 749.
[92] *Historia,* pt. 2, p. 2. *Bybel, II,* 710. With respect to parasiticism and "order" or regulation in generation, see also *Bybel, II,* 633, 713.
[93] *Historia,* pt. 1, p. 23; pt. 2, p. 40. *Bybel, II,* 710–12, 737.

ing even that deviations in the behavior of the caterpillar (eating too little or pupating too soon) could result in malformed imagoes, he wrote, was to make the unalterable foundation of insects and their development wholly accidental.[94] Chance was offensive both because of its irreligious implications and its threat to natural order, and whereas, in one instance, observed natural order was marshalled for the struggle against chance, in the second, that struggle was waged for the sake of order.

The order he seized upon in the life cycles of insects had several levels of significance for Swammerdam. At a very general level, it confirmed that a reassuring order prevailed everywhere. If the repeated generation, rearing, growth, and metamorphosis of the ant and the fly and other insects were regulated and fixed, he wrote, who would dare deny the fitting government of all the parts of the universe?[95] And who would hesitate to place himself calmly under the Creator, he continued, accept His arrangements with contentment, and humbly bow down beneath His Hand?

A subtly converse significance, however, the affirmation, rather, that the insect world was not *excluded* from that universal order of the rest of nature, may have been of greater moment to Swammerdam the naturalist. He had protested the distinction drawn between the larger animals, called perfect, and the small *ondieren* (literally, "monsters") born from accident and putrefaction.[96] It was a generally accepted distinction, basic as well to Harvey's treatment of generation.[97] But Swammerdam declared that if we carefully observed the order of changes and the order of growth in the limbs of both insects and the other animals, it was evident that all God's works were founded on the very same rules.[98] So he wrote in the *Historia,* and he reaffirmed the point in the final conclusion of the *Bybel.*[99] Emphasizing the order in the generation and development of lower animals, Swammerdam consequently did away with the divided world of living creatures, just as the architects of the scientific revolution had done away with the divided cosmos, and he similarly brought all the living world under the rule of a common, all-embracing order.

---

[94] *Historia*, pt. 1, pp. 23, 43, 48–9. Swammerdam was protesting assertions by Goedaert that, according to Swammerdam, seemed to bury all Goedaert's other discoveries in obscurity (ibid., pt. 2, p. 5).

[95] Ibid., pt. 2, p. 40.    [96] Ibid., pt. 1, p. 3.    [97] See note 79.    [98] *Historia*, pt. 2, p. 40.

[99] *Bybel, II,* 867.

Perhaps even more crucial to Swammerdam as a naturalist, however, were implications of the order in insect life cycles that were not so exclusively cerebral. The *Historia* was in fact a preliminary exposition of a new system of insect classification, a system that provided the basic framework of the *Bybel* as well. It not only served as the unifying context of his studies in natural history, consequently, but was in itself, in Swammerdam's mind, an original and important new contribution to the understanding of insect life. It was no small thing to have discovered such rules and foundations (*gronden*) in the nature of things, he wrote.[100]

The system was constructed around four classifications ("orders") determined by four different life-cycle patterns. The viability of this system that gave coherence, direction, and a fair amount of meaning to his lifelong research rested, then, on the inviolability of those patterns, on the order, that is, in insect development. So also did the worth of years of invested time, labor, and skill, as well as the prospect and promise of a vast new arena of continuing and significant discovery.[101] His preoccupation with the order embodied in the rigidity of insect life cycles was intimately linked, consequently, to the significance, both personal and scientific, of his lifelong research effort as a naturalist.

He wrote in the *Historia* that his system laid a firm and unshakable foundation that left no room for chance,[102] but, given the crucial importance of the consistency of insect development to the scientific activity that absorbed him, I am arguing that Swammerdam waged his ceaseless campaign against chance in part, at least, to affirm the order which underlay that foundation. As he himself remarked, moreover, his system of classification also rested on the nature of the pupa,[103] and the most profound importance of his doctrine of metamorphosis as a barrier against accidental generation, I propose, was its implicit significance in the struggle against indeterminacy in insect generation and development.

The threat of that indeterminacy was not just a specter of

---

[100] *Historia*, pt. 1, p. 5. He also pointed out that what he offered in the *Historia* came after 2,000 years of the blind ignorance of even the sharpest minds; ibid., pt. 2, p. 10.

[101] Indeed, countless wonders were yet to be discovered regarding insect development, he wrote, and he offered his own explications and illustrations of the development of selected insects as examples to open the way to further discoveries (ibid., pt. 1, p. 131; pt. 2, p. 1).

[102] Ibid., pp. 56–7.    [103] *Bybel, II*, 873.

Swammerdam's private conjuring. Deriving from classical as well as popular tradition and further embellished by many seventeenth-century observers, the common lore of spontaneous generation was cumulative rather than systematic and characterized by a dis-arming flexibility. There was, it is true, an element of specificity and predetermination granted by that lore. Nearly every plant produced its particular insect, wrote Ulisse Aldrovandi, and most animals generated their own distinctive lice.[104] Nonetheless, whether lice derived from meat, corrupt blood, or from bodily filth and sweat might still be argued, and John Jonston recorded that fleas were said to arise not only from dust but in dog fur from rotting moisture, from drops of sweat on slaves in the "Perienna regione," or also, after all, from copulation.[105] Caterpillars grew from the leaves of cabbages and other plants, or from the dew on leaves, or from eggs butterflies had laid, or, perhaps, from all of these, and Aldrovandi wrote that flies came from worms that came in turn from either sex or putrefaction (though Jonston denied that flies produced by flies were ever worms or that worms from putrefaction turned very often into flies).[106]

Harvey formally linked spontaneous generation and chance, and the mechanists, despite the firm determinism which their philosophy imposed on nature at large, also emphasized the unpredictability of spontaneous generation. Henricus Regius, who had first launched the Cartesian assault in the Dutch universities, explained that, since the finer particles of the earth were almost infinite in their variety and the motions of heat and subtle matter no less so, the variety of animals produced by spontaneous generation was limitless.[107] No matter the number of different animals we might imagine, he wrote, the countless shapes and arrangements of par-

---

[104] Aldrovandi, *De animalibus insectis* [see note 67], pp. 5, 545. See also Athanasius Kircher, *Scrutinium physico-medicum contagiosae luis, quae pestis dicitur* (Rome: typis Mascardi, 1658), p. 40.

[105] Aldrovandi, *De animalibus insectis* [see note 67], pp. 546–7. John Jonston, *Historia naturalis de insectis* (bound with *Historiae naturalis de quadrupedibus libri I* [Amsterdam: apud Ioannem Iacobi Fil. Schipper, 1657]), p. 92.

[106] Johannes Goedaert [& Johannes de Mey], *Métamorphoses naturelles ou histoire des insectes* (Amsterdam: George Gallet, 1700), I, 159. Aldrovandi, *De animalibus insectis* [see note 67], p. 354. Jonston, *Historia naturalis* [see note 105], p. 45.

[107] Henricus Regius, *Fundamenta physices* (Amsterdam: apud Ludovicum Elzevirium, 1646), p. 219.

ticles could produce many more. In effect, the naturalist could never know what might come next.

Less inclined to assume the spontaneous generation of complex insect structure, Robert Hooke nonetheless still marveled at the irregularity of insect generation.[108] Not only was the same kind of creature produced in several ways, he emphasized with mechanist analogies of his own, but the very same creature might itself produce different kinds of progeny. In insisting on rigid inflexibility in insect generation and development, Swammerdam was in fact struggling against a general tolerance of indeterminateness in insect life, and he had to lay his own new foundations for that order that was central to his scientific enterprise.

It was not the fabrication of the parts of the insect body that Swammerdam was ultimately most determined to withhold from transformation, but rather something essential but undefined in the insect's internal substance. Although he insisted that the external parts had taken shape before metamorphosis, he did not concern himself with how or precisely when these intricate constructions had previously developed. The most fundamental point he wished to make was that even the dramatic transformation of metamorphosis was no more than a change in appearances, that the pupa was the animal itself, which hence endured throughout, that the visible change on the outside was not a change on the inside. Only by its external parts was the animal known, however, and the parts of the butterfly discovered beneath the skin of the caterpillar were not significant in themselves, therefore, but as testimony to the persisting presence of the insect and, thus, to continuity within. When his developing skills in insect dissection compromised that evidence, however – so that, faced with the metamorphosis of the internal anatomy of the soldier fly, he now abandoned the analogy with the simple growth of parts in larger animals[109] – he clung nonetheless to the assumption of the persistence of an unchanged internal and essential core, something he identified with the animal itself.

By insisting on that persisting core, Swammerdam was pushing against the constraints of those epistemological imperatives that

---

[108] Robert Hooke, *Micrographia* (1665; reprint, New York: Dover, 1961), pp. 193–4.
[109] *Bybel, II*, 681.

limited knowledge to effects alone and stressed observation over speculation about the unseen interior of things. The discovery of the parts of the butterfly under the skin of the caterpillar was exhilarating precisely because it offered observable, empirical evidence of what had been only an assumption before, supported until then by no more than the fact that one given insect form always succeeded another given form and no other. He challenged his epistemological commitments, I believe, because of a need to provide the order of insect development with a deeper physical foundation. Keeping his speculative elaboration to a minimum, however, he assumed the least that he needed for a persisting corporeal entity to associate with the preservation of predetermined order. He did not ask how that order was embedded within that physical core.

By this physical, that is, *natural,* foundation for the order of insect development, Swammerdam was keeping God as well as chance at a distance. Among the qualities of God that Swammerdam praised in His works were not only wisdom, glory, and goodness but omnipotence and inscrutability,[110] and, no less than chance, inscrutable omnipotence jeopardized the predictable order crucial to Swammerdam's researches in natural history. Leeuwenhoek wrote that the spontaneous generation of an animal would be a miracle and, if spontaneous generation were true, millions of miracles every second would give rise to new kinds of creatures every day.[111] Echoes of the expected rejoinder, that God himself was the source of natural order, are not lacking in Swammerdam's writings, and, in his many troubled moments, Swammerdam doubtless found comfort in reassuring himself of a divinely established order to which he could and should submit.[112] But to explicitly bind God Himself even to His own rules and by His own choice and goodness was a presumption Swammerdam declined to venture; to diminish His incomprehensibility and narrow the

---

[110] See for instance; *Historia,* pt. 1, pp. 28, 147. *Ephemeri* [note 39], p. [3] of the "Ernstige Aanspraak," 79, 92. *Bybel, II,* 505, 625, 680, 713.

[111] Antoni van Leeuwenhoek to the Royal Society, 23 June 1699, in Leeuwenhoek, *Sevende Vervolg der Brieven* (Delft: Henrik van Krooneveld, 1702), p. 107. Leeuwenhoek in fact writes only that new creatures would appear every day, but the context makes it clear that, given the fact that new creatures *do* appear every day, he means new *kinds* of creatures.

[112] Swammerdam, *Historia,* pt. 1, p. 21; pt. 2, p. 40. He speaks of the order in nature as "wise" as well; ibid., pt. 1, p. 23.

infinite chasm between divinity and mortal man would appear to have been fundamentally alien to Swammerdam's religious sensibilities; hence, we may surmise, Swammerdam's unwillingness to leave the metamorphosis of the soldier fly to the almighty arm of God.

I believe that it is within the context of these concerns underlying Swammerdam's commitment to the enduring insect in metamorphosis that his commitment to preexistence is to be understood as well. A peripheral idea that Swammerdam probably never clearly thought out, preexistence appears as a further extension of his efforts to keep the subject matter of natural history the handiwork of God and, at the same time, to lay a natural foundation for the rigid regularity of insect generation and development.

It is perhaps worth recalling that Swammerdam first broached the idea of preexistence while discussing insect metamorphosis, and, judging from the text of the *Historia* itself, studies pertaining to insect generation may have provided the seminal inspiration. Swammerdam was a prominent member of a brilliant circle of anatomists in the Netherlands that was responsible in the 1660s and 1670s for several celebrated advances in anatomy and physiology, not the least of which was the redefinition of the female "testicles" in higher animals as ovaries.[113] But his studies in natural history increasingly dominated his scientific interests.

The preoccupation that immediately preceded the publication of the *Historia* was his determination to resolve the dense confusion that had surrounded the generation of bees since at least the time of Aristotle.[114] Although it figures less prominently in the histories of science than does the redefinition of the mammalian ovaries, Swammerdam's discovery of the eggs within the queen bee was an exciting breakthrough, which ended centuries of ignorance and misguided speculation. It was one of several discoveries that must have sharpened Swammerdam's sense of a new direction or field of research regarding insect generation, a field, moreover, that was relatively free of the competition and the complexities of shared and contested discoveries that entangled research in anatomy.

---

[113] Swammerdam would later write that the discovery of eggs in the ovaries of women first gave him the idea that eggs were to be found in all animals; *Bybel, I,* 305.

[114] Brown, *Scientific Organizations* [see note 2], 280. See Aristotle, *De generatione animalium,* III, 759$^a$8–761$^a$2; Aldrovandi, *De animalibus insectis,* pp. 49, 54–60.

It was during the years preceding the *Historia* that Swammerdam discovered not only the butterfly or moth itself, but also the beginnings of the eggs of *Orgyia antiqua* within the caterpillar, a discovery that, judging from the text of the *Historia,* may well have been the specific source of his idea of preexistence.[115] As he approached the idea in the *Historia,* indeed, he declared that the generation of insects was so clear that we might climb thereby to the true principles of the generation of all animals.[116] His studies of insect generation and metamorphosis would appear, then, to have provided the immediate background for his formulation of the concept of preexistence, and it is not unreasonable to assume that the concerns underlying those studies played a role in that formulation as well.

Like his reduction of metamorphosis to a simple unveiling of parts, preexistence was also conceived by Swammerdam as a barrier against chance. After asserting, as he denied generation in the *Historia,* that nature revealed no more than a propagation and growth of parts, he continued that this left no place at all for chance.[117] That the egg, like the pupa, involved no more than a growth of parts, he added elsewhere, destroyed accidental generation just as the similar nature of the pupa uprooted the idea of a radical metamorphosis.[118]

The chance fended off by preexistence was surely, in part, that which would deny the divine authorship of all God's creatures. Like the persisting physical core in metamorphosis, however, the preexisting animal also precluded that aspect of chance that would have called into doubt the reliable order of insect generation and development. As was the case with metamorphosis, Swammerdam's concern was not the question of how complex structure could come to be; passages suggest that, as the parts characteristic of the butterfly grew or developed under the skin of the caterpillar before metamorphosis, so parts of the embryo developed within

---

[115] *Historia,* pt. 2, pp. 30–1. *Bybel,* II, 565, 568. In both works, Swammerdam refers with respect to the "infinite use" of this observation to the location of his earlier statement on preexistence. See also Brown [note 2], *Scientific Organizations,* p. 281.

[116] *Historia,* pt. 1, p. 51. *Bybel,* I, 34.

[117] Ibid. *Miraculum,* p. 21.

[118] *Historia,* pt. 1, p. 103. *Bybel,* I, 273–4. Because of these implications with respect to both metamorphosis and generation, Swammerdam spoke of the recognition of the common process of development of the egg and pupa as a discovery of "infinite use." Compare note 113 above.

the egg in the ovary,[119] but Swammerdam showed no interest in the processes of that development. His concern was rather the persistence, as in metamorphosis, of a physical core of minimal definition, the "animal itself," that did no more than provide a natural vehicle – fending off omnipotence as well as chance – for the enduring pattern of the life cycle of the species.

Some years ago, F. J. Cole emphasized the difference between the terms *predetermination* and *predelineation* as alternative characterizations of doctrines of preformation, and more recently Peter Bowler has proposed that Swammerdam's conception of generation entailed essentially a preexistent design in the form of a material system programmed to develop into a living organism.[120] I doubt that Swammerdam's thought went so far as to adumbrate any kind of material "system," and we must be careful of anachronism when speaking of "programming" with all its twentieth-century associations. There is obviously much that Swammerdam may have vaguely sensed but could not articulate, but his allusions to Levi in the loins of his father and all mankind in its first ancestors imply a concept of preexistence that could entertain the prospect of the preexistence of the individual even in a moral sense.

Nonetheless, the analogy of programming is a useful one, for whatever the physical nature of the preexisting core of the individual animal, of critical concern to Swammerdam, I have argued, is that it bore within it not only the blueprint for the construction of the animal, but the pattern for the unfolding of both the behavior and the physical changes of its specific life cycle as well. Together with the preclusion of spontaneous generation and the affirmation thereby that even insects were examples of the handiwork of God, a deeper foundation for the predetermination and continuity of life cycles was perhaps the most compelling implication of preexistence for Swammerdam. It defended that natural order essential to his creative effort as a naturalist.

---

[119] In addition to passages cited above in the text, see *Historia,* pt. 2, p. 26, where he also speaks of the parts of the insect within the egg developing from a singular "internal principle" (*inwendige begintsel*).

[120] Cole, *Early Theories* [see note 3], p. 206, Bowler, "Preformation and Pre-existence" [see note 1], p.237.

# Part III

## HISTORIOGRAPHY AND THE SOCIAL CONTEXT OF SCIENCE

# 10

## What is the history of theories of perception the history of?

STEPHEN M. STRAKER

> The *Action* of an *Extended Thing* as such, is nothing but *Local*
> *Motion* . . . ; but it is certain, that *Cogitation*, (*Phancy, Intellection,*
> and *Volition*) are no *Local* Motions; nor the meer *Fridging* up and
> down of the Parts of an Extended Substance . . . .
> Ralph Cudworth, *The True Intellectual System of the Universe*
> (1678)[1]

In this essay I wish to advance the idea that the writing of histories
of theories of perception and cognition is far more central to the
professional goals of the historian of science than has heretofore
been supposed, and this for two reasons. First, since perception
itself is properly understood, whatever else may be said, as the
bridge between "subject" and "object," a causal interrelation between the "self" and the "world," any theory of perception is
forced explicitly to advance or to adopt an understanding of the
subjective "self" as well as of objective "nature." The scientific
problem of how it happens that we have perceptual awareness of
an "external world" simply is a problem which rational self-consciousness puts to itself about itself, and the "subject" must appear
in some way. Second, it is now more than ever possible to argue
that questions of normative epistemology, those concerning the
status of scientific knowledge, will require for their resolution a
prior framework of judgement constructed out of knowledge of
the "self," the "knower," who claims and bears knowledge of the
natural world. A "science of human nature," a "science of mind"

---

[1] Quoted by John W. Yolton, *Thinking Matter: Materialism in Eighteenth-Century Britain*
(Minneapolis: University of Minnesota Press, 1983), pp. 8–9.

on which all else is founded, was correctly identified some years ago by R. G. Collingwood as "history."[2] If these linked propositions are even nearly correct, then the history of theories of perception is uniquely positioned with respect to many of the important questions now on the agenda for the history and philosophy of science.

## KOYRÉ'S RIDDLE

Above all else, Richard S. Westfall has urged us to study the history of the sciences not only for its own sake but also as potentially *the* fundamental mode of understanding our situation in the modern world. Following the example of his mentor, Alexandre Koyré, Westfall has persevered to maintain that scientific work, including historical writing, is a philosophical enterprise of the highest order, one that deals with real issues in a morally serious way. Quite properly, it seems to me, he has joined Koyré in deep skepticism about the fashionable proposal that "science," *qua* "natural philosophy," should be comprehended as nothing other than the ideological expression of "interests" (however sophisticatedly defined). Koyré and Westfall seem to have recognized that the equation of science with "interests" is itself an expression of the modern scientific mind – stated in one of its classic forms, perhaps, by Max Weber – according to which science and, of course, scientific revolutions, have finally no meaning or significance beyond that of "utility."[3]

---

[2] R. G. Collingwood, *The Idea of History* (London: Oxford University Press, 1961), pp. 205 ff.

[3] "Who – aside from certain big children who are indeed found in the natural sciences – still believes that the findings of [science] . . . could teach us anything about the meaning of the world? . . . Tolstoi has given the simplest answer, with the words: 'Science is meaningless because it gives no answer to our question . . . "What shall we do and how shall we live?" ' That science does not give an answer to this is indisputable." Max Weber, "Science as a Vocation," *From Max Weber: Essays in Sociology,* ed. H. H. Gerth and C. Wright Mills (New York: Oxford University Press, 1958), pp. 142–3. Later in the essay, Weber says in superlative fashion that any attempt to answer "our question" introduces a "personal value judgement" (p. 146). The ways in which the "scientific transformation of the life-world" transformed science itself to produce its Weberian embodiment is discussed in the chapter "Science" in Herbert Schnädelbach, *Philosophy in Germany, 1831–1933,* trans. Eric Matthews (Cambridge: Cambridge University Press, 1984), pp. 66–108. The effects of this transformation in philosophy can be gauged by a reading of Hans Reichenbach's *The Rise of Scientific Philosophy* (Berkeley: University of California Press, 1966) and its descendents. For a provocative and severe critique of these developments see Alasdair MacIntyre, *After Virtue: A Study in Moral Theory* (Notre Dame, Ind.: University of Notre Dame Press, 1981).

Behind Westfall's view that it is necessary to conceive of "an autonomous realm of the spirit"[4] – whereby he opens himself to charges of practicing idealist, internalist history – lies, I believe, an awareness born of experience that difficult conceptual necessities are presented in any attempt fully to understand human conduct, intellectual work, or a historical subject's experience of his or her own rational self-consciousness; it becomes especially clear that it is an error to try to comprehend human thought exclusively in the pragmatic categories of the pleasant and the useful.

The arrogance of ideological explanation of the thought of others lies in the claim to understand another's thinking more deeply than he does himself, without being in a position to provide true descriptions of almost any of it. It is a routinised claim to authority where routinised claims must be false, where all authority must be earned in detail and where the mode of its earning is by explaining persons (and their situations) more lucidly to themselves.[5]

Thus Westfall's moving admission, after twenty years of labor, that the past has become less familiar, that the "more I have studied him, the more Newton has receded from me . . ." and become "wholly other."[6]

What applies to individual thinkers certainly holds for groups of them, if not entire cultures, and so the question of the "meaning" or significance of "science" and "scientific revolutions" is a real one, not easily dismissed in the name of rejecting idealism in order to be more scientific.[7]

---

[4] Richard S. Westfall, "Reflections on Ravetz's Essay," *Isis, 72* (1981), 405.

[5] John Dunn, "Practising History and Social Science on 'Realist' Assumptions," in Christopher Hookway and Philip Petit, eds., *Action and Interpretation* (Cambridge: Cambridge University Press, 1978), pp. 145–75, p. 169. "*Justitia est constans et perpetua voluntas suum cuique tribuendi.* If we claim to know about other men, we must try as best we can to give them what is their due, their right. This is a simple moral duty, not a guarantee of epistemological prowess." Ibid., p. 174. I find that Dunn's words explain eloquently Westfall's integrity as a teacher and historian.

[6] Richard S. Westfall, *Never at Rest: A Biography of Isaac Newton* (Cambridge: Cambridge University Press, 1980), p. ix.

[7] Many advocates of a social (science) understanding of "scientific knowledge" fail to realize the degree to which they are themselves captive of this (historically local) conception of science, from the inside of which their attempts to explain science proceed. If this line of argument is correct, then it is the proponents of the "Strong Program" in the sociology of scientific knowledge who are the *real* internalists. One frontier of "sociological" research appears now to be the struggle by thinkers such as Jurgen Habermas and Paul Ricouer to break out of this scientistic prison by way, in part, of a moral redefinition of "interest." See Ricoeur, "Science and Ideology," in *Hermeneutics and the Human Sciences* (Cambridge: Cambridge University Press, 1981), chap. 9, pp. 222–46, and also the two

When Koyré asked the question, what is "the significance of the Newtonian synthesis"? he signaled his intention to deal with the "meaning" of science in the modern world. He concluded that the proper genre for understanding the scientific revolution is "Tragedy." Newton's successes, he tells us, revealed "the tragedy of the modern mind, which 'solved the riddle of the universe,' but only to replace it by another riddle: the riddle of itself."[8] Koyré's proposal is located precisely and insightfully by Edwin Arthur Burtt in *The Metaphysical Foundations of Modern Science*.[9] Modern science presents for our belief a world split in two, a conception of the "real world," composed of dead, "mechanical" matter, a world in which human beings, as we experience ourselves, cannot be located, for it is a world in which persons are wholly "other." As Burtt argues in the eloquent and strangely neglected "Conclusion" to his great work, when such a science turns to the question of understanding persons, it opines that "man is a machine," leaving the intractable riddle of giving an account of "mind," consciousness, self-awareness, perception, intentionality, and "meaning" solely in terms of "matter in motion" (in whichever one of its modern physical guises).[10]

The irony of this situation has not, I think, been sufficiently appreciated. To put it simply and tendentiously, modern empirical science is unable to say anything coherent about its own empiricalness, for the attempt to give an objective account of the "subject" seems irretrievably flawed. Not only is there no persuasive account of how the world (as science understands it) gives rise to "observations" and the observer's awareness of them, there is not even a plausible plot-line for such a story. On the frontiers of "Cognitive Science," the "observer" is treated like a neurophysiological "black box," whose inputs and outputs are taken to be well described, so that the best theory of the human "observing

extremely important discussions by John Dunn, "Practising History" [see note 5] and his *Western Political Theory in the Face of the Future* (Cambridge: Cambridge University Press, 1979), esp. pp. 75–117.

[8] Alexandre Koyré, "The Significance of the Newtonian Synthesis" in *Newtonian Studies* (Chicago: University of Chicago Press, 1968), p. 24.

[9] Edwin Arthur Burtt, *The Metaphysical Foundations of Modern Science*, 1924. The most recent reprinting of the second (latest) edition of 1932 (first edition, 1924) is by Humanities Press (Atlantic Highlands, N.J., 1980).

[10] Ibid., pp. 303–25. Cf. Koyré, "Newtonian Synthesis" [see note 8]: "Thus the world of science – the real world – became estranged and utterly divorced from the world of life, which science has been unable to explain – not even to explain away by calling it 'subjective' . . . Two worlds: this means two truths. Or no truth at all." (pp. 23–4).

instrument" is one cast in the same discourse as the theory of the objective world "external" to the instrument. The thoroughly verified, empirical *fact* (true for all observers) of the experiential *awareness* of meaningful observations is simply and straightforwardly not comprehended. Deep faith is routinely expressed in the idea that by some kind of transubstantiation, "a computer running the program my brain is running is thinking my thoughts and having my experiences," but the mystery remains.[11]

Without knowing it, Thomas Nagel agrees with Burtt that "surely we have somewhere run off the track of sane thinking,"[12] for in his penetrating essay, "What is it like to be a bat?"[13] Nagel declares

Consciousness is what makes the mind–body problem really intractable. Perhaps that is why current discussions of the problem give it little attention or get it obviously wrong . . . [We] have at present no conception of what an explanation of the physical nature of a mental phenomenon would be. Without consciousness the mind–body problem would be much less interesting. With consciousness it seems hopeless.

The very fact that we are permitted to ask the scandalous question, "How do we know we're not just brains in a vat?" is symptom enough of the depth and intractability of the problem.[14]

---

[11] For a perfect display of this assumption at work, see I. Rosenfield, "Seeing Through the Brain," a discussion of David Marr, *Vision: A Computational Investigation into the Human Representation and Processing* in the *New York Review of Books* 11 October 1984, pp. 53–6. The *locus classicus* for a sustained examination of "tragedy" is, of course, the famous "mind–body problem," whose literature is enormous. For a start, at least on the philosophical literature, see Kathleen Emmett and Peter Machamer, *Perception: An Annotated Bibliography* (New York: Garland Press, 1976).

[12] Burtt, *Metaphysical Foundations* [see note 9], p. 316.

[13] Thomas Nagel, "What Is It Like to Be a Bat?" *Philosophical Review*, 83 (1974), 435–50, reprinted in Nagel, *Mortal Questions* (Cambridge: Cambridge University Press, 1979), pp. 165–80. Passage is cited from pp. 165–6.

[14] In an extremely interesting analysis, Hilary Putnam shows how upside-down things are by arguing that the idea we might be "brains in a vat" is "physically possible" but philosophically impossible *and* false. See his *Reason, Truth and History* (Cambridge: Cambridge University Press, 1981), pp. 1–21 and passim. John A. Wheeler remarks that "never so deep has been the mystery in which we live," calling it "the heart of darkness" in "The Universe as Home for Man," *The Nature of Scientific Discovery*, ed. Owen Gingerich (Washington, D.C.: Smithsonian Institution Press, 1975), pp. 261–96, p. 291. To show how intentionality just goes unexplained is the essential point of John Searle's "Minds, Brains, and Programs," *The Behavioral and Brain Sciences*, 3 (1980), 417–24, followed by an "Open Peer Commentary" involving 28 philosophers and scientists, and the "Author's Response," 424–57. For some agonizing testimony from the world of the neurologist, see Jerome Bruner's moving review of Oliver Sack's *A Leg to Stand On*, in *The New York Review of Books*, (27 September 1984), pp. 39–41.

Furthermore, the *structure* of the contemporary debate faithfully mirrors the very subject/object split that is at issue. The proponents of an objective, hence scientific, view and the unscientific philosophical defenders of the "subject" converse past each other, in Kuhnian fashion, across a mutually perceived metaphysical gulf.[15] It is important to realize the degree to which such incoherence typifies a discourse in disarray. Consider, for example, a recent and typical attempt to define "consciousness" in such a way that it could become the object of scientific study.

By way of approximation, I would suggest that if an organism is capable of responding differentially to the same elements [in the environment or in its own "mentation"] in different contexts and hence with different behaviors, that organism is "conscious" of something.[16]

The author has built into the definition those features ("organism," "mentation") that hope to forestall its use to award consciousness to plants or to thermostats. But the attempt to take an "objective" view of the "subject" produces tacit adherence to the proposition that "behavioral equivalence" implies "subjective equivalence." Since this notion is remarkably akin to the now untenable logical empiricist doctrine that "observational equivalence" implies "theoretical equivalence," the conceptual difficulties faced by a "behaviorist psychology" are similar to those already offered as critiques of logical empiricism. Theories that express very different pictures of reality may have identical sets of obser-

---

[15] "In the eye we have on the one hand light falling on this wonderful structure and on the other we have the sensation of sight. We cannot compare the two things; they belong to opposite categories. The whole of metaphysics lies like a great gulf between them." James Clerk Maxwell, "On Colour Vision," in *Sources of Colour Science,* ed. D. L. MacAdam (Cambridge, Mass.: M.I.T. Press, 1970), p. 82. I want to thank Thomas Vinci of Dalhousie University for bringing this reference to my attention. G. Politzer is famous for having remarked that "psychologists are scientific in the manner that evangelised savages are Christian" (in *Critique des fondements de la psychologie,* 1928), quoted by Georges Thines, *Phenomenology and the Science of Behaviour: An Historical and Experimental Approach* (London: Allen & Unwin, 1977), p. 98. In a recent exemplary comment, Douglas R. Hofstadter refers to a paper by John Searle in these words: "this religious diatribe against A. I., masquerading as a serious scientific argument is one of the wrongest, most infuriating articles I have ever read in my life." "Reductionism and Religion," *The Behavioral and Brain Sciences,* 3 (1980), 433.

[16] E. A. Lunzer, "The Development of Consciousness," in Geoffrey Underwood and Robin Stevens, eds., *Aspects of Consciousness,* Vol. I: *Psychological Issues* (London: Academic Press, 1979), 1–2, offering the best "mean" of two extreme positions.

vational consequences.[17] The mistaken assumption of such equivalence is obvious enough in M. E. Levin's claim that

the very fact that a machine solves problems men solve is evidence that what is going on inside it is similar to what goes on inside human problem solvers. Workers in AI [Artificial Intelligence] CS [Cognitive Science] seem to agree that no watertight barriers separate them.[18]

Yet the "materialism" of the reductionist may already be dated, for Levin threatens to become the Lord Kelvin of Cognitive Science when, in his defense of "the ancient thesis that man is a piece of matter," he says,

Modern science brings with it a picture of the universe from which everything but matter is excluded. So far as we know, everything, except possibly the psychological states of sentient beings, is physical. The laws of nature, whose predictive and explanatory power has been re-repeatedly vindicated, are couched wholly in non-psychological terms. As the evidence accumulates, it becomes ever harder to suppose that future discoveries will ever radically alter this world picture.[19]

The instrumentalist equation of predictive with explanatory power goes hand in hand with the heroic but blinkered view that the "story" of the progress of science contains no episodes of radical conceptual change or, perhaps, that such disruptions occur only in the "prehistory" of a mature science.[20] Perhaps Levin would dismiss as metaphysical (or unfalsifiable) the neoromantic, "anthropic principle" of the cosmologists Brandon Carter and John Archibald Wheeler, which posits a cosmic link between mind and

---

[17] See, for example, Clark Glymour, "Theories: An Examination of the Logical Empiricist Philosophy of Science" (Ph.D. dissertation, Indiana University, 1969), where the matter is dealt with rigorously.

[18] Michael E. Levin, *Metaphysics and the Mind–Body Problem* (Oxford: Oxford University Press [Clarendon Press], 1979), p. 189. John Searle has noticed, ironically, that the AI project is unwittingly committed thereby to dualism, for their aim is "to reproduce and explain the mental by designing [machine] programs" which mimic the output of human problem solvers; "but unless the mind is not only conceptually but empirically independent of the brain," so that it doesn't matter what the "matter" is, the AI project has to fail ("Minds, Brains, and Programs" [see note 14], pp. 423–4). Searle is convinced we have good reasons to believe that "brains cause consciousness," even though we have no idea how.

[19] Levin, *Metaphysics* [see note 18], pp. vii, 87; see also his discussion on pp. 60–3.

[20] It is often said that "psychology is still in its infancy" – or words to that effect. This reveals the ways in which a conception of a "science" is already embedded in narrative structure even before the historian looks at it.

matter such that "the universe, through some mysterious coupling of future with past, required the future observer to empower past genesis."[21] But such is the stuff of which conceptual transformations are made.

If we compare objective definitions of "consciousness" with one offered by Nagel, the polarity of opposition is clear.

But fundamentally an organism has conscious mental states if and only if there is something that it is like to *be* that organism – something it is like *for* the organism. . . . If mental processes are indeed physical processes, then there is something it is like, intrinsically, to undergo certain physical processes. What it is for such a thing to be the case remains a mystery.[22]

Nagel is pointing to the simple impossibility of giving an account of the "subject" (of awareness, meaning, and intentionality) with a theory whose adequacy is precisely that of giving a nonsubjective view from the outside.

The recent controversies about the nature of "language" and the question of "talking animals" displays the very same structure of disagreement. What is finally at stake in these debates is whether the animal ("Koko" the gorilla or Herbert Terrace's "Nim Chimpsky") is just *mimicking* American sign language – some-

---

[21] Quoted by Stephen G. Brush, "The Chimerical Cat: Philosophy of Quantum Mechanics in Historical Perspective," Social Studies of Science, 10 (1980), 393–447, p. 431. This paper is extremely valuable for its overview of 150 years of European natural philosophy. Brush posits a "cyclic oscillation between 'Romantic' and 'Realist' periods in science and culture" (p. 393), whose frequency is approximately 30 years.

[22] Mortal Questions [see note 13], p. 166. Nagel states the issue succinctly and articulately as follows: "In a sense, the seeds of this objection to the reducibility of experience are already detectable in successful cases of reduction; for in discovering sound to be, in reality, a wave phenomenon in air or other media, we leave behind one viewpoint to take up another, and the auditory, human or animal viewpoint that we leave behind remains unreduced. . . . But while we are right to leave this point of view aside in seeking a fuller understanding of the external world, we cannot ignore it permanently, since it is the essence of the internal world, and not merely a point of view on it. Most of the neobehaviorism of recent philosophical psychology results from the effort to substitute an objective concept of mind for the real thing, in order to have nothing left which cannot be reduced. If we acknowledge that a physical theory of mind must account for the subjective character of experience, we must admit that no presently available conception gives us a clue how this could be done. The problem is unique." Ibid., p. 175. This view is remarkably like that discussed by Hannah Arendt, in The Human Condition (Chicago: University of Chicago Press, 1958), esp. secs. 36–41, where she discusses modern science's perspective of "the Archimedean point," the point of view of a disembodied, displaced, solitary mind, which has no reference other than itself and everything it can think – a point inside a man who is outside of the universe.

times perhaps in response to certain extremely subtle cues rather like "Clever Hans," the famous horse who could do addition – or is actually *using* the language, whether the gestures or noises have any *meaning* for the ape. As the controversy reveals, one can get rid of "meaning" and all the antireductionist inconveniences that go with it, if one simply adopts some version of the theory that language is nothing but "information processing and transfer." On this instrumentalist conception, human language is regarded as just another means of "communication," on the same continuum as dogs marking hydrants, bees dancing, and ants swapping molecules. All too often, the objection that language is unique in its ability to create and to express "meaning" (and therefore to create "reality" in the form of meaningful "representations") is dismissed as a sign of weakness, a desire to cling to an essentially "religious view" of man.[23]

Such denunciations reflect the fact that this recent dispute is just another battle in a long campaign waged between two rival and incompatible conceptions of language. In the view associated quite properly with modern science, language is *designative,* an instrument of information transfer. As in Hobbes's nominalist theory, for example, language is a means of "communication," and as such allows for the instrumental control of ideas; it is the sort of thing that permits machines to speak and to think, and us to think of ourselves as machines. On the competing view, language is understood to be *expressive* of our selves and of our condition, the means whereby the language-using creature becomes what he is and constitutes his being. In this basically Aristotelian theory of language, as found, for example, in the accounts given in the German "Romantic" tradition beginning perhaps with Herder, language is on the same continuum with art, not just another instrument of survival (as in Darwin's anesthetic musings) but a necessary part of our self-realization. In the fully "Romantic" vision, it is through art and literature that man expresses his true self, realizes himself (in the Aristotelian sense of "becoming"), and in so doing serves as an essential agent of the actualization

---

[23] For an overview of the recent discussion, see: H. S. Terrace et al., "Can an Ape Create a Sentence?" *Science, 206* (1979), 891–902; Martin Gardner, "Monkey Business," *New York Review of Books* (20 March 1980), pp. 3–6, and the correspondence that followed (9 October, 4 December 1980, 2 April 1981); and Nicholas Wade, "Does Man Alone Have Language?" *Science, 208* (20 June 1980), 1349–51. The "unscientific" answer to the last question is yes, by definition. If whales can talk with each other, and therefore with us, then we shall have discovered a race of aquatic, legless men.

of Nature.[24] Human consciousness in this Romanticist view does not simply apprehend the order of nature, it completes and perfects it; the objective and external world is really the "externality" of man's own self-awareness.[25] Here is resolved, I believe, the mystery of the "two cultures": two understandings of language grounded fundamentally on rival accounts of the status of the "subject" in an "objective" world, two different accounts of the "self." Hence it is, as they say, no accident that objections to the designative account of language are taken to be evidence of a nostalgia for a "religious view" of man, sentimental attempts to retain a belief in the immortality of the human soul.

A mature and heroically realistic position is set forth by Paul Churchland in his *Scientific Realism and the Plasticity of Mind.* Implicitly articulating the thesis that "self-knowledge" underlies our theoretical account of perception and cognition, Churchland insists that there is an intimate connection between "our self-conception and the body/mind problem," a problem which, he tells

[24] Charles Taylor, *Language and Human Nature (The 1978 Alan B. Plaunt Memorial Lectures)* (Ottawa: Carleton University Information Services, 1978). See also his *Hegel* (Cambridge: Cambridge University Press, 1975), pp. 3–29.

[25] Owen Barfield, "Participation and Isolation," *Dalhousie Review, 52* (1972), 5–20 (p. 8). See also Taylor, *Language and Human Nature* [see note 24], pp. 14–21, especially, and idem, *Hegel* [see note 24], passim. Clearly, the history of our understanding of the nature of language is relevant to our purpose; one could begin with Hans Aaarsleff, *The Study of Language in England, 1780–1860* (Princeton, N.J., Princeton University Press, 1967); historians of science might find a congenial starting place in Murray Cohen, *Sensible Worlds: Linguistic Practice in England, 1640–1785* (Baltimore: Johns Hopkins University Press, 1977). Perhaps the situation can be summed in one passage from Richard Bernstein's "Why Hegel Now?": "Neither language nor art are to be interpreted from the point of view of how they represent or refer to the world, but rather from the perspective of how they express and realize human life. The thrust of this expressivist orientation is of primary importance for understanding Hegel for three reasons: it is strongly anti-dualist, it makes freedom as authentic self-expression a central value of human life, and it contains an inspiration toward a union and harmony with nature that was jeopardized by the Enlightenment" (in *Review of Metaphysics, 31* (1977), 29–60, quoted by John Tietz, "Davidson and Sellars on Persons and Science," *The Southern Journal of Philosophy, 18* (1980), 237–249 (p. 248). In view of the centrality of language to the whole issue, it is puzzling that there has been so little discussion of Julian Jaynes's argument in *The Origin of Consciousness in the Breakdown of the Bi-Cameral Mind* (Boston: Houghton Mifflin, 1976). The nub of Jaynes's book, if I read the complex and interesting account correctly, is that consciousness arose out of the contingent, social need to express, to realize a "self," and that the vehicle of its birth was *"metaphor."* The metaphor "me" and its analogue "I" are held in the world by narrative. (See pp. 48–66 especially.) On metaphor and the ways in which it populates, if not creates, "the world," see Stephen C. Pepper, *World Hypotheses: A Study in Evidence* (Berkeley: University of California Press, 1960).

us, "will require for its solution, or even for its significant advancement a revolution in our self-conception." Furthermore, Churchland imagines, all the issues of a "normative epistemology . . . will not be solved short of an intellectual revolution in our conception of ourselves as intellectual beings."[26] As he conceives it, in this revolution the "life–world" is transformed into Richard Rorty's "physiological utopia,"[27] and people thereafter experience the world exactly as it is portrayed by physical science; the "two tables" problem disappears, for we are supposed to perceive a table as it really is: the particles themselves, the electromagnetic disturbances in the "aether" as well as in our own heads (neurons firing), and so forth. All this because, for Churchland, perception is so thoroughly theory-laden (and reference so meaning-laden) that we can "exchange the *Neolithic legacy now in use* [as our mode of perceiving and knowing] for the conception of reality embodied in modern-era science."[28] Churchland's and Rorty's fantasies are remarkable for the imagined projection into the realm of self-awareness of the abstracted objective view "from the outside," making it "inner."

There is something deeply wrong with such an attempt to "recognize" ourselves in the image of a successful experiment in Artificial Intelligence. It is as if, to take Nagel's example, we found out that "what it is like to be a bat" is to *be* an expert bat physiologist. Could "I" discover that "I am not really a person"?[29] Putnam has argued that such a possibility is philosophically false. Perhaps it could be said without sentimentality and in the strictest rigor that such a reductionist utopia is an "inappropriate vision from without," one that "disturbs our natural vision," violates an

---

[26] Paul Churchland, *Scientific Realism and the Plasticity of Mind* (Cambridge: Cambridge University Press, 1979), pp. 89, 3–4.

[27] Richard Rorty, *Philosophy and the Mirror of Nature* (Princeton, N.J.: Princeton University Press, 1979), "The Antipodeans," pp. 70–8. The term is Levin's (in his *Metaphysics*).

[28] Churchland, *Scientific Realism* [see note 26], p. 35 (italics mine). Churchland surveys a number of "possible" theories of "the self" that might emerge as a result of scientific work, taking our ordinary experience of ourselves for comparison. He calls our self-awareness "P-theory," claiming that it is inadequate and has not changed in millenia; certainly this is a historically blinkered view (ibid., pp. 107–15). A challenge to the theory-ladenness of perception is argued by Jerry Fodor, "Observation Reconsidered," *Philosophy of Science,* 51 (1984), 23–43.

[29] See John Tietz, "Davidson and Sellars" [see note 25], p. 240: "But this kind of elimination of my status as an agent requires action from *within* the framework of persons which leads to my 'recognition' or to my 'conclusion' that I am not 'really' a person."

authentic sense of who and what we are, and is incompatible with appropriate preferences about how we want to live within a framework of historical self-understanding.[30]

### BURTT'S INJUNCTION

> Possibly the world of external facts is much more fertile and plastic than we have ventured to suppose; it may be that all these cosmologies . . . are genuine ways of arranging what nature offers to our understanding, and that *the main condition determining our selection between them is something in us rather than something in the external world.* This possibility might be enormously clarified by *historical studies* aiming to ferret out the fundamental motives and other human factors involved . . . and to make what headway seemed feasible at evaluating them, discovering which are of more enduring significance and why.
>
> E. A. Burtt (1924)[31]

What stands out in the discussions I have cited is the persistent presence of a notion of *the self*, a "subject" who perceives, observes, understands, uses language, and is always able to recognize itself. In Churchland's more explicit proposals, the history of science we are offered is a "Romance" in which heroic selves sternly face "reality" and leave their disabled "Neolithic" compatriots behind. We have seen this story before. It is precisely the tale told by proponents of the scientific revolution of the seventeenth century about their own transformation of worldview and self-understanding. The "self" persists just because it is impossible to eradicate "meaning," because we can always imagine ourselves asking, What is "the significance of the Newtonian synthesis"?[32]

Those who prosecuted the scientific revolution had, very often,

---

[30] The words are Goethe's, quoted by Erich Heller, "Faust's Damnation: The Morality of Knowledge" in *The Artist's Journey Into the Interior and Other Essays* (New York: Random House/Vintage, 1968), pp. 30–1. "Mayhap we must wait for the complete extinction of theological superstition before these things can be said without misunderstanding. . . . But in these two-sided considerations is bared the terrific difficulty of the modern problem of metaphysics. An adequate cosmology will only begin to be written when an adequate philosophy of mind has appeared . . ." E. A. Burtt, *Metaphysical Foundations*, p. 323–4. The problem should take the form: "could a true and coherent history of theories of perception be written from the point of view of the triumph of the reductionist vision?" Churchland sees that historical guidance is relevant but fails to appreciate the depth and richness of the question.

[31] Burtt, *Metaphysical Foundations* [see note 9], p. 307 (italics mine).

[32] How does it come about that an event or situation has meaning? If it is an event that has no meaning in or for itself – such as a happening in "nature" according to our current understanding of the universe – then it can have meaning only if it has meaning for

clear ideas about what they were doing, and they expressed them nearly as often in the "political" language of liberation, maturation, restoration, and progress. Indeed, Bacon, Hobbes, and to some degree Galileo wrote their *own* histories (histories which they understood themselves to be enacting) in just these terms. If we look at the scientific revolution from the point of view of the "subjects" who gave meaning to their own activities – rather than from the external point of view of what they discovered about the nature of things – we can see that characteristically they understood *themselves* to be authors of and actors in such a "Romance." Engaged in a struggle to banish darkness and superstititon, they took themselves to represent a "coming of age" of the race and saw themselves as "new men."

As epistemological innovators, the moderns of the seventeenth century directed their scorn and polemics against Aristotelian science . . . . From the modern point of view, these earlier visions betrayed a deplorable if understandable weakness of men, a self-indulgence wherein they projected on things the forms which they most desire to find, in which they feel fulfilled or at home. Scientific truth and discovery requires austerity, a courageous struggle against what Bacon called the "Idols of the human mind."

Having introduced a reading of the issue in terms of the "subject" who engaged the revolution, Charles Taylor provocatively suggests an alternative answer to Koyré's question about the meaning of modern science.

Instead of seeing the issue between Galileo and the Paduan philosophers, between modern science and medieval metaphysics, as a struggle between two tendencies in the self, one deploying comforting illusions, the other facing stern realities, we might see it as *a revolution in the basic categories in [terms of] which we understand self.*[33]

someone or something else; it must have meaning for a "subject." The "discovery" that the universe is meaningless (has no intrinsic meaning), however, was not an event in nature; it was a human achievement, a set of actions which, taken together, constitute what we call the "scientific revolution." The discoveries of the new science are the intentional pronouncements of human actors who, as such, have the capacity for intentionality, representation, and signification. See Hayden White, "The Historical Text as Literary Artifact" in R. H. Canary and H. Kozicki, eds., *The Writing of History: Literary Form and Historical Understanding* (Madison, Wisc.: University of Wisconsin Press, 1979). pp. 41–62.

[33] Taylor, *Hegel* [see note 24], pp. 4–5 (italics mine). Taylor goes on to suggest, "The essential difference can perhaps be put in this way: the modern subject is self-defining, where on previous views the subject is defined in relation to a cosmic order." (p. 6)

On this reading, the "self" who has to *face* such struggles is himself a fundamental creation of the new scientific age.

Taylor's suggestion – that we see the scientific revolution not as "an epistemological revolution with anthropological consequences" but instead as a revolution in self-understanding with epistemological consequences – provides the idea I wish to exploit in the rest of this essay.

I want to urge that the time has come for historians of science to confront Koyré's "riddle" of the modern mind directly, to answer Burtt's call for historical studies that see themselves as necessary prerequisites of an adequate philosophy of mind, without which philosophy runs the risk (to which it now succumbs) of merely objectifying the mood of the age.[34] This historical investigation of the "natural philosophy of mind" is, of course, better known as "the history of theories of perception and cognition"; what I shall henceforth call, for short, "the history of theories of perception." I wish, in brief, to examine the desirability and the consequences of adopting the hypothesis that the history of our changing conceptions of nature is the *disguised* form of the real history being enacted, the history of changing conceptions of the "self." If this hypothesis is pursued, the writing of such history will display certain historical and epistemological consequences which would more than justify the attention that ought to be given it.

If the modern scientific worldview came into being because of underlying historical transformations of "self-understanding" ("revolutions" in the "science of human nature"), if the story of nature that resulted is, rather, a story *we* have told about ourselves, then we must acknowledge at the outset that the history of modern science ought to be written not as Romance or Tragedy, but as Irony.[35] That is to say, the history of theories of perception

---

[34] Burtt, *Metaphysical Foundations* [see note 9], p. 304. The project on the part of some philosophers of science "to naturalize epistemology" seems a likely candidate for such unhistorical objectification. For an extended discussion of why "progress" requires historical understanding, without which there is only "change," see R. G. Collingwood, *The Idea of History* [see note 2], pp. 315–34 and also the excellent paper by Stephen Toulmin, "Conceptual Change and the Problem of Relativity" in Michael Krausz, ed., *Critical Essays on the Philosophy of R. G. Collingwood* (Oxford: Oxford University Press [Clarendon Press], 1972), pp. 201–21.

[35] On the question of genre and historical writing, see Hayden White, *Metahistory: The Historical Imagination in Nineteenth-Century Europe* (Baltimore: Johns Hopkins University Press, 1973) and the discussion by Christopher Norris, "Some Versions of Narrative," *London Review of Books*, 6 (1984), 14–16.

should locate itself as agnostic with respect to the findings and the program of modern natural science. It need not take for granted – as does Koyré's tragic mode – the necessity of the worldview which science presents for our belief. There are independent, and stronger, grounds for doing this, which have important consequences for our question about the status and nature of the history of theories of perception.

These lie, essentially, in the problem forcefully brought to our attention by T. S. Kuhn, but in fact present throughout the history of modern philosophy, namely, that no one knows whether we have grounds to take as demonstrated (or as demonstrated to be "superior" to all other possibilities) the truths or the research programs that modern science presents for our belief. Indeed, as is well known, much current discussion is concerned to identify what features such a demonstration would have to possess and which would allow for the identification of what counts as "progress" in science. These epistemological matters connect to our purpose in the following way. To put it compactly: *If* there are always, in principle, at least two "theories of nature" (cosmologies), which are incompatible with each other but equally adequate to the "evidence," then the question of which theory to "hold" admits of some kind of genuine choice. But the name of considerations that are given in a situation of genuine choice is "ethics." Therefore, epistemology becomes ethics.[36]

The dominant contemporary response to this conclusion is pragmatism, almost always involving instrumentalism with respect to theories. Thus epistemology is "reduced" to science itself, and ethics goes utilitarian. This is because the notion of " 'believe' what is useful to believe" quickly dissolves (in the solvent of a robust skepticism) into " 'use' what is useful to use."[37] Because I

---

[36] "The fact that almost any philosophical doctrine may find realization either in a cosmology, that is, in a theory of the universe that is capable of sensual representation and/or in a *theory of man,* which may also be sensually realized in a corresponding society . . . makes it very clear that . . . philosophical positions . . . may be said to be blueprints for possible, *and realizable,* universes. These positions admit of a *genuine choice* . . . and . . . ethics is, therefore, the basis of everything else." Paul K. Feyerabend, "Problems of Empiricism" in Robert Colodny, ed., *Beyond the Edge of Certainty* (New York: Prentice-Hall, 1965), p. 219.

[37] The most vigorous spokesman for such thorough pragmatism at present is Richard Rorty. Rorty always writes provocatively, as in his *Philosophy and the Mirror of Nature* [see note 27], but finally appears as a charming, because complacent, Thrasymachus, for which see his *Consequences of Pragmatism (Essays: 1972–1980)* (Minneapolis: University of Minnesota Press, 1982). Charles Taylor is properly alarmed by Rorty's position, for which

wish to retain the Aristotelian prejudice that we are not under constant illusion concerning our frequent expressing of judgements about what is good and what is true – indeed, that the fact we do talk about such things is our distinguishing attribute[38] – I am more attracted to the philosophical quest for an absolute frame of reference in which to view human choices. Our "Kuhnian" premise is that a true account of nature is not directly available to us by any known canons of scientific method. If epistemology does "reduce" to ethics, then – if only in a realist's abhorrence of the meaningless void of an instrumental pragmatism – I would want to take seriously the possibility that there *is* an absolute frame, namely, knowledge of "the self," what Collingwood called the "science of human nature or of the human mind," known by its proper name as "history."[39] To pursue this possibility is to notice important differences between the natural and the social (or human) sciences and to reverse their priority as authoritative with respect to normative epistemology. In such a resolution of relativism, the "subject" is more objective than the "object." More to

see Taylor's, and Hubert Dreyfus's remarks on pp. 52–3 of "A Discussion," *Review of Metaphysics*, 34 (1980). See also Rorty's "Reply to Professor Yolton," *Philosophical Books*, 22 (1981), 134–5. (Cf. the compelling case against such pragmatism made by Leo Strauss in the "Introduction" to his *Natural Right and History* [Chicago: University of Chicago Press, 1953], pp. 1–8.)

[38] Aristotle, *The Politics*, ed. and trans. Ernest Barker, (London: Oxford University Press, 1958), pp. 5–7 (1253$^a$). See also Robert Bellarmine, *De laicis or The Treatise on Civil Government*, trans. Kathleen E. Murphy (New York: Fordham University Press, 1928), p. 21. In the Aristotelian tradition, the existence of language is taken as evidence that, unless the universe is absurd and human beings are unlike every other naturally existing thing, human beings have some kind of access to the good and the true.

[39] R. G. Collingwood, "Part V. Epilegomena. I. Human Nature and Human History," *The Idea of History* [see note 2], p. 220. "Thus history occupies in the world of to-day a position analogous to that occupied by physics in the time of Locke: it is recognized as a special and autonomous form of thought, lately established, whose possibilities have not yet been completely explored . . . [W]hereas the right way of investigating nature is by the methods called scientific, the right way of investigating mind is by the methods of history." Ibid., p. 209. "It might be said that to describe the rational activity of an historical agent as free is only a . . . way of saying that history is an autonomous science. Or it might be said that to describe history as an autonomous science is only a disguised way of saying that it is the science which studies free activity. For myself, I should welcome either of these two statements, as providing evidence that the person who made it had seen far enough into the nature of history to have discovered (*a*) that historical thought is free from the domination of natural science, and is an autonomous science, (*b*) that rational action is free from the domination of nature and builds its own world of human affairs . . . [and] (*c*) that there is an intimate connexion between these propositions." Ibid., p. 319.

the point, a historical understanding of transformations in our comprehension of "self" – in which the history of theories of perception and cognition is a key element – would then properly play a more central role in our attempts to settle epistemological issues. The test of a theory would then be something about ourselves – "history" or the story of "man's world" (to which it might be said we have special access) – rather than something about "nature" or "God's world" (which we cannot know directly).[40]

Such a proposal has been recently approached and forcefully argued by Alasdair MacIntyre in two papers discussing the significance of Kuhn's *Structure of Scientific Revolutions*. MacIntyre proposes that it is through the writing of histories, the constructing of true coherent narratives, that epistemological evaluations are accomplished, and "practices" in disarray and incoherence rescued.

The criterion of a successful theory is that it enable us to understand its predecessors in a newly intelligible way. It, at one and the same time, enables us to understand precisely why its predecessors have to be rejected or modified and also why, without and before its illumination, past theory could have remained credible. It introduces new standards for evaluating the past. It recasts the narrative which constitutes the continuous reconstruction of the scientific tradition.[41]

[40] See the extremely interesting argument by Shirley Robin Letwin, "Nature, History and Morality" in R. S. Peters, ed., *Nature and Conduct*, Royal Institute of Philosophy Lectures, Vol. 8: 1973–1974 (London: Macmillan Press, 1975), pp. 229–50, in which she urges that the issue becomes one of preferring to live according to an understanding of what it is to be a human being, on the obligation, therefore, to stand firmly on certainties known to be fragile (pp. 245–6, 250). Similarly, consider Alan Ryan's observation that in the social sciences "when we elucidate concepts we are elucidating the possibilities of social life, and conversely when we explain social life we elucidate the concepts available to members of that society." Consequently, it is "the task of social science to reflect on the concepts with which we make social life intelligible . . . to show what would be lost were certain concepts not available . . . to inquire into the rationality of life understood in terms of various conceptual schemes." (*The Philosophy of the Social Sciences* [London: Macmillan Press, 1970], p. 145.)

[41] Alasdair MacIntyre, "Epistemological Crises, Dramatic Narrative, and the Philosophy of Science," *The Monist*, 60 (1977), 453–71, reprinted in G. Gutting, ed., *Paradigms and Revolutions* (Notre Dame, Ind.: University of Notre Dame Press, 1980), pp. 54–74, p. 73. See also his "Objectivity in Morality and Objectivity in Science," in Engelhardt and Callahan, eds., *Morals, Science and Society* (Hastings-On-Hudson, N.Y.: The Hastings Centre, 1978), pp. 21–39; and Charles Taylor's excellent paper "Interpretation and the Sciences of Man," *Review of Metaphysics*, 25 (1971), 3–51.

If we applied these ideas to our question, our task would be the interpretation and construction of histories of theories of perception that admit

the possibility of constructing an intelligible dramatic narrative which can claim historical truth and in which . . . [such] theories are the subject of successive episodes. It is because and only because we can construct better and worse histories of this kind, histories which can be rationally compared with each other, that we can compare theories rationally too . . . It is only when theories are located in history, when we view the demands for justification in highly particular contexts of a historical kind, that we are freed from either dogmatism or capitulation to scepticism.[42]

Of course, such "histories" have usually been composed by or with guidance from the practitioners themselves. However, the transformation of the writing of history into a self-conscious enterprise is, as Collingwood observed, a recent development, one that has its own history, a history that makes intelligible (even if not persuasive) the claim that a true historical understanding of theories of perception would stand epistemologically prior to, and therefore in a position to judge, the truth of theories of perception themselves. "An understanding of the concept of the superiority of one physical theory to another requires the prior understanding of the concept of the superiority of one historical narrative to another. The theory of scientific rationality has to be embedded in a philosophy of history."[43]

I want to discuss MacIntyre's point with an example both he and Taylor mention, the case of Galileo. For MacIntyre, Galileo succeeds because "he, for the first time enables the work of all his predecessors to be evaluated by a common set of standards. . . . The history of late medieval science can finally be cast into a coherent narrative."[44] In just these terms, but in view of the attention to the "self" I am here urging, the narrative that Galileo permits seems somewhat less coherent than MacIntyre imagines. A significant part of the new Galilean account is a story about the kinds of people who believe Aristotelian theories and, by contrast, those who do not. In the *Dialogue Concerning the Two Chief World Systems,* Galileo notices that the Aristotelian philosophers, Simplicio and his ilk, understand nature through the category of "perfec-

---

[42] MacIntyre, "Epistemological Crises" [see note 41], 62, 73–4.
[43] Ibid., p. 70.    [44] Ibid., p. 61.

tion." They cling to their beliefs, Galileo opines, because of "their great desire to go on living . . . and the terror they have of death."[45] With such observations, Galileo tells his readers that the question of deciding between "world systems" is charged with the issue of deciding (or facing) the question *who we are going to be*. In claiming that Aristotelianism is a pathology, Galileo is saying that a particular conception of "the self," or what a "self" ought to be, is mistaken and wrong-headed. In saying that Aristotelians persist in their beliefs chiefly because of their fear of death – which is, in fact, to call it a "superstition" – Galileo reveals that he finds it incomprehensible how any self-respecting adult could believe Aristotelian theory on its own terms, without a prior and underlying need for reassurance and comfort, for the heart of a heartless world. On Galileo's view, the Aristotelians have embedded their autobiographies in the framework of a larger story about a redeeming nature.

Unless we are to adopt the Romantic or Tragic mode and say (with Galileo) that the story of man and nature is now definitively told, we should recognize that Galileo's own case is properly vulnerable to analysis in just the same terms. What kind of a "self" is it who can understand (and constitute) his own condition, give meaning and sense to his own conduct, by reference to a "Book of Nature" written entirely in "mathematical characters"? We have to imagine that in the last analysis, Galileo can comprehend his own biography (of which *he* is the author) as embedded in a story about Nature, a story in which he is able to recognize himself.[46] For Galileo, a story in which nature is effectively a neutral field for the deployment of practical human industriousness cannot be far from wrong or even particularly original. What is new, it seems to me, is that in Galileo's time such a view of nature and of theory actually threatened to become legitimate and thereby displace existing authorities.

[45] Galileo Galilei, *Dialogue Concerning the Two Chief World Systems,* trans. Stillman Drake (Berkeley: University of California Press, 1962), p. 59. It is extremely difficult to know whether Galileo really believes this or whether it is "rhetoric." But this is not just a historian's problem; it is also a problem for Maffeo Barberini, Benedetto Castelli, and perhaps even for Galileo himself. Even if it is only rhetoric, it could not be judged effective unless judged persuasive and believable.

[46] At least this is a goal to which he must aspire. See Taylor, "Interpretation and the Sciences of Man" [see note 41], 16, 47 especially, and Hayden White, "The Historical Text as Literary Artifact" [see note 32], pp. 60–1.

An approach along similar lines suggests ways of resolving the discussion between myself and David C. Lindberg concerning the status of Kepler's optical work. It will also be clear that the point is fully generalizable to all such disagreements concerning interpretation. In a number of publications, Lindberg has carefully brought support to the conclusion that Kepler's work is best understood as a set of events in the narrative that is the history of medieval optics. Kepler's successes even complete and fulfill that tradition, representing its mature form.[47] I have held that Kepler represents a decisive break with the medieval tradition, a break characterized by what I have understood to be Kepler's "mechanization of light and vision."[48] As is often the case in such discussions, Lindberg and I do not seem to disagree about the "facts." At some level of description, what Kepler did is, in fact, indistinguishable from the activities of a medieval theorist. If we are to take seriously the subject (the "self"), however, we have to ask whether a "history" at that level of descriptive adequacy would be properly intelligible *to Kepler* or, indeed, to John Pecham as an account of what they actually did. Their understanding of themselves as optical theorists are irreducibly present, it seems to me, and must be considered in any account that aspires to historic truth.

Kepler dramatically associates himself with the tradition of *perspectiva pingendi,* represented for him most outstandingly by Albrecht Dürer, and turns to this set of practices for authoritative guidance in optical matters. He also expresses his preference for such people as Jost Bürgi, unlettered instrument maker at Rudolf II's court in Prague, in comparison to the scholars. It is extremely difficult, consequently, to imagine that he sees himself in his activities as a theorist of vision in ways similar enough to the medieval authors of optical texts that we can legitimately insert him in a story about the successful extension and continuation of their work in the seventeenth century. The outstanding feature of Kepler's "solution" to the problem of how vision takes place is his asser-

---

[47] David C. Lindberg, *Theories of Vision from Al-Kindi to Kepler* (Chicago: University of Chicago Press, 1976), esp. pp. 178–208. See also his "Laying the Foundations of Geometrical Optics: Maurolico, Kepler, and the Medieval Tradition" (William Andrews Clark Library Publication), forthcoming.

[48] See S. Straker, "The Eye Made 'Other': Durer, Kepler, and the Mechanisation of Light and Vision" in L. A. Knafla, M. S. Staum, and T. H. E. Travers, eds., *Science, Technology, and Culture in Historical Perspective,* University of Calgary Studies in History, No. 1, (Calgary: 1976), pp. 7–25.

tion that we see by means of a "picture" "painted" by "little brushes" of light on the opaque back wall of the eye. Kepler's theory shares strongly the perspectivist painter's focus of attention on a physical (behaviorist?) simulation of a visual experience by means of colors arranged on a surface. With Kepler, as with Galileo, it seems to me we are witnessing the rise to intellectual prominence of what is essentially a "craft" approach to theoretical knowledge and a change in the context of optical, and therefore psychological, theorizing. Although Kepler does plainly subscribe to a "faculty" theory of perception, he does not elaborate it, declaring, rather, that the equipment of the optician takes us only as far as that "painting" on the back wall of the eye. Thus he is able to overcome the previously daunting problem of how it is we see things right-side-up, even when the picture in the eye is inverted.

That Kepler's understanding of himself at both levels – the "self" as theorist as well as the "self" who is represented *in* those theories – is more like that of the artists than the academic theorists is reinforced by Svetlana Alpers's recent and forceful interpretation of "Northern" (Dutch) art in the seventeenth century, which she understands in large part to be a series of scientific experiments in *The Art of Describing*.[49] She sees these artists as engaged in an activity best described as an investigative craft of "seeing," in which painting is the manifest exponent. In this explication of "seeing" by "painting," this frank exploration of just what can be observed, the painter operates out of a craft tradition, which has deep experience of mirrors, lenses, and other optical devices (such as the glass urine bottles that Kepler used to observe and judge refractions). In a loving devotion to the painting of what can be seen, Alpers argues, the painter is inviting us to resist the demands of a spurious (and verbal) knowledge that goes beyond and tries to penetrate the *appearances*. If Alpers is right, then there are grounds for imagining how deeply indeed Kepler may understand the eye, literally, as a device for making pictures which then, by the action of the spirits and faculties of vision, we see.

My suggestion is that Kepler effectively decides that his "project" of understanding light and vision is most authentically engaged by inserting himself into the tradition of the artist–engi-

---

[49] Svetlana Alpers, *The Art of Describing. Dutch Art in the Seventeenth Century* (Chicago: University of Chicago Press, 1983). See also the extremely thoughtful review by Andrew Harrison in *Art International* (January–March, 1984), pp. 53–6.

neers. In relocating the project, he changes for himself and those who follow the categories in terms of which we think about visual perception. If this is right, he terminates the tradition of medieval optical theory essentially by abandoning it.

An approach along these lines would put the history of science (and the history of theories of perception in particular) squarely back into the history of Western philosophy and would resist all attempts to lift science out of history. For my argument simply follows a train of thought that begins with Kant's explication of Newtonian science and acknowledges that "nature" exists for us only by virtue of certain *categories of thought* possessed by minds.[50] That such categories are historically mediated is not news to historians of science, for "historicized Kantianism" is, after all, just another name for the position argued somewhat uncannily (after studying Burtt and Koyré) by Thomas Kuhn in his famous book. In closing his consideration of the problematic posed by science, Kuhn said that the question his book raises is, "What must the world be like in order that man may know it?"[51] The history of theories of perception calls for a set of companion volumes to Kuhn, Burtt, and Koyré which inverts and amplifies Kuhn's question by asking, What must "mind," the "self," be like in order to have understood the world in the ways that it has?

The critical literature generated by Kuhn's work opens the further possibility of replying directly to any suggestion that the undertaking I propose is necessarily an "idealist" history. An examination of the place and necessity of acquiring "cognitive authority" in the sciences makes it more than plausible that "paradigms" are to be comprehended as *historically* created and sustained. As the program I suggest requires, such a history of categories of understanding would include the historical creation and maintenance of conceptions of "the self." On this subject, existing historical literature is vastly informative, and some of it – for example, Marvin Becker, R. H. Tawney, or E. P. Thompson – brilliantly eloquent.[52]

[50] See Louis Mink, *Mind, History, and Dialectic: The Philosophy of R. G. Collingwood* (Bloomington, Ind.: Indiana University Press, 1969), pp. 146 ff and R. G. Collingwood, *Essay on Metaphysics: Philosophical Essays* (Oxford: Oxford University Press [Clarendon Press], 1940), II, 245–281.

[51] T. S. Kuhn, *The Structure of Scientific Revolutions* (Chicago: University of Chicago Press, 1962), p. 173.

[52] See M. D. King, "Reason, Tradition, and the Progressiveness of Science," *History and Theory, 10* (1970), 3–32 and S. Straker, "A 'New Physiognomy of Servitude': Some

## NATURE AND HUMAN NATURE

Though words be the *signs* we have of one another's *opinions* and
intentions, because the *equivocation* of them is so *frequent according
to the diversity of contexture,* and of the company wherewith they
go, which, the presence of him that speaketh, our *sight* of his
*actions,* and *conjecture* of his *intentions,* must help to discharge us
of; it must be *extremely hard* to find the *opinions* and meanings of
those *men* that are *gone from us long ago,* and have left us no other
signification thereof than their books, which cannot possibly be
understood without *history,* to discover those aforementioned
circumstances, and also without great prudence to *observe* them.

Thomas Hobbes, *Human Nature* (1640)[53]

I wish to close by surveying what I think are some important issues in the history of theories of perception and by placing these
matters in the context of the preceding discussion.

It is worth noticing first of all, that in the Scholastic or "classical" theory of perception, there is no mind–body problem. Were
a good Aristotelian to be put the question, how can we have
knowledge of the external world?, the best reply would be, what
do you mean by "external"? The problem of consciousness and its
awareness of the world is successfully and seriously addressed in
medieval optics precisely because the Scholastics posit as a theory
of knowledge the hypothesis that "mind" and the immaterial forms
("ideas") that constitute mind are taken to be objectively existing
entities. The key to Aristotelian epistemology is the theoretical
understanding that "Nature" exists as an intentional act. In Plato,
Aristotle, Augustine, and Thomas and throughout the tradition,
it is understood that the fully actualized Mind (that is, God) creates
and maintains the world as an embodiment of Its own thought.
All creation is in this sense an expression of God. Just as human
ideas are clothed externally as *words,* in language, so the thought
of God, the *Logos,* the *Verbum,* is deployed externally as "Nature."[54] No wonder perception and cognition are characteristically

Comments on A New Philosophy of Science," *4S Review, 2* (#3, 1984). See Marvin
Becker, *Florence in Transition,* 2 vols. (Baltimore: Johns Hopkins University Press, 1967–
8, esp. the "Epilogue to Vol. II; R. H. Tawney, *Religion and the Rise of Capitalism* (New
York: Harcourt, 1926), and E. P. Thompson, "The Moral Economy of the English
Crowd in the Eighteenth Century," *Past and Present,* No. 50 (1971), 76–176.

53 In William Molesworth, ed., *The English Works of Thomas Hobbes* (London, John Bohn,
1840; reprint ed., 1966), *IV,* 75.

54 Charles Taylor, *Language and Human Nature* [note 24], p. 12. See also his *Hegel* [note
24], pp. 3–29. John W. Yolton quotes a late seventeenth-century writer referring to the

veridical, for both man and nature express themselves in "impressions" or "images" whose material embodiments constitute the individual naturally existing things found in the material world. Similarly when they are actualized in the mind, they constitute the act of perceiving and knowing.[55] A. Mark Smith has expressed this overall causal theory elegantly in explaining that for the Aristotelian,

the act of apprehension or sense cognition is bilateral, contingent not only on the potential of external objects ultimately to express themselves intelligibly, but also on our power, as appercipient beings, to accept those expressions intelligibly, or literally, to *realize* them.[56]

The knower and the known become one. In gaining knowledge of nature, "mind becomes what it thinks and may be said to know itself."[57]

With the scientific revolution, the revival of ancient materialism, medieval nominalism, and Stoic philosophy, and the "mechanization of the world picture," the intentional cosmos was replaced with the world-machine, and the mind–body problem came to the forefront of modern consciousness. Or, so our common understanding would have it. Several recent studies have, however, given grounds for some measure of skepticism about this portrait of the birth of the modern world in the seventeenth century.

Scholastic theory of "a Material Phantasy offered to the Intellect, a Species impressed upon it, or expressed in it." (*Thinking Matter* [see note 1], p. 154)

[55] I am grateful to Katherine Park of Wellesley College for sharing with me material that arises out of her work with Charles Schmitt on a forthcoming translation of Gianfrancesco Pico's *De imaginatione*.

[56] A. Mark Smith, "Getting the Big Picture in Perspectivist Optics," *Isis, 72* (1981), 568–89, p. 575. Smith's claim that "the ulterior concern of the perspectivists was epistemology" (p. 569) does not follow from what he does demonstrate, namely, that the Aristotelian "participation theory" is the *context* in which they understood themselves to be working. Unless further evidence of a different kind is offered – evidence distinctly about their own conception of what they are doing – it could still be the case – and what Lindberg's work suggests – that the medieval perspectivists' own view of their activity is that they are are studying the "perspectival" preconditions characteristically accompanying normal vision. That much optical theorizing has this character seems to be implied in Stephen Gaukroger, "Aristotle on the Function of Sense Perception," *Studies in the History and Philosophy of Science, 12* (1981), 75–89.

[57] Owen Barfield, *Saving the Appearances. A Study in Idolatry* (New York: Harcourt, Brace & World, 1965), p. 101. Barfield offers a history of theories of perception that is radically different from those in the mainstream of the history of science. It might be regarded as a "rival theory," with roots in the Romantic tradition, a theory well worth historical investigation.

John Yolton has recently argued, in two important books, that we have been engaged in a systematic misunderstanding of seventeenth- and eighteenth-century texts about perception and cognition.[58] If we properly understand the writings of early modern theorists, Yolton suggests, we shall find that they are *not* advancing a "representation theory" in which "ideas" stand in for objects perceived; nor are these thinkers prey to the skepticism that is supposed to accompany a renunciation of authentic acquaintance with the world of things. Rather, they endorse a theory of direct perception whereby the perceiver is immediately cognitively acquainted with things perceived. Perception *is* cognition, in these views, and we directly apprehend the world, there being "no cognition at a distance." Furthermore, the notion of being "present to the mind" is apparently not intended in a spatial or physical sense but rather as a "cognitive presence," evidence of which is the immediacy of meaning and signification to cognitive awareness. The notion of perception involving the "natural signs" of a visual "language of God" or of "nature" is often adopted.[59] If Yolton's reading is even nearly correct, there is much more continuity with the Scholastic "participation theory" in early modern thought about perception than we are used to supposing. There are actually independent grounds for suspecting that such is the case.

One is John Dunn's discovery that the political theory of John Locke is irreducibly grounded in a theological explanation of human "rights." It is understood by Locke that each of us must re-

[58] John Yolton, *Perceptual Acquaintance from Descartes to Reid* (Minneapolis: University of Minnesota Press, 1984), see, e.g., p. 5. The other study is idem, *Thinking Matter* [see note 1]. Unfortunately I did not see these two books in time to use them constructively in the present argument. There is no doubt, however, that Yolton's studies are taking up what I have called Burtt's injunction. That such rereading of the texts about "perfection" is required is underlined by John J. MacIntosh's careful discovery that there is no such thing as thing as *the* distinction between "primary and secondary qualities." His study of the literature since the sixteenth century reveals "some twenty odd ways of making a distinction." MacIntosh concludes, "We should ask, with respect to each primary/secondary quality distinction: what is (or was, for past needs need not be present ones) the point of making it?" "Primary and Secondary Qualities," *Studia Leibnitiana, 8* (1976), 88–104, quoting from pp. 88, 103.

[59] For a survey of these conclusions, see Yolton, *Perceptual Acquaintance* [see note 5], pp. 3–17, 204–23 and *Thinking Matter* [see note 1], Preface, pp. ix–xiv. After reflection on the import of his reinterpretation of these texts, Yolton says, "If we could understand how physical movements, as well as physical objects, become meaningful, we might be able to give up the talk of causal interactions between mind and matter." *Thinking Matter*, p. 204.

spect each other's "rights" to "life, health, liberty," and "property" because each of us is *God's property*.[60] Take away this theological fact, and Locke has no theory of political obligation whatsoever, modern or otherwise.

If what we recognize as "liberal politics" and a "modern" sense of the self is not particularly articulated by Locke, we have strikingly similar theological reasons for doubting that Isaac Newton is the author of a "materialist conception of nature." In the unpublished manuscript *De gravitatione et aequipondio fluidorum*, Newton proposed that God continually maintains the existence of "matter" "by the sole action of thinking and willing" a volume of space (His *sensorium*) to present such appearances to us. Newton goes on to say that his theory "clearly involves the chief truths of metaphysics and thoroughly confirms and explains them," for "atheism" is thereby defeated and the existence of "mind" and its "union with body" is made both intelligible and possible. Thus, Newton's preferred theory of matter also requires an actively present God.[61]

Locke seems to have considered a similar task for God in order to make perception and cognition comprehensible. In the *Essay Concerning Human Understanding* he suggested that, for all we know, God could give "to some system of matter, fitly disposed, a power to perceive and to think" and that such a "super-addition" of the power of thought to matter is no less conceivable than that God "should superadd to it another substance with a [its own] faculty

---

[60] John Locke, *Two Treatises of Government*, ed. Peter Laslett (Cambridge: Cambridge University Press, 1960), Chap. II, par. 6, p. 311, and chap. V, "Of Property," pp. 327–44. For his exemplary treatment of these issues in intellectual history, see John Dunn, *The Political Thought of John Locke: An Historical Account of the Argument of the 'Two Treatises of Government'* (Cambridge: Cambridge University Press, 1969) and in the same *genre*, James Tully, *A Discourse on Property: John Locke and his Adversaries* (Cambridge: Cambridge University Press, 1979). See also E. J. Hundert, "Market Society and Meaning in Locke's Political Philosophy," *Journal of the History of Philosophy*, 25 (1977), 33–44.

[61] Isaac Newton, *Unpublished Scientific Papers of Isaac Newton*, ed. and trans. A. Rupert Hall and Marie Boas Hall (Cambridge: Cambridge University Press, 1962), pp. 89–156, quoting from pp. 138–9, 141–3. The significance of the presence of God in these matters was revealed by Locke when he revealed what our situation would be like if there is no God. "If man were independent [of God] he could have no law but his own will, no end but himself. He would be a god to himself and the satisfaction of his own will the sole measure and end of all his actions." (John Locke, Ethica B. MS Locke cap. 28, p. 141, p. 141, quoted by Dunn, *The Political Thought of John Locke* [see note 59], p. 1)

of thinking."[62] Apparently, the farthest thing from Locke's mind is the idea that the mind is a "mirror of nature," as Rorty, along with many others, suggests.[63]

Yolton's rereading of these classics in modern perception theory suggests that as seems to have happened to some degree in political theory (Locke and Hobbes) and certainly in the theory of matter (Newton and Boyle), seventeenth-century thinkers are translating a Scholastic "participation" theory into a "materialist" discourse, a move that requires for coherence the introduction of an inscrutable (voluntarist) God to serve actively as the creator and bearer of things and thoughts in the world.[64] The mere possibility of such suspicions strongly vindicates the importance of examining the history of theories of perception as a means of clarifying the historical situation.

It may be that the "modern problem" is not decisively present until God is absolutely absent, a denouement which perhaps only occurs coincidentally with the "scientistic" transformation of the world in the nineteenth century, a transformation described by Weber as the disenchantment of the world. It is easy enough to see that after about 1800, all the powers of expression that in the Scholastic view belong to God or Nature become lodged in "Man," not so much attuned to a cosmos, but creating it, not cosmically defined, but self-defining, as in the Romantic vision. The neoromantic revival now underway primarily in North America, as philosophers of science (prepared by Kuhn and Feyerabend) discover Hegel and the "hermeneutic circle," seems far less celebratory of the intellectual powers of Man than the German Roman-

---

[62] Quoted by Yolton, *Thinking Matter* [see note 1], p. 14.

[63] If Yolton is right, then the mainstream of philosophical discussion about this problem is wrong in supposing that "The wrong turn" was taken in the seventeenth century by Descartes and Locke. See, for example, Richard Rorty, *Philosophy and the Mirror of Nature* [see note 27]. Rorty's theme, the idea that "the mirror of nature" as a metaphor for the "mind" is an invention of the "scientific revolution," seems to me simply a mistake. Consider, just for example, Thomas Aquinas's remark that *"verbum intellectum est tanquam speculum, in quo res cernitur,"* quoted by Owen Barfield (from *De natura verbi intellectus*), *Saving the Appearances* [see note 56], p. 95. Notice, again, the connection between perception, meaning, and language.

[64] For a provocative discussion of the historical relations between atomism in science, voluntarism in theology, and their political involvements, see Francis Oakley, *Omnipotence, Covenant, and Order* (Ithaca, N.Y.: Cornell University Press, 1984). Here also is an energetic defense of Arthur Lovejoy's "history of ideas."

tics. For although the "self" has become hyperactive, it is entirely a subjective self, impotent to claim any real understanding of the world. In the recent writings of Nelson Goodman and Hilary Putnam, the "world" is understood as a "construct" of *our* devising, not just as we please, but with the participation of that world. We are engaged in "world making," in co-constructing the world, "not with hands but with minds, or rather with languages or other symbol systems";[65] "the mind and the world jointly make up the mind and the world."[66] It would appear, however, that such American neoromanticism is in danger of uniting comfortably with the "objectivists," and CS and AI theorists, as all are tempted to espouse a "pragmatism" of some kind whose view of theory and of knowledge will be primarily "instrumental."[67] The point is that such contemporary discussions are properly the object of historical study, and that indeed, only historical studies can "rescue" philosophy from present infelicities.

These considerations reinforce the conviction that the historian of theories of perception cannot take for granted the epistemological success of recent "scientific" theories and thereby grant them special privilege as objects of study. If the ironical reading of Koyré's story is fruitful, we might expect to find the most historically interesting accounts of perception and cognition in just those discussions resoundingly declared to be "antiscientific," namely, the cosmology of the German "Romantics." For the historian to take such material seriously no more requires an acceptance of its presuppositions and conclusions than does the important study of the "hermetic tradition" in an attempt to understand the situation of science in the sixteenth and seventeenth centuries. It would at least be judicious to ask whether among these thinkers, men who consciously set themselves against the Newtonian synthesis, and therefore, for Koyré, against the modern age, there can be found,

---

[65] Nelson Goodman, *Of Mind and Other Matters* (Cambridge, Mass.: Harvard University Press, 1984), p. 42, pp. 34–5.

[66] Hilary Putnam, *Reason, Truth, and History* (Cambridge: Cambridge University Press, 1981), p. xi. Putnam continues, "Or to make the metaphor even more Hegelian, the Universe makes up the Universe – with minds – collectively – playing a special role in the making up." (p. xi)

[67] See Richard Rorty's contribution to the fascinating discussion about hermeneutics and "understanding in the human sciences" (with Charles Taylor and Hubert L. Dreyfus) in "A Discussion" (note 37), 3–55. In his "Reply to Professor Yolton" (note 37), Rorty says, "I don't think that the human understanding has a nature to be enquired into. Understanding is a matter of inventing useful vocabularies." (p. 135)

nevertheless, attempts at a "properly scientific" understanding of perception and cognition. Indeed, the persuasive argument, primarily by Taylor,[68] that these thinkers are adumbrating a much more coherent, and thus scientifically plausible, account of human language than the essentially Hobbesian, modern scientific account lends support to the proposal that the standardly dismissed, "unscientific" theories of the Romantics should have a place in the history of theories of perception. This will not surprise historians of modern physics, who have already found themselves led into the territory of *Naturphilosophie*.

These theoretical considerations and historical possibilities reveal a "network of the self-understanding," a web of connections that figure into any serious account of perceiving and knowing. These are the structural members of a bridge that spans the metaphysical chasm between the knower and the known. In proposing that the history of theories of perception is particularly fundamental in our attempts to understand our condition and our capacity to know, I am suggesting that we see such a history as the play within the play of our changing conceptions of nature. The history of theories of perception is the *inside* story whose outside is the history of natural science.

[68] Taylor, *Language and Human Nature* [see note 24].

# 11

## Tycho Brahe as the dean of a Renaissance research institute

VICTOR E. THOREN

The perspective of time has not been kind to the reputation of Tycho Brahe. The manifold achievements that loomed so large in the seventeenth century now seem to have considerably less than the overwhelming significance they had then. To modern generations, who have no memory of the grasp of the Aristotelian worldview on Renaissance thought, the 300- and 800-page tomes[1] that Tycho wrote to initiate the challenge to it a generation before Galileo's work seem hopelessly pale in comparison with the dramatic revelations of the telescope and the brilliant polemics of the *Dialogues*. To generations who have learned by rote in their formative years that the earth is moving, the Tychonic System[2] appears at best timid, and at worst simpleminded, no matter how much influence it had in the seventeenth century. And to generations, finally, who have been educated in the glorious generalities of Kepler's laws and Newton's mechanics, the particularities – no matter how elegant – through which Tycho advanced so spectacularly the accuracy of the solar[3] and lunar theories[4] of the day have seemed, likewise, to hold little interest. So Tycho has been left with his

---

[1] Tycho Brahe, *De mundi aetherei recentioribus phaenomenis* (1588) and *Astronomiae instauratae progymnasmata* (1602).

[2] For a recent discussion of the background of the Tychonic system, see V. E. Thoren's "Tycho Brahe's System of the World and the Comet of 1577," in *Archives internationales d'histoire des sciences, 29* (1979), 53–67.

[3] Y. Maeyama, "The Historical Development of Solar Theories in the Late Sixteenth and Seventeenth Centuries," *Vistas in Astronomy, 16*, (1974), 35–60.

[4] For Tycho's contributions to the lunar theory, see V. E. Thoren's "An Early Instance of Deduction Discovery" (*Isis, 58* [1967], 19–36) and "Tycho Brahe's Discovery of the Variation" (*Centaurus, 12* [1967], 151–66).

observations, and all too frequently, even in this respect, with credit more for being the person who happened to collect the data used by Kepler, than for having conceived the ideal of minute-of-arc accuracy, and having invented the methods of constructing and utilizing instruments[5] that allowed him to obtain it. There is one aspect of Tycho's work, however, for which current historians should have more appreciation than the seventeenth century did: A generation before Bacon sketched for a generally unappreciative audience an outline of his Utopian establishment for the advancement of learning, Tycho had one in operation on Hven.

From at least as early as the planning of a "garret" with accommodations for eight, Tycho envisioned having assistants to work with him at Uraniborg. Already by the beginning of 1578[6] he had his first one, a fellow Dane named Peder Jacobsen, who hailed from the village of Flemløse on the island of Fyn. In the early 1580s, Flemløse was gradually joined by others, until a group numbering eight to twelve members[7] of varying degrees of permanence and competence came to be established. Thanks to the curiosity of an anonymous inmate who compiled a fragmentary list of his fellows[8] toward the end of the 1580s, it is possible to obtain a glimpse of the spectrum of people involved.

For the most part, naturally, Tycho's students were Danes, drawn from the highways and byways of the realm as far away as Norway and Iceland. But there were several from Holland and Germany, and at least one three-month visitor from England (John Hammond, who would later become physician to the family of James I).[9] With the exception of two or three who were acquaintances from Tycho's own student days, and an occasional "older" travelling scholar, the students were twenty-year-olds. But they

---

[5] See V. E. Thoren's, "New Light on Tycho's Instruments" in *Journal for the History of Astronomy, 4* (1973), 25–45.

[6] *Tychonis Brahe Dani Opera Omnia*, ed. J. L. E. Dreyer (Copenhagen, 1913–27), X, 59, 69. [Hereafter cited in the form: X, 59, 69.]

[7] See the seventeenth-century biography of Tycho written by Pierre Gassendi, *Tychonis Brahei Vita . . .* (Paris, 1654). Since it is most readily consulted in a modern translation (Swedish) by Wilhelm Norling (*Tycho Brahe, Mannen och Verket* [Lund, 1951]), having different pagination, I shall cite it under the name of Gassendi, but with the page numbers of Norlind. Thus: Gassendi, p. 45.

[8] *XIV*, 44–5. The list is also printed as an appendix to J. L. E. Dreyer's well-circulated *Tycho Brahe* (1890; New York: Dover, 1963) [hereafter cited as Dreyer], pp. 381–4.

[9] Wilhelm Norlind, *Tycho Brahe, En levnadsteckning med nya bidrag belysande hans liv och verk* (Lund, 1978) [hereafter cited as Norlind], p. 98.

were veterans of the university curriculum, and were, in theory, at least, striving for excellence in some form of higher learning just as Tycho had been after his first trip abroad. Of course, in practice, there were, inevitably, enough mere curiosity-seekers and dilettantes, so that already by 1589, Tycho was more than a little bit jaundiced, and in no mood to mince words to applicants.[10]

If Victorinus Schönfeld [Professor of mathematics at Marburg] wants to send his son here to stay with me, that is all right with me. But as soon as his son or I feel that it is appropriate for him to leave, either of us must be free to bring it about. Whether he has obtained the M.A. degree, or not, is immaterial to me. I would prefer that he really be a master of arts, rather than just have the degree. But that is no easy matter, so it will suffice if he is a serious student.

As would only be expected under any such circumstances, many of the three-score souls who must have rotated through Urani-borg by 1597 got enough of astronomy, island life, or Tycho to satisfy them in just a few weeks. Already by the time of composition of the list of 1588 or 1589, eight of the thirty-two enrollees could not be remembered by name, and several others were explicitly credited with tenures of half a year, or less. On the other hand, no less than ten people are known to have stayed with Tycho two and a half years or longer, and two of them stuck it out for terms approaching ten years.

As nearly as can be determined, the most important technical assistant Tycho had during his long tenure on Hven was not one of his students at all, but rather one of the artisans from his shop, Hans Crol. Crol was a goldsmith, lured from Germany (or some Slavic land beyond) sometime prior to 1585, when his name first appears beside an observation in Tycho's log.[11] All we know about his term on Hven is that when he died and was buried there on December 4, 1591,[12] he had "had charge of Tycho's instruments for many years, and had even built some of them himself."[13] If this description suggests that he probably did not arrive early enough to participate during the crucial period of the early 1580s, when the bulk of Tycho's most important instruments were conceived and fabricated, he clearly provided other services that removed

[10] *VI*, 198.    [11] *X*, 373.
[12] *IX*, 106-7. A son of Crol's had died on the island a year and a half earlier (*IX*, 85).
[13] *VI*, 299, 371.

him from the status of mere "technician." By 1590, he had done enough observing to establish a reputation for very keen eyesight, and accordingly, to be singled out by Tycho as the one to make a series of important measurements of the angular diameters of the various planets.[14] And if it is correct to ascribe to him the compilation of the above-mentioned list of Tycho's students, he was obviously occupied with the program in a way that made it something more than just a "job" for him.[15]

Among those who were, more properly speaking, Tycho's students, the best known was one of the later entrants on the scene, Christian Sørensen. Longomontanus, as he later styled himself in Latin, labored at Uraniborg from 1589[16] until Easter of 1597, when Tycho closed its doors forever. That he had by that time made himself practically indispensable to Tycho appears more from the importunate requests he subsequently received to rejoin his exiled master, than from the rather restrained letter of recommendation he received. But the fact that it and a later one for him are the only such letters to survive,[17] combined with the fact that Longomontanus was the only one of Tycho's students to attain a professorship in astronomy (University of Copenhagen, 1605–47), suggests that Tycho managed to convey his respect for his student's abilities to someone besides Longomontanus himself. On the basis of Longomontanus's *magnum opus,* the ponderous but well-circulated *Astronomia Danica* of 1622, 1643, and 1663, it seems safe to assume that most of his work for Tycho was of a theoretical nature. He says that he did some considerable work (data reduction, probably) for Tycho's star catalog,[18] which was compiled in the early 1590s. Most likely he did much of the planetary work that was done during that time, too, because, when he eventually yielded to Tycho's pleas and rejoined him at Prague in 1600, he was immediately set to work on the theory of Mars. Shortly thereafter,

---

[14] *XI*, 292.

[15] The identification has been based on handwriting – always a difficult undertaking. Longomontanus is in every respect a more logical candidate.

[16] A document cited by Norlind (p. 102) sets Longomontanus's arrival in the summer of 1590. The traditional 1589 date stems from Tycho's letter of recommendation in 1597 (*VIII*, 384), which refers to eight years of service. It is not at all impossible that Tycho either misremembered or counted the portions of both 1590 and 1597 as "years."

[17] *VII*, 384; *VIII*, 335. Tycho also wrote a very nice letter for Kepler at their (temporary) parting of the ways (*VIII*, 324–5).

[18] C. S. Longomontanus *Astronomia Danica,* (Amsterdam, 1622), p. 201.

he made the final alterations and adjustments to the theory of the motion of the moon, on which he had also previously bestowed some considerable labors.[19]

Almost equally favored with the opportunity to do independent technical work was another Dane from Longomontanus's neighborhood in northwest Denmark, Elias Olsen Morsing. The earliest evidence of Olsen's presence on Hven is from April, 1583, when the daily weather notations in Tycho's meteorological diary begin to appear in his hand. From what can be gathered concerning the arrivals of the people before and after him (see note 8) on the list of Tycho's students, it is doubtful that he had been there more than a few months prior to that time. He would remain at Uraniborg until his death in 1590.[20]

The best documented of Olsen's contributions was a solo trip to Poland in the spring of 1584. The purpose of the expedition was to confirm Tycho's hunch that, through ignorance of the effects of refraction, Copernicus had underestimated the latitude of his observatory at Frauenberg by more than 2′, with concomitant effects on his determination of the obliquity of the ecliptic.[21] Olsen made the requisite observations and returned with results that apparently satisfied Tycho in every respect. As with Longomontanus, we may assume that Olsen, too, played some kind of role in the construction of Tycho's star catalog, but probably one involving more observation than computation – if only because Tycho's star catalog was still in its early, observational stages when Olsen died. His participation in two other projects is more appropriately described below in the context of the activities themselves.

The longest termed of all Tycho's associates was the already-mentioned Peder Jacobsen Flemløse, his first known student. Flemløse's acquaintance with Tycho dated from at least 1574, when he dedicated to Tycho a Latin poem which argued that, although

---

[19] For biographical information on Longomontanus, see *Dictionary of Scientific Biography*, *XIV*, 445–8.

[20] *IX*, 82. Because of a diary reference to an(other) Elias Olsen in 1596, Dreyer (123) believed that the reference to his "obiit" intended to convey his "abiit" (departure). However, the exact recording of the hour (11-1/2 noct.), and the unlikelihood that anyone would leave at midnight for a 2 or 3 hour boat trip to the mainland, argue strongly for "obiit."

[21] Although Tycho mentions 23°28′, Copernicus actually settled on 23°2/5′ (*De revolutionibus*, book III, chaps. 2, 6, 10).

eclipses were naturally occurring phenomena, the solar eclipse of late 1574 foretold the second coming of Christ.[22] Since the pamphlet was published in Copenhagen, it seems reasonable to assume that its author was among the auditors of the lectures Tycho gave at the university that fall. Flemløse's presence at Hven is documented by references to him in Tycho's observation logs from the beginning of 1578 right up through the fall of 1587,[23] when he is credited with the responsibility of supervising a crew in the detailed recording of a lunar eclipse. And when he left Tycho's service shortly thereafter, to become physician to the Governor of Norway (Axel Guildenstern, Tycho's mother's cousin), he continued to contribute by conducting surveys for Tycho in Norway.[24] Such a person would obviously become thoroughly familiar with Tycho's instruments over the years. Thus, when Tycho encountered, in 1586, a situation in which he feared that his priority in certain features of his instruments might have been compromised, he could send Flemløse to the Landgrave of Hesse to look into the matter in a subtle way.[25]

For the most part, however, Flemløse probably occupied himself with medical alchemy. Already in 1575, he had published a translation of a medical text; and when he went on the above-mentioned trip to Germany, Tycho's letter of introduction referred to him as being skilled in "pyronomics." Given Tycho's own interests in chemical medicine, Flemløse's eventual employment as a physician, and the fact that when he died suddenly in 1599 he was just about to obtain the M.D. degree from the University of Basle, it is not difficult to credit Longomontanus's statement that Flemløse came to Hven because of the relevance of astronomy to medicine.[26] Nor was he the only one. Already during the winter of 1581/2, he was joined by a medical student named Gellius Sascerides from the University of Copenhagen, who would stay on Hven for over six years without making any notable astronomical contribution. Of course, even he participated sufficiently in the program to earn his spurs as an observer;[27] but it is clear that his major interest, and probably his major occupation at Hven, lay in alchemy. When he left Hven in 1588, he would pursue medical studies to an M.D. degree from Basle in 1593, and eventually

---

[22] Dreyer, pp. 117–8.    [23] *X*, 59, 92, 102, 107, 124; *XI*, 143, 164.
[24] Norlind, p. 94. *V*, 342.    [25] *VI*, 40, 58.    [26] Dreyer, p. 118.
[27] *X*, 107, 140, 155, 373; *XI*, 143, 164. See Tycho's introduction of him in *VI*, 104.

serve as professor of medicine at the University of Copenhagen from 1603 until his death in 1612.[28]

By virtue of their aggregate forty years of service, and the (not totally independent) fact that their interests coincided so directly with Tycho's, Crol, Longomontanus, Olsen, Flemløse, and Gellius must be ranked in a class by themselves as contributors to Tycho's program at Uraniborg. However, the picture of intellectual life on Hven would be seriously deficient if it did not include glimpses of a considerable number of background figures and activities. One such was the German mathematician Paul Wittich. A fellow student of Tycho's from his days at Wittenberg (1566),[29] Wittich showed up on Hven in the summer of 1580. Whatever their relationship had been before, Wittich made an enormous impact now. Armed with some kind of knowledge of the existence of a way to replace the tedious multiplications and divisions of spherical astronomy with much less troublesome additions and subtractions, he managed (with some unspecified collaboration from Tycho) to develop in a very short time some new trigonometric identities,[30] which promised to be of great utility in astronomical data-reduction processes. Unhappily, however, Wittich stayed less than four months. Apparently, Wittich was primarily a mathematician at heart, and so excessively theoretical in his orientation, that he just could not work up a full-time interest in the numerous practical pursuits, which constituted the prerequisite for the renovation of astronomy.[31] At any rate, despite Tycho's hope that they would have a long and fruitful relationship, Wittich soon packed his bags for a "temporary" absence to collect an inheritance from a rich uncle, and disappeared from Hven forever. As an unduly nervous Tycho would find out later, Wittich took with

---

[28] J. R. Christianson, "Tycho Brahe's Cosmology from the *Astrologia* of 1591," *Isis* 59 (1968), 312–18.

[29] Norlind, p. 101.

[30] The bases of the so-called method of prosthaphaeresis were the two formulas $\sin a \sin b = \frac{1}{2}(\cos[a-b] - \cos[a+b])$, and $\cos a \cos b = \frac{1}{2}(\cos[a-b] + \cos[a+b])$. For the background and development of the method, see *Dictionary of Scientific Biography* (*XIV*, 470–1).

[31] According to Tycho (*IV*, 453–5), Wittich was with him about four months, and proved to be very good at mathematics, particularly trigonometry. But he did not even know the constellations; and even went so far as to proclaim that such information was no more necessary for the astronomer than a knowledge of herbs was for a physician. In fact, he actually left Hven, making derogatory remarks about Tycho's extreme emphasis on observation.

him – and passed on to the Landgrave of Hesse – some impressions of Tycho's new instruments. But he left behind him trigonometric formulas that would give Uraniborg the most advanced methods then known for handling its computations.

Equally interesting but for different reasons, was a Dutch student named Willem Janszoon Blaeu. Blaeu was about twenty-five years old when he wintered on Hven in 1595/6, and seems already to have been committed to a technical career as a producer of maps and globes. Such commercial interests were by no means beneath the dignity of royal mathematicians or even of professors of mathematics. But they could also be indicative of background or interests which were more technical than academic in character, and that seems to have been the essence of Blaeu's situation. Although he was sufficiently fortunate in his family circumstances to have had access to university education, he spent the years immediately prior to his tenure on Hven as a general apprentice in a family mercantile office.[32] Whatever the deficiencies of his formal education may have been, however, Blaeu made the most of his time with Tycho. Displaying the entrepreneurial instincts that would propel his firm to great prominence in the seventeenth century, he sought out and carried to completion a project of his own: a study of the path of the comet of 1580 from observations made by Tycho, Wittich, and Flemløse many years earlier.[33] It was not the exhaustive analysis that Tycho had conducted for the comet of 1577, but it was a nice student's exercise, relevent to the work Blaeu would be doing for the rest of his life. And the entire experience was obviously very stimulating for him. Fifty years later, his celestial globes were still advertising the stellar positions of Tycho's catalog, and his *Atlas* was including a map of Hven, conveying descriptions of the activity there and imparting anecdotes of Tycho unknown in other sources.[34]

There can be little doubt that the common denominator of schooling and service for all people who set foot on Hven was astronomy, the raison d'être of Uraniborg. So rare were exceptions to this generality that the above-mentioned list of students contains a notation to the effect that one Sebastian, a German who

[32] *Dictionary of Scientific Biography* (II, 185).    [33] *XIII*, 325–31.

[34] Blaeu also did enough observing to discover (and depict on his globes in 1600) the variable star 34 Cygni (III, 407). On his maps and globes, see Ernst Zinner. *Deutsche und Niederländische Astronomische Instrumente des 11.–18. Jahrhunderts* (Munich, 1956).

was on Hven for a month or two, did *not* study mathematics.[35]
What this most probably meant, however, was that he had done
nothing at all in the sciences of arithmetic, astronomy, geometry,
and music beyond the smattering of introductory concepts involved in the curriculum for the Bachelor of Arts Degree – and
that he had no intention of doing anything further. For, in fact,
there is no good reason to believe that any significant number of
Tycho's students were accomplished mathematicians or astronomers. That, presumably, was one of the goals that motivated ap-
prenticeship on Hven. Certainly, command of spherical trigonometry, which constitutes the basis of all astronomical
computation, was so far from routine that there was institution-
alized instruction in it at Uraniborg: A notebook containing ex-
ercises in it is still extant in Copenhagen.[36] But if computational
aid was something that could be expected only from advanced
students, observational help was another matter.

Prior to the granting of Hven, Tycho made his observations
either by himself, or with occasional aid from a servant. With the
maturation of Uraniborg, however, circumstances conspired to
reduce considerably his role in gathering his observations. Many
of his more sophisticated instruments required two people just to
sight them. Nor can it have taken Tycho long to discover that
having someone other than the sighter come with a lantern to read
positions off the instrument would facilitate the next observation
(by preserving, as we would now say, the dark adaption of the
observer's eyes). In due time, in fact, Tycho was probably using a
recording crew of three: one to hold the lantern and read out sight-
ings, another to sit by (or carry) the log and make entries, and a
third to stand by the clock and call out times as readings were
recorded.[37] And since such a crew could have worked much faster
than any instrument could be sighted, it could easily have serviced
several sighting crews at once. When combined with the fact that
good observing conditions seem not to have been something which
could be taken for granted,[38] and that Tycho had plenty of things
to do with his time, it was probably inevitable that the observing

---

[35] See number 26 on the lists cited in note 8. See also Norlind, p. 97.
[36] Norlind, pp. 101–2.
[37] See Tycho's oft-reproduced depiction of the mural quadrant (*V*, 28).
[38] See *X*, 231, and Tycho's meteorological diary (*IX*, 5–146 passim), for numerous references to days as gray or cloudy.

would come to be relegated to the assistants. How long Tycho even continued to man an instrument himself is open to question. In later years, there are several notations in the log stating that he has performed or verified a particular alignment himself,[39] which suggests conversely, that he was not doing so with very many of the others. In fact, even the appearance of his handwriting in the log, which at least documents his presence at the observing sessions, becomes less and less frequent after 1585. The implications of these facts for the nature of Tycho's contribution to his observing program are ambiguous. In all likelihood, he was in attendance at each session – but working on his own writings, providing occasional supervision, and just generally exercising a constraining influence. Even that would have been no mean achievement: Through most of the twenty-year existence of Uraniborg, there were about 185 observing sessions a year – on average, one every other night (or day) of the year. But if Tycho had had to do all the work himself, *some* aspect of his renovation of astronomy would have made notably less progress.

Although astronomy was clearly the primary activity at Uraniborg, it was neither an exclusive nor even an overriding concern of Tycho and his collaborators. Among the earliest records of business for example, pre-dating even the building of the manor house, are some papers labeled "Geographic observations made on the island of Hven." Collected, apparently, before the end of 1579,[40] they consist basically of triangulations between various landmarks on Hven and the surrounding mainland. Angular measurements to outside points allowed Tycho to establish the location of Hven relative to prominent nearby cities, whereas triangulations (and pacings, of course) on the island itself provided the data necessary for mapping Tycho's new domain. By 1585, Tycho was talking with the king about providing an improved map of the whole kingdom, and had obtained permission to use all the old maps and charts of the realm preserved in the state archives.[41]

---

[39] X, 127, 243, 362; XI, 58 103, 163, 279, 287; XII, 18, 50, 72, 121, 282, 301, 345.

[40] The observations (V, 294–300) were bound into Tycho's log of astronomical observations in a group, following the observations for the year 1579 (V, 338), indicating that the last of them was made before the beginning of 1580.

[41] VII, 102. A crude, hand-drawn map of unknown vintage (but postdating 1584, when Stjerneborg was built) is reproduced by Dreyer as V, 293. Something like it may have been sent to Ranzov, for the map published by Braun (Civitatis Orbis Terrarum, IV [1588], no. 27) shows a marked similarity, in its poorly rendered outline, at least, to the drawing.

It is more than likely that the map was associated with the project of Tycho's childhood tutor and friend, Anders Vedel, to write a history of Denmark, for nothing seems to have been done on it until 1589, when Tycho sent Elias Olsen with Vedel on a tour of Denmark.[42] In the same year, and again in 1590 and 1592, Peder Flemløse made surveys in Norway.[43] There was even some discussion of compiling a general table of geographical locations of prominent European cities.[44] But the work was never extended southward to Tycho's satisfaction. The few eclipse observations he could solicit from various correspondents, and the more readily obtained reports of latitudes, were eventually compiled into a table of longitudes and latitudes. But the result was not something that Tycho was willing to include in his catchall *Progymnasmata,* and the table remained in manuscript until Longomontanus published it.[45] The mapping got only as far as an improved chart of Hven, which Tycho had printed by 1592 and eventually published in his *Mechanica.*[46] Although these results were not enough to establish any major reputations in geographical studies, they were still far from negligible. Tycho's map was the first one made of any part of the north that was based on actual measurement, and, to one authority,[47] at least, "makes a wholly modern impression," sufficient with his use of the method of triangulation to earn him

[42] In fact, Hven is displayed twice. Plate 26 (vol. *IV*), featuring Frederick's Kroneborg castle, but covering the whole north end of the Øresund, shows Hven very prominently and even names Uraniborg. The house, grounds, and instruments for the main map (plate 27) were done (recut in reduced size for the insets) from woodcuts that Tycho had already had printed on his press, and was circulating to his friends (*VII*, 96, 104). They were eventually published in the *Mechanica* (*V*, 60, 72, 138, 142). The presence of what was essentially a country estate among plates which generally depicted cities made Hven very conspicuous, and Tycho seems to have been very pleased (and perhaps surprised) with the results (*VII*, 386).

[43] *V*, 300–4, 342; *VII*, 219.    [44] *V*, 342.

[45] *V*, 309–13. Dreyer notes (*V*, 343) that the place names are written in Dutch, rather than Danish, without suggesting that the list might therefore have been compiled by Blaeu, who was at Hven (as we have seen) over the winter of 1595–6 and did at least one other project more or less independently. Tycho's reference to the list as being essentially unfinished in 1597 is on *V*, 116. Although Kepler's "Catalogue of Principal Places of Europe" in the *Rudolphine Tables* (33–6) must surely be drawn from Tycho's list, it is sufficiently altered and extended to constitute a new work: See V. Bialas, "Data Processing in the Rudolphine Tables," in *Vistas in Astronomy,* ed. A. Beer, *18* (1975), 749–67.

[46] *VI*, 295; *V*, 150.

[47] See H. Richter. "Wilhem Janszoon Blaeu with Tycho Brahe," *Imago mundi 3* (1939), 53–60. Richter shows, incidently, that all of the interesting geographical work of Tycho was done before Blaeu [see note 45] came to Hven.

recogniton "as one of the pioneers in the technique of cartographic measurement."

Another enterprise in which Olsen and Flemløse figured prominently was Tycho's program in astrological meteorology. As with so many of his contemporaries, Tycho's view of the universe pivoted around the presumption that there was some kind of link between the cyclings of the heavens and the vicissitudes of earthly life. The most obvious way of exploring this relationship was through study of the weather; and already before the appearance of the New Star in 1572, Tycho had begun to generate almanacs based on some combination of astrological and astronomical principles, and to keep track of their performances.[48] From October of 1582, he resumed at least part of this project by having one or another of his students – Olsen for several years – keep daily records of the weather on Hven. In addition, someone may well have been generating trial almanacs, for Tycho is known to have been providing some kind of annual prognostication to King Frederick.[49] Presumably, Olsen was associated with this work in some way, for when Tycho issued the first book from his printing press on Hven, a meteorological calendar for 1586, he published it under the name of "Elias Olsen of Denmark, apprentice in the practice of astronomy to the nobleman Tycho Brahe."[50]

Precisely what role Olsen played in the actual production of the calendar is impossible to say. The fact that it bears a strong resemblance to one that Tycho appended to his little booklet on the New Star printed in 1573 is only natural under the circumstances,[51] and all one can say is that if Tycho was merely attempting to dissociate himself from an enterprise, which, after the publication of literally thousands of similar items during the previous

---

[48] I, 38–9. The originally unpublished charts survived to be printed by Dreyer in I, 75–130.

[49] See discussion later in the text and note 60.

[50] "Diarium astrologicum et metheorologicum anni a Christo 1586. . . . Per Eliam Olai Cimbrum, Nobili viro Tychoni Brahe in Astronomicis exercitiis inservientem. Ad Loci Longitudinem 37 Gr. Latitudinem 56 Gr. Excusem in Officina Uraniburgica."

[51] The book is described briefly by Dreyer (p. 125). The account of the comet was largely astrological. But extant manuscripts in Tycho's hand accord with it so well that Dreyer felt (reluctantly) obliged to publish it as part of Tycho's collected works: IV, 399–414, 512. In fact, Norlind (p. 354) sees enough similarity between the 1586 volume and Tycho's unpublished calendar for 1573 (I, 73–130) to assume (with Hayek; VII, 102) that it was essentially Tycho's work.

century, was in very low repute,[52] he did not work very hard at it. Between printing it on his own press, and appending to it his observations of the comet of 1585 (under his own name), he certainly did not fool his old teacher and long-time correspondent, Brucaeus, who chided him for sponsoring it by saying that it reminded him of Cato's query as to whether astrologers could keep from smirking at each other when they met.[53] Tycho, himself, had discovered a new and fundamental difficulty with the project around 1580, while comparing observations made from various sites in Europe on the comet of 1577: Stations that were quite close together, astronomically speaking, could nevertheless be subject to quite different meteorological conditions on a given night.[54] He would, therefore, seem to have been the first person to note publicly just how limited in extent local weather patterns could be. But although this finding impressed him sufficiently to get mentioned both in *De Mundi* and in his later correspondence with Rothmann, it does not seem to have destroyed his faith in the basic idea of astrological meteorology – at least not privately. For he followed it up in 1591 with a list of 399 aphorisms for predicting changes in the weather, published at Hven in both Danish and German, under the name of Peder Jacobsen Flemløse.[55]

Given the existence of the one type of astrological study of Hven, it can scarcely be surprising that investigations of the more usual judicial astrology were being carried out there, too, particularly because it seems more than likely that expectations in that area

[52] J. G. G. Hellmann, *Versuch einer Geschichte der Wettervorhersage im XVI Jahrhundert* (Berlin, 1924).

[53] *VII*, 100. Brucaeus had objected in advance to Tycho's permitting such "trite and vulgar things" (*VII*, 91) to be printed on Hven, and in a second letter expressed his fear that Tycho's sponsorship would "confirm the vanity of the art" (*VII*, 92). He promised to try to keep an open mind toward it, but, as we have seen, was unable to restrain himself when he saw the results. Even Hayek, who was much more receptive to astrology, expressed his opinion that Tycho had been well advised to leave his name off the calendar, even though the section on the comet elevated the tract above the general run of such things (*VII*, 104).

[54] *IV*, 113. From the distance of a planet – even the very modest distances assumed by the sixteenth century – places such as Hven and Prague would be indistinguishable, and could therefore scarcely be supposed to undergo different influences from the planet. The passage setting out this difficulty was probably written before 1580, and certainly before the end of 1584, when Tycho finally abandoned the obliquity and latitude used in the computations involved. In 1588, he adverts to the same problem in correspondence with Rothmann (*VI*, 142).

[55] For a précis of the book, see Dreyer, pp. 118–9.

were a considerable element in Frederick II's original decision to found and fund Uraniborg. And, like the weather studies, the researches into horoscopes were conducted in a way that elevated them above the connotation of the term "occult science." So rigorous was the empirical element in Tycho's study of horoscopes, in fact, and so aggressively did he pursue it, that he eventually exhausted his ingenuity and was forced to face the possibility that what he wanted to do simply could not be done. His youthful naiveté went early, hastened no doubt by peer ridicule for posting a horoscope (1566) predicting the demise of a Turkish sultan who turned out to have been dead for some weeks before Tycho cast his horoscope,[56] but probably caused primarily by numerous less spectacular and unconfided misjudgments. By the time of his "Oration on Astrology" at the University of Copenhagen in 1574, he was making hedging references to the need for correct procedures and correct planetary positions,[57] and suggesting that human will could counter the influence of the stars.[58] After being installed on Hven, he was called upon to provide horoscopes for the sons born to Frederick in 1577, 1579, and 1583,[59] and it is clear that it was a task that became progressively less congenial. Someone had to do the rather considerable background computation required for the one-hundred-page horoscopes cast for each princeling, and insofar as skilled assistance was available in the early days of Uraniborg, Tycho doubtless availed himself of it – even if, for these documents, there was no dodging the final responsibility of signing Tycho's own name to them. Naturally , the spectacular comet of 1577 required an extended interpretation by Tycho, and something of the same nature may have followed the brief appearance of comets in 1580 and 1582 (the comet of 1585 was written up for Olsen's published calendar). Tycho was also compiling annual prognostications for Frederick, and, since Tycho's sole allusion to them makes it obvious that he generated them

---

[56] I, 135, and X, 13.

[57] There could be other reasons for caution, as well. In 1573, Tycho foresaw astrologically what he thought might be the death of Frederick II, and accordingly disguised his prediction of the situation (I, 58–64, 132–40).

[58] I, 38–9. For later, stronger statements, see I, 185, 196, 205. On the last page, Tycho adds the warning (to the king) that if the time of birth is erroneous by as little as four minutes, the whole judgment would be different.

[59] They have been published in I, 179–280, and summarized at some length by Dreyer (144–54).

only under duress,[60] one can imagine a creative role in this enterprise for some student, even if Tycho undoubtedly retained editorial control. After 1588, when the necessity for such productions ended with Frederick's death, Tycho's sister, Sophie, continued to provide a stimulus for astrological studies. A bona fide intellectual in her own right, noted for her learning by Tycho and other contemporaries, she maintained an active collection of horoscopes for the specific purpose of comparing predicted and observed fates of various friends and acquaintances.[61] Newly independent after the death of her husband in 1588, she appeared at Uraniborg four or five times annually for visits that ranged from a few days to a few weeks, and would certainly have encouraged and advised any students who shared her interests.

Astrology was not Sophie's only intellectual interest. Alchemy, too, absorbed many hours of her time, and provided a subject for continuing discussion with her brother. Tycho's claim late in life to "have been occupied by [chemistry] as much as by celestial studies since [his] 23rd year"[62] was no doubt somewhat enthusiastic, but

---

[60] By 1587, Tycho was expressing more than mere reservations. The occasion arose out of an inquiry from a German nobleman who was married to one of Tycho's cousins, and was consulting Tycho on behalf of his liege, the duke of Mecklenburg. The problem was that the duke had obtained two prognostications for 1588, which conflicted so radically that one had the year governed by two beneficent planets, the other by two malevolent ones. Tycho, of course, was able to account for the opposition by pointing out that one judgment was based on Ptolemaic and the other on Copernican tables, and was willing to state further that since neither set of tables was accurate, neither set of predictions could be relied upon. In fact, he went on to say, he did not willingly involve himself in astrological matters. He sent a prognostication to King Frederick every year, but only because the king expressly demanded one; for he really did not like to be associated with such doubtful predictions. So far, this could be a matter of pure pragmatism: a statement that astrology would not be worth doing until its astronomical basis was reformed. But Tycho went on to say that even if all astrologers used the same tables, very few pairs among a hundred prognostications would agree with each other because astrologers used many different bases and procedures to generate their judgments. That was why Tycho himself never placed any trust in them, and wanted to restrict himself to astronomy, where one could attain real truth (*VII*, 116–9). What is all the more remarkable about this frank document is that its ultimate addressee, the duke of Mecklenburg, was King Frederick's father-in-law! Tycho even went so far as to tell him that if he wished to see what he had predicted for 1588, he would have to ask the king because Tycho himself had not bothered to keep a copy of his prognostication. It can scarcely be doubted, therefore, that Tycho had already expressed at least some of these opinions to his royal patron.

[61] Dreyer, p. 201.

[62] *V*, 118. In 1588, he volunteered a similar assertion to Rothmann. (*VI*, 144–6). See also *VII*, 94–5.

it is clear that the *"ars spagyrica,"* as he termed it, absorbed more time and energy at Uraniborg than did any other study except astronomy. The original construction of Uraniborg included a very large basement laboratory with sixteen built-in alchemical furnaces; and, within three to four years of the completion of the building, Tycho had the winter dining room on the main floor fitted out with a small lab so that extended distillations could be observed frequently without the trouble of running back and forth to the basement.[63] Exactly what form these experiments took is unlikely ever to be discovered. In marked contrast to the reams of material accumulated in the pursuit of astronomy, the alchemical researches are not documented by a single note recording any experiment or observation. There can be little doubt, however, that the general object of the researches was to produce medicines, rather than gold.[64] Tycho's circle of close associates and correspondents included at least four physicians who treated him as a fellow professional,[65] and a fifth who criticized him for dispensing free medicines from Hven.[66] The medical pursuits of Flemløse and Gellius, whom we may assume were the principal participants in the experiments, confirm this orientation. And although they had to go elsewhere to obtain degrees, they would find no university with even the research ethic, let alone the research facilities, that they enjoyed at Uraniborg.

One of the more frequent visitors Tycho entertained over the years was his erstwhile preceptor, Anders Vedel. By one of those twists of fate that seem to throw great men together with supra-statistical frequency, Vedel was in the process of earning a reputation in Denmark's literary history that is essentially the equal of Tycho's in astronomy. In addition to having established himself as a theologian of some stature, he had completed an epoch-making Danish translation of Saxo Grammaticus's chronicles of medieval Denmark, and was embarked on an ambitious attempt to

---

[63] *V*, 142.

[64] *VII*, 75–6. In 1580, Tycho wrote that Uraniborg had been built with a special laboratory "pro chymicis exerciius, praesertim quantum ad Medicinem Spagiricam spectat" (*VII*, 59).

[65] Han Frandsen Ripensis and Johannes Pratensis were professors of medicine at Copenhagen; Brucaeus, professor at Rostock; and Hayek, personal physician to three Hapsburg emperors.

[66] Peder Sørensen, royal physician at Copenhagen, was the critic (Gassendi, pp. 307–8).

write a complete political history of the realm.[67] Some of his interests seem to have penetrated Tycho's circle. Tycho's publication of a Latin poem in 1575, exhorting the women of Denmark to contribute rags for the paper to print Vedel's translation of Saxo, need not represent anything more than support for a friend.[68] And the assignment of Flemløse and Olsen to mapping activities associated with Vedel's history might, likewise, have been merely the pursuit of a project requested of Tycho by the king (who, however, was by that time dead). But the fact that two of Tycho's students would occupy themselves sufficiently with historical writing in an era when history was not yet really recognized as an academic discipline, to win terms as official historiographer of Denmark,[69] suggests strongly that Vedel's interests at least came under discussion at Uraniborg.

There can be little doubt that Vedel's other major interest, theology, came under a great deal of discussion at Hven, too, even though it was probably not "researched" with the same self-consciousness that astronomy was. In the case of this subject, however, the interest was surely independent of Vedel. Its original status as the foundation of European culture and its residual importance as the core of Renaissance education must have rendered it almost as important as astronomy in the spectrum of total intellectual concern at Uraniborg. In fact, given the consistency with which certain doctrinal positions traceable to Tycho's old teacher, Neils Hemmingsen, turn up in the affairs of Tycho and various members of his circle, it seems obvious that the subject must have been of vital interest even to Tycho, himself. For his students, there was additional incentive for taking it seriously. It was the basis of practically every intellectual job in Renaissance society.

As should be clear by now, life was by no means all study and drudge work for Tycho's assistants at Hven. Those who demonstrated their competence were given opportunities for intellectually challenging work that were not available anywhere outside Uraniborg. Yet, there is a limit to the length of time that can be spent in the short-term pursuit of self-satisfaction. Sooner or later it becomes necessary to look to the future. For most of Tycho's

[67] *Dansk Biografisk Lekikon, XXV,* 183–92.     [68] Dreyer, p. 79.
[69] J. R. Christianson, "Tycho Brahe's Facts of Life"; *Fund og Forskning i Det kongelige Biblioteks samlingar, XVII* (1970), 21–8.

students, who were not only free agents, but obviously a highly selected, strongly motivated bunch, some reasonable anticipation of long-term return for their labors was an important consideration. Future employment as an astronomer was not a reasonable expectation – although one of Tycho's students, Longomontanus, would, indeed, attain that goal. But general employment in the "white collar" establishment of the day was a reasonable expectation, and one that was routinely attained by "graduates" of Uraniborg. The usual point of entry to the system after University was a post as parish priest somewhere in the provinces. With the support deriving from Tycho's special status and connections, however, one could not only expect a surer, and probably more favorable, entry into the system, but aspire to rise further in it, as well. Cort Axelson, for example, after two and one half years of service on Hven, was subsidized for seven years of study in foreign universities, by employment as traveling tutor to one of Tycho's brother's sons. After his return, he would become the first Norwegian to secure an appointment as professor (of theology) at the University of Copenhagen.[70] Just a couple of years later, another three-year veteran of Uraniborg, Johannes Isaacson Pontanus, would accompany another of Tycho's nephews on five years of similar travels. Born in Denmark of Dutch parents, Pontanus would find his professorship in Holland.[71] But he would also be one of Tycho's two students to serve a term in Vedel's old post as Royal Danish historiographer. The other one, Johannes Stephanius, got a professorship at Copenhagen.[72] The basis on which Tycho recruited (or, at least, retained) Flemløse was the royal promise of a canonry at Roskilde;[73] he later placed him in the post of physician to the governor of Norway. Through Tycho's recommendation, Gellius got a royal stipend to support his studies.[74] Even those who won no such direct and conspicuous rewards for service seem to have benefitted significantly from participating at Hven; for no less than four of Tycho's students who entered the church rose to the rank of bishop.[75] At the very least, Tycho chose his students very perceptively.

[70] *Dansk Biografisk Leksikon, I,* 545–7. In 1610, Axelson edited the introductory "Oration on Astrology" from Tycho's Copenhagen lectures of 1574.
[71] *Dansk Biografisk Leksikon, XVIII,* 448–50.    [72] See note 69.    [73] *XIV,* 11.
[74] *XIV,* 27.    [75] Dreyer, p. 237.

Concerning the shorter-term pecuniary aspects of study on Hven, there is no documentation whatever. Institutional tuition was not completely unprecedented in Europe, and private tutorial fees were routine at all universities. Yet, it seems very unlikely that either Tycho or King Frederick ever even considered casting Tycho in the economic (and, by implication, social) role of professor. We may presume, therefore, that only actual living expenses could have remained as an issue. A significant fraction of those had already been underwritten by Tycho (or Frederick) in the capital expenditure for the construction of Uraniborg. Free bed was surely assumed by anyone who visited Uraniborg, as an act of noblesse oblige, if nothing else. So, too, doubtless, was board – at least for a day or two. For longer terms, however, some kind of understanding would have been necessary. Again, in order to avoid putting Tycho in the position of haggling over bills like an innkeeper, the understanding was probably that Tycho's royal stipend should cover the boarding costs of his students – at least those from Denmark. The advanced ones, who could provide real, professional help, must even have received some kind of pittance for incidental expenses. For, whereas many of them doubtless had personal resources for such requirements, Longomontanus, for one, almost surely did not.

The task of supervising the instruction and inquiry of his students and assistants imposed another set of concerns, which doubtless assumed annoying proportions at times, but it was surely one from which Tycho realized a considerable net advantage. It is extremely unlikely that Tycho provided any routine instruction himself. Remedial work in astronomy, and even the lessons in spherical trigonometry, must surely have been handled by assistants or senior students. Aside from the responsibility of organizing such activities, therefore, Tycho's teaching activity must have occurred almost exclusively at higher levels, in much the same way that postgraduate scientific research teams operate in the modern university.

The dual role as dispenser of both knowledge and patronage at Uraniborg certainly offered Tycho a decided advantage in his relationships to his students; and the evidence suggests that he was not one to yield it. Kepler would actually leave Tycho's service at one point, because of a feeling that Tycho was treating him as a

menial subordinate rather than as an associate.[76] A German student named Frobenius, who came to Hven in about 1591 with a
Master of Arts and good letters of recommendation, likewise found
Tycho's mien too haughty;[77] and it is very possible that Wittich's
departure was prompted by the same consideration. The pattern
displayed in these interactions is quite consistent with the picture
that emerges not only from various other incidents in Tycho's life,
but from the domineering behavior displayed by the Danish nobility, generally. On the other hand, we have even more evidence
– and apparently Tycho's students did, too – that his autocracy
was essentially benevolent. Those who were at Uraniborg to do
serious work, and did not take Tycho's dominance personally,
clearly got on with him very well. In addition to the several long-
termers already mentioned, one might cite Christian Johannis Ripensis, who, although he eventually made his career in theology
(rising to professor and bishop), was on Hven for four years, and
continued to send observations (comet, solar eclipse) back to Tycho long after he left.[78] Blaeu and Pontanus published fond recol-
lections of Uraniborg and Tycho a generation after the demise of
both institution and founder.[79] In fact, given the extent to which
the institution was a reflection of its founder, it may not be inappropriate to conclude with a vivid counterexample to Dreyer's
depiction of Tycho as a curmudgeon and tyrant. It is found in a
letter written to Hayek at the court of Rudolf II in 1591. At issue
was one Peter Jachinow, who had appeared in Prague soliciting
work as a mechanician (specializing in the building of odometers)
and using Tycho's name as a reference. Upon being queried by
Hayek, Tycho affirmed at considerable length that Jachinow had,
indeed, rigged odometers to the coaches of many nobles (including German princes and the King, himself) and that "when he was
last here he adapted an automaton of this type to my own carriage
in which I regularly travel about this island with my friends. This
machine indicates the whole miles and also their different divisions
and their subdivisions into sixty parts by central pointers, and makes
it clear by striking distinct sounds with two bells." "If you will
recommend this man," Tycho continued, "in such a way that . . .
he may be able to hire out his labor . . . and, by this means acquire

---

[76] See the discussion by Dreyer, pp. 294–6.    [77] Norlind, p. 100.
[78] *Dansk Biografisk Leksikon, xix*, 478–80.    [79] Norlind, pp. 99–100.

a small amount of money, you will thereby be doing me a favor; for I love this man for his evident honesty and trustworthiness, and I am sorry for him because in his advanced age he has to travel up and down to support himself and his family." "Don't let his efforts down – I do not doubt you will do it for my sake – so that he may not complain whenever he returns here that my little recommendation, such as it is, turned out to be fruitless for him."[80]

[80] *VII*, 320.

# 12

## Agricola and community: cognition and response to the concept of coal

JAMES A. RUFFNER

I have omitted all those things which I have not myself seen, or
have not heard or read of from persons on whom I could rely.
Georgius Agricola[1]

These words by Georgius Agricola (1494–1555) are from his work
*On Metals*, which is regarded as a classic of the emerging techno-
logical tradition that depended significantly upon local artisans as
authorities. They apply equally to his earlier works concerning
more generally *Underground Things*, which culminated some fifteen years of work to reform and revitalize the scholarly tradition
to accord with the Greek naturalists who derived authority from
direct observations.[2] The *Subterranea* provided the basis for the
sobriquet "father of mineralogy" and have been deemed his most
important scientific contribution, on a par with the work of Cop-

---

[1] Georgius Agricola, *De Re Metallica*, trans. H. C. Hoover and L. H. Hoover (London,
1912; reprint, New York, 1950), pp. xxx–xxxi.
[2] Georgius Agricola, *De Ortu et Causis Subterraneorum, libri V. De Natura eorum quae Effluunt ex Terra, libri IIII. De Natura Fossilium, libri X. De Ueteribus et Nouis Metallis, libri
II. Bermannus, sive de re metallica dialogus. Interpretatio Germanica Vocum rei Metallicae, addito
Indice Foecundissimo* (Basel, 1546). For English translations, see *De Natura Fossilium (Textbook
of Mineralogy)*, trans. M. C. Bandy and J. A. Bandy, Geological Society of America
Special Paper 63 (New York, 1955). *Bermannus* (important omissions) in Wolfgang Paul,
*Mining Lore* (Portland, Or., 1970), pp. 252–311, 813–43. Brief selections as follows: *De
Ortu* (1558) in Hoover [see note 1], pp. 46–52, 595–6; *De Natura . . . Effluunt* in Bandy,
op cit., pp. 61–2; *De Ortu* and *De Natura . . . Effluunt* (1558) in R. J. Forbes, *More Studies
in Early Petroleum History* (Leiden, 1959), pp. 37–40. For German translations see *Ausge-
wahlte Werke*, ed. Hans Prescher, 12 vols. incl suppls. (Berlin, DDR, 1955– ) [hereafter
cited as *Werke*].

ernicus and Vesalius.[3] Recent scholarship has reached such divergent conclusions, however, that detailed reevaluations are in order, keeping his own stricture in mind.[4]

This study focuses on the development and early diffusion of the concept of coal as a bitumen. This focus permits a detailed analysis of an important but relatively neglected aspect of Agricola's geological writings to determine what was problematic under what circumstances, which persons he trusted, and how he revised what he read or heard in the light of what he probably had seen for himself. Just as significantly, since the essence of science is communication and consensus building, this chapter is concerned with following in detail the immediate fate of the concept among peers who had overlapping interests.

### ZWICKAU: THE PLACE AND ITS COAL

Zwickau, in Agricola's day, was a five hundred-year-old town on the Mulde River in the part of Saxony known as Misnia. For well

[3] Hoover [see note 1], pp. iii, xii. F. D. Adams, *The Birth and Development of the Geological Sciences* (Baltimore, 1938), p. 195. H. Wilsdorf, "Georgius Agricola," *Dictionary of Scientific Biography*, 16 vols. (New York, 1970–80), I, 77–9. J. Eyles, "Georgius Agricola (1494–1555)," *Nature, 176* (1955), 949–50. See also R. Hookyaas, rev. of Hans Hartman, *Georg Agricola, 1494–1555* (Stuttgart, 1953), *Archives internationales d'histoire des sciences, 7* (1954), 198–202; J. Eyles, "Georgius Agricola," essay rev. of *Georgius Agricola 1494–1555 zu seinem 400 Todestag, 21 November 1955* (Berlin, 1955) and Georgius Agricola, *De Natura Fossilium* (New York, 1955), *Nature, 177* (1956), 1144–5; A. Sisco, rev. of Helmut Wilsdorf, *Georg Agricola und seine Zeit* (Berlin, 1956), *Isis, 49* (1958), 368–9; G. White, rev. of Georgius Agricola, *De Natural Fossilium* (New York, 1955), *Journal of Geology, 65* (1957), 113–4; and R. Forbes, rev. of Georgius Agricola, *Ausgewählte Werke. De Natura Fossilium Libri X, die Mineralien* (Berlin, 1958) and *Isis, 51* (1960), 239.

[4] Compare A. G. Debus, *The Chemical Philosophy*, 2 vols. (New York, 1977), II, 540. C. Webster, *From Paracelsus to Newton, Magic and the Making of Modern Science* (Cambridge, 1982), pp. 55–6, 79, 83. R. Halleux, "La littérature géologique française de 1500 à 1650 dans son contexte européen," *Revue d'histoire des sciences, 35* (1982), 111–30. Idem, "La nature et la formation des métaux selon Agricola et ses contemporains," ibid., *27* (1974), 211–22. H. Prescher, "Die Bedeutung Agricola (1494–1555) für die Lagerstattenkunde," *Geologie, 20* (1971), 740–7. W. R. Albury and D. R. Oldroyd, "From Renaissance Mineral Studies to Historical Geology, in the Light of Michel Foucault's *The Order of Things*," *The British Journal for the History of Science, 10* (1977), 187–215 (pp. 189–90). D. R. Oldroyd, "From Paracelsus to Haüy: The Development of Mineralogy in Its Relation to Chemistry," (Ph.D. dissertation, University of New South Wales, 1974), pp. 15–20, 58–9, 102–8 (microfilm available from Center for Research Libraries in Chicago). Agricola never fitted Oldroyd's analysis very well and, for practical purposes, he was dropped from the published papers. See D. R. Oldroyd, "Mechanical Mineralogy," *Ambix, 21* (1974), 157–78 (p. 159); idem, "Some Neo-platonic and and Stoic Influences on Mineralogy in the Sixteenth and Seventeenth Centuries," *Ambix, 21* (1974), 128–56.

over two hundred years, blacksmiths within easy transport of the river had been served with a fuel mined near the town and known in the vernacular as *Steinkoln*.[5] It was the only known source of that fuel in the entire *Erzgebirge,* or "Ore Mountain" region of Saxony and Bohemia until the late 1530s or 40s.

The coal of Zwickau occurs in two basic types.[6] Both types are classified today as highly volatile bituminous coals and are extremely light, with nearly identical specific gravity in the range 1.08 to 1.25 and average values near 1.20. Otherwise, they are as different as can be.[7]

A bed of "soft" coal, upwards of 10 meters thick, rises to within 2 meters or less of the surface along an arc 4 kilometers south of town. Few banded coals have more than 5 percent fusain or "mineral charcoal." The "soft" coal, known latterly as *Russkohle* or "soot" coal, contains in the aggregate of its layers some 25 percent fusain, 60 to 65 percent dull-to-midluster attrital matter, and 10 to 15 percent vitrain and bright attrital matter. So long as they are only a few millimeters thick, these "bright" layers merely add variation in luster and zones of comparative strength to an otherwise dull, somewhat silky block that is quite pulverent and extremely dirty to the touch. A bed of "hard" coal lies 40 to 50 meters deeper, although it also rises to within a few meters of the surface a hundred meters south of the outer limits of the *Russkohle.* The "hard" coal, known locally as *Pechkohle,* is either nearly pure vitrain or a banded coal with multiple layers of vitrain and bright attrital matter. It is generally free of fusain and is clean to the touch. A still deeper bed contains benches of *Pechkohle* interleaved with thin layers of *Russkohle* and considerable "dirt" bands of extraneous mineral matter.

The earliest mines exploited the *Russkohle* near the left bank of

---

[5] In 1348, blacksmiths were prohibited from buring *Steinkoln* within the "lower half" of the walls. Emil Herzog, *Geschichte des Zwickauer Steinkohlenbaues* (Dresden, 1852), p. 3.

[6] For the modern basis of observation and terminology, see J. M. Schopf, "Field Description and Sampling of Coal Beds," *U.S. Geological Survey Bulletin, 1111B* (1960), 25–69, plates 6–27. *Field Description of Coal,* ASTM STP 661, ed. R. R. Dutcher (Philadelphia, 1978). F. T. C. Ling, "Coal Macerals," in *Coal Structure,* ed. R. A. Meyers (New York, 1982), pp. 8–49.

[7] Congrès Geologique International – Commission de Stratigraphie, *Lexique stratigraphique international* (Paris, 1955– ), *I*, fasc. 5ci. (Allemagne-Carbonifère), 235–58. W. Gumz and R. Regul, *Die Kohl* (Essen, 1954), pp. 154–5. O. Stutzer and A. C. Noe, *The Geology of Coal* (Chicago, 1940), pp. 46, 108–9, 206–9, 241. H. Geinitz, H. Fleck, and E. Hartig, *Die Steinkohlen Deutschland's und anderer lander Europa's,* 2 vols. and Atlas (Munchen,

the river.[8] Outcrops across the river in a woods at the base of a hill known as the *Kohlberg* were not exploited because they occurred in a hunting area. These outcrops caught fire in 1479, according to a contemporary tale, from a gun flash during a fox hunt. The mines caught fire in 1505, spreading fear that a new Etna or Vesuvius was about to form. They burned throughout Agricola's lifetime and, indeed, for several hundred years beyond.

The first mines beyond the right bank of the river, and the burning outcrops, were opened in 1530.[9] By 1541, the new mines yielded over two-thirds of Zwickau's coal, and some of them were deepened significantly in the period 1541–6.[10] Increased demand for coal is also indicated by exploitation near Dresden beginning in 1540 and other places in Misnia and Bohemia later in the decade.[11] It is likely that the deep *Pechkohle* and the next deeper composite bed were first exploited during this period of expansion. Thus, Agricola's first thoughts on coal, in 1528, were fashioned in the limited light of *Russkhole,* one of the most extraordinary types found anywhere.

#### BERMANNUS AND THE ISSUE OF COAL'S NAME

One of Agricola's earliest influences came from Erasmus Stella (1455–1521), the leading scholar of Zwickau when Agricola served there as assistant schoolmaster.[12] Stella decried the corruption of mineral terminology since antiquity and composed a little tract on gems organized by color to avoid the problem of alphabetical listing when the names are uncertain. Agricola went a step further in believing that one could begin to understand the problems of nature and of natural power only by relying on the ancient masters who observed things for themselves or on one's own observa-

---

1865), *I,* 6; *II,* 194, 201; Atlas, taf. I–III. Otto Stutzer, "Russkohle von Zwickau," *Zeitschrift der Deutschen Geologischen Gesellschaft, 84* (1932), 222–9.

[8] Herzog [see note 5], pp. 10, 36–8. Petrus Albinus, *Meissnische Land und Berg Chronica* (Dresden, 1589–90), p. 187.

[9] Herzog [see note 5], pp. 11–13. Geinitz, Fleck, and Hartig [see note 7] *II,* 53. Coal of an unspecified type was discovered in the new area between 1524 and 1527; see *Werke,* Suppl. *I,* 510.

[10] W. Dohlen, *Die Ökonomische Lage der Zwickauer Bergarbeiter im Vorigen Jahrhundert* (Leipzig, 1962), p. 9. See also *Werke,* Suppl. *I,* 38, 501.

[11] Albinus [see note 8], pp. 189, 198. Geinitz, Fleck, and Hartig [see note 7], p. 11.

[12] L. Thorndike, *A History of Magic and Experimental Science,* 8 vols. (New York, 1923–58), *VI,* 302–3. For Stella's likely role model, see *Werke, I,* 117–18.

tions. Initially, he was convinced that proper names were critical in this regard, and he challenged others to investigate these questions for themselves.[13]

Another early influence came from Petrus Plateanus (d. 1551), the schoolmaster at St. Joachimsthal when Agricola settled there to serve as town physician and apothecary.[14] Agricola had just returned from Italy, where he worked on the editorial staff of the Aldina editions of the works of Galen and Hippocrates. En route, he had visited other important mining centers including, once more, Zwickau. His intention to prepare Latin annotations for the works of Dioscorides and Galen quickly expanded into a program to elucidate everything about the German mines in the light of all the authentic ancient writings. Plateanus encouraged him to write a preliminary statement and lined up the backing of Erasmus and the celebrated Froben Press of Basel that did much to spread new scholarship. He also prepared a German–Latin mineral lexicon to accompany the resulting work, *Bermannus, sive de re metallica*.

Agricola's treatment of coal illustrates nicely some of his concerns, trusted sources, and initial thought processes. An unacknowledged actor was Albertus Magnus, whose treatment of coal at Liège demonstrably set the problem for Agricola.[15] Albertus had made more first-hand observations than most scholars in the post-ancient world, but Agricola did not always trust his results. In this case, Plateanus, who was born in Liège and educated at a nearby university, probably confirmed the report.

Albertus opposed fiery-red but fire-resistant *carbuncles* (rubies) to various inflammable substances including one genus of *carbones* (charcoal), manufactured from wood, and another genus of *carbones* (mineral coal), formed naturally underground from a moist, fatty earth that is altered by heat. If the original moisture were overwhelming, *calcem* (chalk?) would form. If the original mass of

[13] Georgius Agricola, *Bermannus, sive de re metallica dialogus* (Basel, 1530), p. 1. See also A. G. Debus, *Man and Nature in the Renaissance* (Cambridge, 1978), pp. 7–8 and M. J. Rudwick, *The Meaning of Fossils*, 2d ed. (New York, 1976), pp. 3–4. Compare Rudwick's analysis of a similar concern of Conrad Gesner (1516–1565) to carry out direct investigations and to encourage his readers to do the same, ibid., pp. 9–14.

[14] On Plateanus, see W. Fischer, *Zum 450 Gerburtstag Agricola's* (Stuttgart, 1944), p. 36. *Werke, II*, 312. *Allgemeine Deutsche Biographie*, 56 vols. (1875–1912), *XXVII*, 241–3. See also his forward to *Bermannus*.

[15] For an analysis of the use of Albertus's work elsewhere in *Bermannus*, see *Werke, II*, 235–41. This link was missed.

fatty earth were highly divided and the moisture were overcome so as to yield ample pore spaces for the passage of fire, *carbones saxeos* (rock coal) would form. It retains black-earth-like qualities, but is very heavy (*gravis valdè*). It is found at various unnamed places in Germany and near the city of Liège, where it is eagerly sought by blacksmiths because it makes a very hot fire. He went on to discuss volcanoes.[16]

In a parallel fashion, Agricola used charcoal namesakes to digress from the mineral *minium* to ,coal and Zwickau's famous burning *Kohlberg*.[17] A further digression on how Zwickau received its name provides another clue to Agricola's state of mind.

Zwickau originally was called *Cycnaeam* or "place of the swans." The modern name, according to Stella, stemmed from a Saxon word *"verzwickt,"* which Emperor Henry III allegedly used to express the cut-up street plan after the town was rebuilt in the eleventh century to avoid periodic flooding.[18] The unstated point is that if its name reflects something of its nature as a city, the name of Zwickau's coal ought to reflect something of its nature as a mineral. Consider now the charcoal namesakes Agricola found in the literature.

Minium ore (red lead) had been termed *anthrax*, according to Vitruvius, because clods brought from a mine resemble glowing pieces of charcoal.[19] Rubies and other fiery-red gems, to use Al-

[16] Albertus Magnus, *De Meteoris libri IV*, 4.3.19, in *Opera*, ed. P. Jammy (Lugduni, 1651), II, 199. Magnus, *De causis et proprietatibus elementorum*, 2.2.2–3, in *Opera Omnia*, Borgnet edition (Paris, 1890–9), IX, 644–7. Compare Aristotle, *Meteorologica*, 4.9, 387$^a$18–388$^a$10, trans. H. D. P. Lee (Cambridge, 1952), pp. 352–7. Chronicles and ordinances at Liège dating from 1213 refer to the discovery around 1195 of coal known as *terra nigra* or *terra nigra carbonis simillima*, "the black earth that resembles charcoal." See E. Polain, "La vie à Liège sous Ernest de Bravière," *Bulletin de l'Institut Archéologique Liègois*, 61 (1937), 1–162 (pp. 86–100). *Charbon de roche* is listed in imports to Bruges in the year 1200 A.D. See J. U. Neff, *The Rise of the British Coal Industry*, 2 vols. (London, 1932), I, 6. Albertus would have learned the names, properties, and uses of coal during journeys by foot to religious houses between Cologne and Liège, some of which were in the forefront of coal mining.

[17] *Bermannus* (Basel, 1530), pp. 106–9. I have transliterated Greek terms whenever used.

[18] *Verzwict* is inserted in the German text to indicate Henry's pun that Zwickau had been rebuilt in a most intricate pattern. According to Herzog [see note 5], p. 2, the name *Zwickau* was derived from *Zwicz*, the Slavic God of Fire, because certain black stones caught fire when inadvertently used by early shepherds to build fireplaces. There may be truth to both accounts, and Henry's irony more clever than imagined previously.

[19] The word "anthrax" is not in the standard modern text at the likely place of this reference. Vitruvius, 7.8.1 in *Vitruvius on Architecture*, ed. and trans. F. Granger (Cambridge, Mass., and London, 1962), II, 114. Vitruvius attributed a carroty or red color to minium

bertus's example, had been termed *carbuncles,* according to Pliny, because they resemble little pieces of burning charcoal.[20] Agricola then made a vague allusion to the use of the term *anthrax* by unnamed writers on rural subjects for an unspecified type of earth.[21] Finally, he recalled Theophrastus's use of the term *anthrax* for a certain mineral that also has an earthy nature and is used by coppersmiths in the place of charcoal.[22]

The third allusion provides a further clue to Agricola's problem. The rural subject is viticulture and the material is *terra ampelitus* or "vine earth." After being dissolved in olive oil to produce the desired consistency, it was smeared on grape vines to suppress worm or other pest infestations. Its relation to charcoal is not found in an ancient name, it is found in Dioscorides's description. He described a worthless white kind as "ashy and unmeltable," and a valuable black kind that readily "melts" or dissolves in olive oil when crushed as having the luster and general appearance of charcoal made from the "pitch" of pine tree.[23] Later, this account would be significant in building a theory of coal in relation to the bitumens, using *Pechkohle* and *ampelitus* as important connections. Here, Agricola lacked that connection and was working on quite a different problem. The German name simply did not accord very well with the properties of the soft charcoal-like mineral he knew.

Agricola was certain that coal was exactly the same earthy fuel reported by Theophrastus that ignites and burns just like charcoal. Recalling Zwickau's *Russkohle* (and Albertus's theory of origin), Agricola stated that this mineral fuel is excocted or baked by the terrestrial heat until it becomes black and light just like charcoal. Indeed, it is so porous (*rarae*) that it floats on water. Imagine Agricola's consternation that the vulgar German name for it joined the

ore and then moved on to a discussion of black things. Agricola made a comparable abrupt shift to a discussion of coal.

[20] Pliny, 37.25.92–8; 37.29.10 in *Natural History*, ed. and trans. H. Rackham and others, 10 vols. (Cambridge, Mass., and London, 1938–62), X, 239–43, 246–7.

[21] The identity, according to Prescher, remains questionable. *Werke, II,* 196. But see my analysis below.

[22] Theophrastus, 2.16, in *De Lapidibus,* ed. and trans. D. E. Eichholz (Oxford, 1965), pp. 62–3. Also *Theophrastus on Stones* ed. and trans. E. R. Caley and J. F. C. Richards (Columbus, Oh., 1956), pp. 48, 85–6. The fuel mostly likely was lignite. Agricola and his peers had no doubt it was coal in its local sense.

[23] Dioscorides, 5.181, in *The Greek Herbal of Dioscorides,* trans. John Goodyer (1655), ed. R. T. Gunther (Oxford, 1934), p. 659. The heading that indicates ampelitus is a "bituminous earth" is Goodyer's addition and is one indication of Agricola's influence.

terms for "charcoal" and "stone" as if the original correct Greek term had been *lithanthrax*. Theophrastus called it *anthrax,* to be sure, but characterized it as *geodeis* (earthy), not as *lithodeis* (stony). If, Agricola indicated, the designation "stony" (*lithodeis*) were to be used, it might better be applied to the coal at Liège, which he described as *satis gravis* or "tolerably heavy."[24]

### BERMANNUS AND THE PROBLEM OF COAL THAT FLOATS

It is easy to understand the distinction between light and heavy coal, based on Agricola's limited experience and sources. The claim that the light variety could actually float, however, is difficult to accept. All types of coal, and most types of lignite, have a minimum specific gravity in the range of 1.05 to 1.50. Zwickau's coal falls near the bottom of that range, of course, but no piece of raw coal of significant size can float. As will be seen, Christoph Entzelt (1517–86) altered the claim in an otherwise verbatim copy of the passage. Agricola's modern editors comment only that some kinds of coal are surprisingly light when lifted, as if Agricola had been deceived by that experience alone.[25] A more likely explanation is found in the sources for the discussion.

Theophrastus mentioned a porous stone with the appearance of rotten wood that can be made to burn repeatedly without being consumed if it is soaked in olive oil.[26] Similarly, Albertus mentioned stones embued with naphtha that are left very light and porous, and able to float in water, just like pumice, after they are burned. He also suggested that a volcano would burn for many years or even perpetually if the fuel were regenerated periodically.[27] Albertus did not indicate a relationship between these light, naphtha-imbued stones and the rather heavy, "rock" coals he had just discussed, but Agricola seems to have been less cautious. He had seen a proto-volcano where the fuel was known to be coal of

---

[24] *Bermannus* (Basel, 1530), p. 109. Paul [note 2, pp. 299–300] provides a partial translation, leaving out the crucial passages about light and heavy coal and the coal that can float.

[25] Editor's note in *Werke* [see note 2], *II, 196.*

[26] Theophrastus, 2.17 [see note 22, 1965], p. 63. The section follows immediately the one treating "earthy coal."

[27] Albertus, *De Causis Elementorum,* 2.2.3 [see note 16], *IX,* 647.

a type that resembles charcoal, if not exactly rotten wood or pumice, and he probably conflated the comments of Theophrastus and Albertus with what he had seen. His compulsion to question the aptness of the term *Steinkohle* and the force of his rhetoric carried him beyond actual observations. Until a better explanation is found, the example needs to be set against Agricola's usual reputation as a skilled observer. His understanding and rhetoric, however, changed significantly in the light of new information of *Pechkohle* and influence of unknown proportion from Valerius Cordus (1515–44).[28]

### TOWARD A NEW CONCEPT

By ancient consensus, the category of material known as "bitumen" was comprised of *bitumen* properly called, also known as *bitumen judaicum* or asphalt; pissasphalt; naphtha, also known as petroleum; and either "maltha" or "mumia." Certain stones such as gagates sometimes were said to give off the odor of bitumen when burned but were not explicitly categorized as such. These classical views were reiterated in Agricola's day by no less an authority than Antonius Brasavola (1500–44) and by Walther Ryff (fl. 1543).[29] Another benchmark can be found in Brasavola's understanding that amber, camphor, manna, and sugar are examples of "concrete" juices of vegetable origin.[30]

Cordus, possibly to an even greater extent than Agricola, gained extensive field experience in the Erzgebirge and Harz mountains and nearby districts. He corresponded with Agricola and sent him specimens. Lectures to explicate Dioscorides in the light of his current work were delivered at Wittenberg between 1539 and 1543. Notes indicate that he had extended the category of bitumen to

[28] Agricola's friendship with Cordus is touched breifly in *Werke, I*, 231; *II*, 278. See also "Valerius Cordus," in *Dictionary of Scientific Biography* (note 3) and T. A. Sprague and M. S. Sprague, "The Herbal of Valerius Cordus," *Journal of the Linnean Society of London, Botany, 52* (1939), 1–113 (pp. 3–8).

[29] Antonius Brasavola, *Examen Omnium Simplicium Medicamentorum, Quorum in Officinis Usus Est* (Venice, 1539) pp. 452–7. Thorndike [see note 12], *V*, 445–71. For Ryff, see note 31.

[30] Brasavola, ibid, pp. 363, 397. Agricola's concept of "concrete" mineral juices is examined with translations of key passages by Hoover [note 1], pp. 1–3, 46–52. Related points of interest are touched in K. H. Dannenfeldt, "Ambergris: The Search for its Origin," *Isis, 73* (1982), 382–97.

include amber, camphor, ampelitus earth, Gagate stone, Thracian stone, and mineral coal.[31]

Cordus asserted that a similar origin for such traditionally distinct and apparently different materials, ranging from white (and translucent) to jet black, could be established only after long effort to gather the evidence and fashion the arguments. His notes provide some details for amber. Otherwise, they indicate, without supporting evidence, that the Gagate stone of antiquity is the same thing as German *Steynkolen,* that ampelitus earth is a crustaceous form of gagates with a fissile nature like mineral coal, and the gagates and Thracian stone are quite similar in substance and powers. Cordus also alluded to Agricola's very diligent searching and examination in accordance with his own efforts and announced that Agricola would soon publish extensively on the origin and material of minerals. Cordus's tragic death surely cut short his own contributions.

Agricola added the same members as Cordus to the category of bitumen, plus two other stones from ancient sources. His key assertion indicated that the names Gagate stone, Thracian stone, Samothracian gem, Obsidian stone, ampelitus earth, and stone coal had been applied inadvertently to the *same* earthy bitumen by different authors in different locales.[32] Cordus, of course, had indi-

---

[31] Ryff edited student lecture notes included with his own annotations (1543) in Dioscorides, *De medicinali materia libri sex, Joanne Ruellio interprete . . . Additis etiam annotationibus . . . per Gualtherum Rivium . . . Accesserunt priori editioni, Valerii Cordi annotationes . . .* Franc[ofurti, 1549]. For Ryff, see pp. 44–5, 50, 427; for Cordus, see pp. 459, 472–3, 533. Cordus revised sections on bitumen and amber, but not the coal "family." These revisions together with related field notes on "fossils" were published in Valerius Cordus, *In hoc volumine continantur Valerii Cordi . . . Annotationes in . . . Dioscordi . . . de Medica materia libros V . . . Sylva, qua rerum fossilium in Germania plurimarum, Metallorum, Lapidum et Stirpium . . . Gesneri collecta* (Argentorati, 1561), pp. 16$^r$, 32$^v$, 83$^r$, 84$^v$.

[32] Agricola, *De Natura Fossilium* (1546), letter of dedication and pp. 229, 233–7; (1955), pp. 2, 61, 66–70. Modern editions of the main sources used by Agricola to support his arguments are as follows: Nicander, "Theriaca," lines 35–50 in *The Poems and Poetical Fragments,* ed. and trans. A. Gow and A. Schofield (Cambridge, 1953), pp. 30–1. Theophrastus [see note 22], 2.9–17. Posidonius, quoted in *The Geography of Strabo,* trans. H. Jones, 8 vols. (London and New York, 1917–32), III, 266–7. Anon., *De Mirabilibus Auscultationibus,* 832$^b$26–30, 833$^a$25–28, 841$^a$28–35 in *The Works of Aristotle,* ed. W. Ross, 12 vols. (Oxford, 1908–52), VI. Dioscorides, 5.146–7, 181 [see note 23], pp. 653, 659. Pliny, 33.30.94; 35.51.178–182; 35.56.194; 36.34; 37.67.181 [see note 20], IX, 72–3; IX, 392–5, IX, 404–5; X, 112–15; X, 310–1. Galen, *De simplicium medicamentorum temperamentis ac facultatibus,* 9.10 in *Opera Omnia,* ed. C. Kuhn, 20 vols. (Leipzig, 1826; reprint ed., Hildesheim, 1965), XII, 181, 203–4.

cated only that Thracian stone is similar and had not included Obsidian or Samothracian stone. Who had what idea first and how much each profited through private exchange is a crucial issue that will require considerable study before being mooted or resolved. Here it can be noted that Agricola was the first to have his conclusions in print and supported by substantial arguments. Cordus had a important role in shaping consensus or raising the opposition, but Agricola's much fuller account obscured it at most points in later treatments.

Combining the accounts of the Gagate stone by writers such as Nicander, Dioscorides, and Pliny, and what Galen took to be the same thing, Agricola indicated two distinct types, with properties similar to but assorted differently from the two types of coal at Zwickau. One type is usually black, squalid, crusty, and exceedingly light. The other type is black, smooth, porous, light, brittle, has an appearance little different from wood, and leaves an indelible mark on pottery. Pliny also described a stone from Samothracia as black, very light, and similar in appearance to wood, from which account Agricola concluded it is really the smooth form of gagates. Less convincingly, he concluded that Pliny's account of a volcanic glass known as Obsidian stone reported the same thing, simply because it has the same color, luster, and ornamental uses.

The odor of Thracian stones in burning is similar to gagates because each was said to resemble that of buring asphalt. They must also be light because they are carried by a river current. The paradox, according to most accounts, is that a glowing Thracian stone becomes totally inflamed after it is taken from a blazing fire and sprinkled with water, and is extinguished to the core when oil is thrown on it. The paradox of Gagate stone is that it can be burned without being consumed by even the most violent fire. Pliny, however, muddled the distinction. He indicated that soothsayers used gagates' property of burning without consumption to convince clients that their wishes would come true. Yet, he also attributed the oil/water paradox both to gagates and to Thracian stone, as tradition had it.

Agricola gave equal credence to these claims, and took the convergent points as evidence that Gagate stone, Thracian stone, Samothracian gem, and Obsidian stone are identical bitumens, not essentially different from mineral coal. A factor, I am convinced, was Galen's willingness to consider a stone he found near the Dead Sea or "Lake of Asphalt" to be the same thing as gagates, which he could never find.

Agricola knew (from Cordus?) of a heavy combustible mineral encrusted with pyrites found near Eisleben in the Harz Mountains. He classified it as a complex *mixta* that combines pyrites and a metal to make it heavy and some bitumen to render it combustible. He regarded the stone found in Thrace, known as *spinos,* to be similar, if not the same thing. According to Theophrastus, it bursts into flame when broken pieces are wetted. Agricola decided that it is not likely the same thing as Thracian stone, which has similar behavior because the latter is light, whereas Theophrastus said that *spinos* is heavy.[33] He applied similar considerations to Theophrastus's reports of stones from Binae and Lipari. Agricola's premise was that the bitumens and sulfurs, as simple substances, are light and readily combustible in comparison to other minerals. The weights of complex mineral *mixta* vary inversely with the amount of bitumen and sulfur present, whereas their ease of combustion varies directly thereby.[34]

Combining the accounts of ampelitus earth by writers such as Posidonius, Pliny, Dioscorides, and Galen, Agricola noted that it either has an asphaltic nature or bears a strong resemblance to asphalt, that the valuable kind has the luster and appearance of charcoal made from the pitch tree, and that unlike most other "earths," it comes near to the essence of stone. Agricola concluded from these accounts that black ampelitus is a hard, earthy bitumen of the finest quality and is altogether the same kind of thing as mineral coal and the other "stones." He went on to say that when it is mixed with too much earth, ampelitus acquires a different color and belongs to a different genus – a reference to the worthless, white kind mentioned by Dioscorides.

Agricola also noted that Theophrastus mentioned "coal" but had said nothing of ampelitus, whereas the writers who treated "ampelitus" said nothing of coal. This symmetry was further evidence that "ampelitus" and "coal" were the same thing.[35]

---

[33] Agricola, *De Natura Fossilium* (1546), pp. 233, 370; (1955), pp.67, 215–16. I have not found the passage attributed to Theophrastus in which *spinos* is said to be heavy.

[34] Ibid., (1546), p. 361; (1955), pp. 202–3.

[35] Ibid. (1546), p. 234; (1955), pp. 67–8. Theophrastus [see note 22], sec. 49, did mention a Cicilian earth that was smeared on grapevines to suppress worms, but it was prepared by boiling rather than by mixture with olive oil as was ampelitus.

Finally, Agricola rejected the common notion that sulfur is the primary cause of underground fires because it is quickly extinguished by water.[36] The ancient accounts convinced him that only asphalt and related bitumens can burn readily in the presence of underground water, and indeed have the burning enhanced by it. Coal must be related because blacksmiths frequently watered it, and the mine fires near Zwickau burned relentlessly, despite the presence of underground water.

In sum, Agricola found that coal could be as different as mineral charcoal and mineral pitch. Correspondingly, many light minerals mentioned in antiquity were sometimes said to resemble wood and sometimes to resemble asphalt or tree pitch. The fragmentary ancient descriptions of them dealt with other properties found in coal or the classical bitumens. He concluded that all accounts pointed to the same rather variable material, even though they had been given different names.

### NAME AND NATURE REVISITED

One of Agricola's major concerns when he first wrote about coal in *Bermannus* was that the name of a thing should be related to its essential nature. Then, he had been uncomfortable with the German name for coal. Now, he took a pragmatic approach to names and indicated cultural pride that a name such as *Steinkoln* could be made up as easily in German as in Greek or Latin. He stressed, however, that authors should explain the basis for a name in whatever language they write. Thus, in the particular case of the name for coal in the German vernacular, he suggested that the word for stone connotes hardness only in comparison to charcoal.[37] For Agricola, the true bituminous "stones" were the heavy *mixta*, discussed above, that combine bitumen with metal and stone. He allowed that the true bituminous "coals" could be termed "stone coals" without contradiction so long as the correct denotations are kept in mind. He preferred, however, to use Latin nomenclature that combined the term for charcoal with the term for earth, bitumen, or mineral, as in *carbones terrenos, carbones bituminosi,* or *carbones fossiles.* Names such as *lapis carbones, lithanthrax,* and *Stein-*

---

[36] *De Ortu* [see note 2], pp. 17, 35; Forbes [see note 2], pp. 37–9.
[37] *De Natura Fossilium* (1546), pp. 233–4; (1955), pp. 66, 68.

*kol* were relegated to the index and lexicon.[38] Ironically, by the end of the century, such names had become commonplace when writing about coal.

### ON THE ORIGIN OF COAL

Agricola established the notion in *De Ortu et Causis Subterraneorum* that each mineral is generated from water and earth that has been cooked to yield a specific juice that is subsequently congealed.[39] His elaboration of this view contains possible traces of similitude and the use of organic analogies. His conclusions owe more to Aristotle and a picturesque style of writing, however, than to any of the newer intellectual fashions, which he opposed in other contexts.[40]

Agricola believed that, in general, combustible minerals form from juices "squeezed" by the force of subterranean heat out of a rich, fatty earth, or the destruction of an existing deposit and the regeneration of its fatty juices. When the heat is gentle, juices trickle from the fatty earth like resins from the larch, fir, or similar tree, and they will be white and have a sweet oily taste. Evidently, he had camphor and ambergris in mind. Under the action of increasing heat, the juices turn from white to red, from red to yellow, from yellow to green or black, just like the different phlegms and biles produced in animal bodies under moderate heat. Here, he must have had the variegated forms of amber in mind. When the force of heat is very great, black juices gush from the precursor fatty earth just like pitch flows from pine wood being made into charcoal or used as a torch. These black juices, presumably, give rise to petroleum, asphalt, and the other black bitumens.

The further process of transformation into the solid earthy forms is hinted in *De Natura Fossilium:* "Bitumen either inside or outside the earth may be made dense and changed until it becomes to some

---

[38] *Interpretatio Germanica Vocum rei Metallicae, addito Indice Foecundissimo* (1546) [see note 2], p. 476; Index, unpaged.

[39] See Hoover [note 1], pp. 46–9, 52.

[40] See Halleux [note 4] and Oldroyd [note 4, diss.], pp. 15–16. Albury and Oldroyd [see note 4], pp. 189–90, include Agricola among Renaissance scholars such as Gesner and Aldrovandi who relied on a complex web of resemblances or "similitude" in compiling "histories" of minerals. Key elements of that mode of thought, such as the macrocosm-microcosm relationship and the doctrine of "signatures," are missing from Agricola. See Michel Foucault, *The Order of Things* (New York, 1970), pp. 25–30. Compare Rudwick's analysis of Neoplatonic thought in works by Gesner and Cardano [see note 13], pp. 18–21. See also Debus [note 13], pp. 11–15.

degree as hard as stone."[41] The statement follows a brief discussion of *bitumen* in the particular sense of asphalt, so it may be assumed he saw *that* material, rather than, say, petroleum, as the intermediate precursor to coal formed in the earth, to gagates formed in the earth or in the sea, and to the "same" minerals formed similarly but known under other names in antiquity. Notably missing is any direct expression that earthy bitumen achieves its final form by "cooking," unless that action is implied in the phrase "made dense and changed."

A slightly revised edition of *Bermannus* was published with the other *Subterranea*. There, the notion that coal is "cooked" was retained with the added proviso *attamen pingue* to make it clear that the final product retains the fattiness of the precursor.[42] It was the only significant change in his early treatment of coal. In the new works, he silently dropped the distinction that coal is either light and earthy or heavy and stony. It was not really consistent with his newer idea. Given the apparent reaffirmation in the revised edition of *Bermannus,* however, it is not surprising that some successors continued to offer the simple but outmoded twofold distinction in their Agricolan-influenced writings and borrowings.

### ON THE DESCRIPTION OF COAL

The external properties of the earthy bitumens were found to be highly variable with regard to hardness, luster, and graininess or other evidence of structure. Agricola's most complete description was of ampelitus and merely reiterated that given by Dioscorides.

Agricola's original contributions were limited to brief descriptions of the types of coal found at Zwickau and their strata.[43] The very thick uppermost bed contained a type he described only as

---

[41] *De Natura Fossilium* (1546), p. 232; (1955), p. 68. My translation.

[42] *Bermannus* (1546) [see note 2], p. 458.

[43] *De Natura Fossilium* (1546), p. 236; (1955), p. 69. See also U. Horst, "Gedanken über Georgius Agricola," *Geologie,* 4 (1955), 599–614 (p. 603). Horst is pleased Agricola made a distinction between the two basic types that is still valid. He blunts the point, however, with the suggestion that the German expressions *Weichsteinkolen* and *Pechsteinkolen* probably were in use long before Agricola's time. He provides no evidence for that conjecture. Such older uses would not necessarily undercut my understanding that the bed of rather pure pitch coal was discovered only in the 1530s, if the miners recognized the thin layers of *Pechkohle* that pervade the soft coal bed, more or less as trace amounts of pitch may remain on charcoal. Prescher [see note 4], pp. 744–5, sees this passage as initiating stratigraphy, but the next contributions I have seen occur nearly a century later. More work needs to be done to fill or anlayze that "gap."

"soft coal" (*mollium carbo*). This type is listed in the lexicon as *bituminosi carbones molles* and *Weichsteinkolen*. The coal in the most significant lower vein below a very dense rock was described only as being so hard it has been given the name "pitch" (*pice*), which it resembles in color and luster. This type is listed in the lexicon as *bituminosi carbones duri* and *Pechsteinkolen*. The still deeper layers were said to contain "bituminous cadmia" and finally aluminum pyrites, pure copper, and "coal" (*carbones*), without further modifier or description. These terms and descriptions can be compared with Cordus's field notes of "black hard bitumen" found in the Swiss Alps that resembles "bituminous or fossil coal," of "black hard bitumen" found at three places in the Elbogoner region of Bohemia that are "crustaceous, light, burn violently and emit copious vapors (smoke)," and sometimes contain a "hotch-potch of useless copper pyrites," and of "bitumen just like burned out ashes" from the same region.[44]

## ON THE PLACES AND USES OF COAL

Agricola mentioned other locations in which earthy bitumen was found. The likely references to ampelitus, gagates, and obsidian specify no more than a general region or country of origin with an ancient text as authority. The likely references to coal mines always specify a specific locale, although the authority sometimes remains obscure. Albertus, of course, was the source of information that coal was found at Liège, but he merely alluded to other distant places in lower Germany where it was found. Agricola's specific references to Aguigranus (Aachen), two other remote places elsewhere in Germany, and the Alps, and three places in the Elbogoner almost certainly came from Cordus. Many persons, including himself, might have provided information of coal mined near Dresden. Agricola's reference only to the burning mine at Dysert Heath near Edinburgh probably came from an obscure printed source concerned mainly with underground fires. It is likely an oral source would have also provided information about English and Welsh mines and the considerable trade with the continent.

---

[44] Cordus, *Sylva* (1561) [see note 31], p. 220ᵛ. Cordus may have seen rock asphalt as well as coal and lignite.

Agricola listed uses of earthy bitumen as a substitute for charcoal and wood fuel (coal), as a materia medica and beauty aid (gagates and ampelitus), as a substance to guard grapevines from worms and rodents (ampelitus), and as a material for making beads and doublets (gagates and obsidian). The passage was written as if one material could be put equally to any of the uses.

He explained the brittleness of iron worked at a coal-fired forge in terms of the coal's "fattiness." Even charcoal had a "fatty" component or else it could not have burned, but Agricola sensed there was something special about the "fattiness" of coal coincident with what Albertus had termed its very hot fire that tended to corrupt the final product.[45] Thus, coal was avoided for fine iron work unless charcoal was not available. Coal, by implication, was acceptable and possibly even preferred for working many rough iron products where the cross sections were large, and brittleness was not a major problem.

Agricola came full circle, of vastly increased compass. His initial interest in coal grew out of contrasting accounts by Theophrastus and Albertus of mineral fuel used by metal workers in place of charcoal. His mature thoughts ended with a discussion that highlighted uses and limitations of coal in metal working, but also included other uses traditionally associated with different minerals. How much he owed to Cordus remains to be developed. The effect was to place coal squarely in a novel context drawn

---

[45] Albertus [see note 16], *II*, 199. The idea that coal can serve blacksmiths better than wood or charcoal is found in folklore attributing its discovery at Liège to an old man in white, an angel, or Jesus Himself. The truth of the matter may involve Cistercian or "white" monks who were pioneers in coal mining from the north of England, through Liège and Aachen, to Zwickau. See Polain [note 16], Herzog [note 5], p. 7, and L. J. Lekai, *The Cistercians, Ideals and Reality* (Kent, Oh., 1977), pp. 321–3.

John Leland (c. 1506–1552) saw metal workers in parts of Yorkshire using "yerth cole" or "se cole" despite a local abundance of wood. See *The Itinerary of John Leland in or about the years 1535–1543*, ed. L. T. Smith, 5 vols. (Carbondale, Ill., 1964), *IV*, 14–15.

Vannoccio Biringuccio (1480–1539?) mentioned metal workers and lime burners in Tuscany who also used a "stone" that burns despite an abundance of wood and went on to indicate criteria for selecting suitable wood and proper methods of making good charcoal. By implication, the metal workers had trouble working some iron products with the charcoal commonly available. See *The Pirotechnia of Vannoccio Biringuccio*, trans. C. S. Smith and M. T. Gnudi, (New York, 1943; reprinted, 1959), bk. III, chap. 10, pp. 173–9.

It is said frequently that blacksmiths used mineral coal only when charcoal was scarce and very expensive. The full story of the use of coal for certain metal working operations in *preference* to charcoal remains to be developed.

largely from ancient texts, which were not to be venerated but rather selected, synthesized, emulated, and updated. More generally, his work and that of Cordus quickly achieved their intended status on a par with ancient authority.

### MINING AND REFINING AGRICOLA

Girolamo Cardano (1501–76) modified Agricola's list of bitumen by treating the earthy bitumens as conceptually distinct types, by omitting obsidian, and by distinguishing between *succinum* or "yellow" amber and *ambra* or "gray" amber (ambergris). The list now became *asphaltum, pissasphaltum, napta, gagates, ampelitus, maltha, Thracian lapis, carbones fossiles, succinum, petroleum,* and with uncertainty, *ambra* and *camphora.*[46] Agricola had not made the genus/species distinction in his discussion. Indeed, no writer of the period made it consistently. Cardano did so partially with the indication that coal, gagates, and asphalt belong to the same genus. With a touch of Neoplatonism, he indicated that asphalt and gagates have the purest and most abundant fattiness, whereas coal has a lower *virtu,* stemming from its origin in a black, light, and fatty earth.

Christoph Entzelt (1517–86), the Lutheran pastor of Saalfield in the Thuringer region of Saxony, and Severin Goebelio (1530–1612), a physician from Gedansk in the Baltic amber district, cribbed or reworked various sources to place mineralogy on sound philosophical bases. Entzelt provided a comprehensive treatment, built largely on the sulfur/mercury theory. Goebelio limited his scope to a naturalistic discussion of amber, petroleum, and the congealed bituminous juices, preceded, however, by a religious allegory in which the Christian mystery is depicted in images that range from tenuous and subtle heavenly spirits down to crass, earthbound juices. For both authors, most ideas on bitumen came straight from Agricola and Cordus, although Goebelio also drew on Entzelt's formulation.[47]

---

[46] Girolamo Cardano, *De Subtilitate libri XXI* (1550), in *Opera Omnia,* 10 vols. (Lyon, 1663; reprint ed., Stuttgart, 1966), *III,* 442–4.

[47] Christoph Entzelt, *De Re Metallica* (Francofurti [1551]), pp. 107–8, 125–32, 143–4, 157, 180–99. See also W. F. Parish, "Distillation of Petroleum," *Franklin Institute, Journal,* 203 (1927), 806–9. *Gagates lapis* is translated anachronistically as "lignite." The concept of "lignite" belongs to the eighteenth century. Severin Goebelio, *De Succino libri duo* (1558) in Conradus Gesnerus, *De Omni Rerum Fossilium Genere, Gemmis, Lapidibus, Metali . . . (Tiguri, 1565),* separately paged, pp. 27ᵛ–30ʳ.

They differed most significantly on *succinum*. Entzelt excluded it because it is extinguished by water, whereas gagates was said to be inflamed by water. Goebelio indicated that it forms from clear and very pure bituminous juices. Both omitted or excluded camphor and obsidian. Both accepted Thracian stone as the same thing as gagates. But both, like Cardano, included coal only as a member of the same genus and not as the same mineral.

In substantial agreement with Agricola, Entzelt saw gagates and coal as originating from naphtha or petroleum that has been excocted and hardened by the terrestrial heat. The generally higher quality of gagates was in accord with the riverine environment, where it takes its final form, in contrast to the underground environment, where coal forms. Goebelio accounted for variation in quality in terms of thick, black bituminous juices that might be mixed with other mineral juices. Entzelt relegated ampelitus to the section on the "Earths," but indicated, in a passage possibly drawn from Cordus, that it was sometimes considered a bitumen. Goebelio included ampelitus simply as a type of bitumen in the same genus as *Steinkol,* pissasphalt, and gagates.

Entzelt's concept of gagates, with a possible assist from Cordus, included "yellow" varieties of agate that occur in yellow, yellow-green, black- and yellow-striped, and red-spotted forms. The basis for including these quartz minerals with the bitumens certainly was not combustibility or oil solubility. It must have been little more than a name and uses similar to gagates, although the variegated colors may have conjured images of films of congealed petroleum. Similarity, if not precisely similitude, was philosophically important for him.

Entzelt understood Agricola's charcoal namesake minerals in *Bermannus* to include minium, rubies, turf, and coal. The group was useful because all were red or could turn red. He recast Agricola's information into a chapter "On Cinnabars," in which he reiterated the distinction between light and heavy coal, including the claim the former could float. In the Chapter "On Lithanthrax," however, he qualified the claim to indicate that the light variety only floats sometimes. Goebelio's section on the four earthy bitumens consists largely of extracts from *De Natura Fossilium,* but his use of the same qualification indicates he relies very selectively upon Entzelt's work as well.

As seen, Entzelt conjectured that the unnamed earth in *Bermannus* (which I have shown to be ampelitus) was *Thorff,* that is, *torf*

| | Yellow | *Gagetsteyn, aagetsteyn, bornsteyn, bernsteyn, birnsteyn,* the color usually approaches green, like the skin of a lion. | Stones hardened from naphtha or bitumen, i.e., petroleum in Cicilia, and in the Gagate River where it falls into the sea, by heat that obviously is its own or rather that of water. |
|---|---|---|---|
| *Gagas* or *gagates* | Black | Thracian stone, *Schwarz aagetsteyn.* | |
| | *Lithanthrax, Steynkole.* | A stone from naphtha or bitumen hardened by the terrestrial heat. It is a species of gagates. | |

Figure 12.1.    Entzelt's schematic summary of gagates. Adapted from Christoph Entzelt, *De Re Metallica* (Frankfurt, 1551).

or turf. He also conjectured that this earth undoubtedly is a type of bitumen and that it is the "mother of coal." He did not attempt to relate this view to the more prominent theory of origin directly from petroleum, but, as will be seen, a later writer made just this connection.

Entzelt minimized Agricola's distinction between coal as a light, simple substance and the bituminous *mixta* of Islebia as a heavy, complex one. This mineral, from Entzelt's own district, was known locally as *der Schyffer*. He regarded it merely as a species of *Steinkol* with similar "dryness" to permit its burning, but he also reverted to Agricola's view and asserted that it is a vareity pyrites made of bitumen.

Despite some inconsistencies, Entzelt's most important contribution, for the topic at hand, was his simplification, superior organization, and pioneering use of schematic arrangements, such as that in Figure 12.1, to present Agricola's idea in summary fashion. Goebelio's treatment presented many of the same points in usefully compact form, but without the diagrams.

A quite different response was made by Pierre Belon (1517–61).[48] Singularly, he accepted (without credit) the notion that obsidian is a bitumen. He also accepted the conclusion that gagates

[48] Pierre Belon, *De Admirabili Operum Antiquorum et Rerum Suspiciendarum Praestantia* (Paris, 1553), bk. III, chaps. 2–5 and 7 (there is no chap. 6), pp. 41$^r$–48$^r$. Chapters 5 and 7 are translated in R. J. Forbes, "Pierre Belon and Petroleum," *More Studies* [see note 2], pp. 3–13. See also P. Delaunay, "L'adventureuse existence de Pierre Belon du Mans," *Revue du seizième siècle*, n.s., IX (1922), 251–68.

is a bitumen and offered a quotation to that effect which he attributed to Galen. In fact, the passage was copied from *De Natura Fossilium* where Agricola added his own comment to a passage he cited from Galen without using quotation marks to set it apart. Belon did not check the original and inadvertently lifted the composite passage saying he had it from Galen.[49] The example is only one of several indications that Belon closely followed the appropriate parts of Agricola's *Subterranea* as he prepared his own manuscript on related topics. Even though he had traveled extensively with Cordus, the main conclusions Agricola and Cordus reached about coal in relation to the minerals named in antiquity somehow did not attract his interest. The related claim that amber is a bitumen, however, was dismissed out of hand and the ancient idea of Dioscorides reiterated.

Pietro Mattioli (1500–77) raised strong points against new views, especially, of "earthy bitumen," followed by even more traditionalist reactions of Andrés de Laguna (d. 1560).[50] Both criticized Pliny for confounding gagates, a famed materia medica, with Thracian stone. Mattioli indicated that no such uses had been reported for the latter, and that most likely the stories of it are fables. Laguna indicated only that Pliny's view is against all writers and that Thracian stone is unknown in the present. Both indicated a medical formulation that, following Aetius (sixth century), involves quenching burning gagates with wine. Mattioli argued this effect proves it is not a bitumen, or else it would be inflamed (implying thereby that wine is a kind of water), all of which reminded him, in an aside, of agate, which also has medical uses. Laguna used Aetius's example more simply to differentiate gagates from Thracian stone. He also argued that Gagate stone is quite different from agate.

Mattioli accepted reports of unknown origin that gagates is burned instead of wood in Flanders and Brabant, but he argued

---

[49] Forbes [see note 2], p. 13. Compare Belon, p. 48$^r$ with Galen (note 32), *XII*, 203 and *De Natura Fossilium* (1546), p. 234.

[50] Pietro Andreas Mattioli, *Pedacii Dioscoridis de Materia medica libri sex* (Venice, 1554), pp. 77–80, 630–1, 638–9. For English translation, see R. J. Forbes, "Matthiolus and Petroleum," *More Studies* [note 2], pp. 16–36. See also J. Stannard, "P. A. Mattioli: Sixteenth Century Commentator on Dioscorides," *University of Kansas Libraries Bibliographic Contributions*, *I* (1969), 59–81, (pp. 67–9). Andrés de Laguna, *Pedacio Dioscorides . . . acero de la materia medicinal . . .* (Salamanca, 1566; reprint ed. in 2 vols., Madrid, 1968), *II*, 560. The first Spanish edition was 1555.

directly against Agricola's view that mineral coal is the same thing as gagates. Agricola (and Theophrastus) had indicated "coal" ignites and burns just like charcoal. Mattioli probably had not seen coal in use, but he knew that charcoal does not emit flame unless urged by bellows and does not emit the odor of bitumen. He assumed from Agricola's account that coal behaves in exactly the same way, whereas what he considered to be the same thing as gagates (from Innsbruck) is so full of bitumen it produces flame at once without bellows and emits copious fumes. Moreover, the distillation of gagates, following Mesue (tenth century), yields a large amount of medically useful oil, although Mattioli denied this possibility for coal.[51] Laguna simply opposed all persons who confounded gagates with the "black rock" burned in Flanders, which neither emits the odor of bitumen nor yields oil. Mattioli agreed with "some" persons who see no great difference between gagates and ampelitus, thereby denying indirectly the related claim that ampelitus is another name for coal. Laguna merely treated ampelitus traditionally as an unrelated earth.

The difference with Agricola (and Cordus) on the category of bitumen could be extended, but these examples reveal the range of contention. Without putting too fine a point on the matter, it was essentially Cardano's list of bituminous species, with amber and the earthy bitumens added, and camphor and frequently ambergris omitted, that was adopted by key writers of the seventeenth and eighteenth centuries, such as Sennert, Barba, LeFebvre, Woodward, Boerhaave, Hill, and Diderot.[52]

---

[51] The claim probably was based on finding no report rather than on actual distillation. By 1641, it could be claimed "everyone" knew of medical uses for an oil "drawn" from "Carbo Petrae," which was differentiated from gagates using another of Mattioli's arguments. In general accordance with Cardano's list, coal was classed, nevertheless, as a bitumen. See Johann Schroder, *The Compleat Chymical Dispensatory*, trans. W. Rowland (London, 1669), pp. 53, 277. For a recipe see Johann Glauber, *Pharmacopea Spagyrica* in *The Works of . . . Glauber*, trans. C. Packe (London, 1689), pt. II, pp. 125–6.

[52] Daniel Sennert, *Epitome Naturalis Scientiae*, 3rd ed., rev. (Oxoniae, 1632), pp. 352–61. Albaro Alonso Barba, *Arte de los metales* (Madrid, 1640); English trans. in *A Collection of Scarce and Valuable Treatises upon Metals, Mines, and Minerals*, 2nd ed. (London, 1740), chap. IX, p. 25. Nicaise LeFebvre, *A Compleat Body of Chymistry*, trans. P. D. C. (London, 1670), pt. II, pp. 74, 291–2, 311. John Woodward, "Fossils," in *Lexicon Technicum*, ed. John Harris, 2 vols. (London, 1704; reprint ed., New York and London, 1966), I, unpaged. Hermann Boerhaave, *A New Method of Chemistry*, trans. Peter Shaw, 2 vols. (London, 2d ed., 1741), I, 117–9. John Hill, *A General Natural History*, 3 vols. (London, 1748–52), I, 408–21. *Encyclopédie, ou Dictionnaire raisonné des sciences, des arts et des métiers*, ed. D. Diderot et al., 39 vols. (Paris, 1751–80), "Bitume," and "Charbon minérale."

Cardano's knowledge of coal was extended during a trip to Scotland and Liège in 1552, including a stop at Lyon during the Milanese carnival.[53] He noted the coal there was used exclusively by blacksmiths because it makes a very strong fire useful for working iron, and he mused about the way it retaliated with bad, oppressive fumes. In Scotland, like many other travelers of the period, he was struck by the use of coal for all purposes because of the scarcity of wood, and not simply for rough, smithy work.

All too briefly, Cardano mentioned the English and Scottish coal trade with France. He also indicated with disappointing terseness that "the light ones are pure and lustrous, while the others have a more earthy look" (*leves sunt qui puriores, ac splendent: alii magis terrei*), adding he had seen some of these coals reduced to very small pieces in transit "not so much because of their weight as their lack of strength" (*non tam illorum gravitatis quàm horum imbecillitatis*).[54] He did not relate these observations to the contrary observations of Agricola about bright coal that was hard and the more earthy coal that was soft. Also, Cardano was probably the only author widely read over the next century who actually had a chance to see the coal from Liège. Unfortunately, he did nothing to corroborate or refute the notion that the coal at Liège might be markedly heavier or more stonelike than the rest. He also failed to advance the debate by ignoring differences with Belon and Mattioli concerning the category of bitumen, even though he referred to their work in his discussion of plants.

Conrad Gesner (1516–65) was sufficiently impressed with Agricola's *Subterranea* to name him as the Saxon Pliny when that attribution was still high praise. He intended to produce a more comprehensive work and published various short tracts that he hoped would elicit further information of the same kind. Most notably for the topic at hand, his preliminary volume included contributions from Johannes Kentmann (1517–74), a physician from Dresden, and that of Goebelio.

This list is neither complete nor inclusive of contrary views. All the works, however, were pivotal for their fields and are indicative of the wide range of works through which Agricola's ideas flowed at third, fourth, and more remote hands. The key secondary sources, I believe, were those of Cardano and Entzelt. See also C. S. St. Clair, "The Classification of Minerals: Some Representative Mineral Systems from Agricola to Werner," (Ph.D. diss., University of Oklahoma, 1965).

[53] Jerome Cardan, *Book of my Life*, trans. J. Stoner (New York, 1930), pp. xxi, 97–101.
[54] G. Cardano, *De Rerum Varietate libri XVII* (1557), in *Opera Omnia* [see note 46], *III*, 47.

Kentmann had assembled a collection of several hundred mineral specimens and organized it in classes similar to, but more elaborate than, those introduced by Agricola. As published by Gesner, the classified list of bitumens included mineral coal along with camphor, petroleum, "mumia," and asphalt; amber specimens were listed in a parallel category.

The coal collection included one sample from Bohemia (Elbogoner?), two from the St. Joachimsthal Valley (unknown in other sources I have seen), and three dug not far from Dresden. These were listed as "the black Bohemian mineral called stone coal, so consumed by earth that it cannot be polished" (*Bohemisch Steinkolen*), "soft friable bituminous coal in an ash-colored stone" (*Steinkolen in einen gemeinem Gravenstein*) "bituminous coal of the same type, but with a white efflorescence" (*Steinkolen in weissen Flussen*), "soft and fissile bituminous coal" (*Weichsteinkolen*), "hard and fissile bituminous coal . . . almost like pissasphalt" (*Pechsteinkolen*), and "hard and fissile bituminous coal with alumino-pyrites" (*Steinkolen mit Alumstein*).[55] The latter three names are the same used by Agricola. Based on Agricola's very terse description, Kentmann probably assumed what he had from Dresden to be the same types, and he adopted the same terminology. In any case, it is clear that Kentmann actually had samples in hand because he provided information on encrustations and on the laminated nature of the hard and soft coals provided by neither Agricola nor Cordus.

Unfortunately, Kentmann had no samples from Zwickau, the Low Countries, or the British Isles. Equally unfortunate, Cardano or Kentmann could not compare notes with Agricola or Cordus. Each used a slightly different approach to description. Each provided very terse accounts that provided less information than might have been forthcoming if they had been encouraged. The community of interest could not be built because Agricola was soon dead, Cordus had died long before, and there were no adequate mechanisms to stimulate the others as they moved on to other interests. Gesner's network of scientific correspondents did not survive him. His untimely death cut short both his efforts at com-

---

[55] Johannes Kentmann, *Catalogus rerum fossilium*, pp. 22$^r$–24$^v$, in Gesner [see note 48]. My translation of Latin descriptors.

munity buidling and his own research program on minerals. Kentmann's efforts were quoted but not extended for many years.

The *Lexicon Alchemiae* of Martin Ruland the Younger (1569–1604) is noted as a key to the meaning of various arcane terms used by Paracelsus and his followers.[56] That is half the story. Many of its long articles are verbatim extracts from Entzelt's *De Re Metallica*. The chapters on cinnabar, coal, and the other bitumens, including the schematic diagrams, for example, were cut into articles and printed in dictionary order without credit to Entzelt. Kentmann's related information on coal also was incorporated without credit to the source. Hence, Ruland's dictionary was a major conduit for ideas rooted in Agricola's work and put into a more popular format by his successors. John Webster (1610–82), for example, relied on Ruland for information on coal rather than on Agricola or direct experience.[57]

### CARRYING THE CONCEPT OF COAL TO "NEWCASTLE"

The first use of these ideas on coal in an English language text was provided by William Fulke (1538–89). Other influences could be shown, but his work is a fitting place to end this study of the development and early reception of the concept of coal as a bitumen. Fulke provided a unique blend of current and ancient ideas devised for the enlightenment of the "common" reader. As such, it gives us a good gauge of the assimilation of ideas first put together in detail by Agricola less than twenty years before. The ideas on coal almost certainly came indirectly through Cardano. Unlike, Agricola and Cardano, or even Entzelt, Fulke relied on the increasingly popular sulfur theory of combustion.

Brimstone, according to Fulke, is generated from earthy and airy vapors, fumes, and exhalations, and is the chief and most notable kind of combustible substance. The other "fat and oyly" minerals, such as "seecoles" or ordinary mineral coal of English fame and "gette" (gagates), are "generated of such lyke *vapors* as brym-

---

[56] Martinus Rulandus, *Lexicon Alchemiae* (Francofurti, 1612; reprint ed., Hildesheim, 1964). Idem, *A Lexicon of Alchemy* trans. A. E. Waite (privately printed, 1893; reprint ed., London, 1964). The text and translation are not always reliable.

[57] John Webster, *Metallographia* (London, 1671).

stone is, but then they are diversely mixed."[58] Following Cardano and quite possibly his own observations of residues, Fulke said that coal consists of much earth mixed with brimstone and gagates seems to be the same thing, but better concocted.

Agricola touched a point of popular fear by relating his ideas to volcanoes and other burning hills. Fulke addressed another point of popular fear by relating his version of these ideas to comets and the other fiery meteors still considered part of the terrestrial realm. He explained all such phenomena in terms of earthy and airy exhalations drawn from the "fatness" of the earth and lofted into the highest regions of the air by the heat of the sun. In the extreme case of comets, the loss of matter is so extensive as to leave the soil dry, barren, and unproductive and to so fill the air with corrupt exhalations as to infect the bodies of people and animals.[59] In one stroke, Fulke explained why the appearance of a comet heralds famine, pestilence, or other disaster.

Fulke said nothing of the reduced soil fertility when its "fatness" is drained only enough to form sulfur, bitumen, coal, or an oil spring. It is interesting to note in this regard Gabriel Plattes's advice some years later to seek coal in grounds that are neither very fertile nor extremely barren, but of indifferent fertility.[60] He also suggested that the precise spot to drop a shaft could be found by placing a new, white, wooden bowl or a clean piece of white tiffany on the ground about the middle of May, when the subterranean vapors are strong, in order to catch the telltale signs of an underlying coal deposit. Coal itself was seen as the sort of black fatty earth cut for fuel (turf) that has been impregnated and hardened by the vapors of other combustible substances such as asphalt or sulfur. The treatment offers well-assimilated ideas provided by Agricola and his closest followers. Agricola, with notable assistance especially from Cordus, Cardano, and Entzelt, had established coal as a standard subject that bore his general imprint with little advancement until the end of the eighteenth century. Silent

---

[58] *A Goodly Gallerye, William Fulke's Book of Meteors,* (1563), ed. T. Hornberger (Philadelphia, 1979), p. 111. Hornberger (p. 108) is wrong on the dates, but probably right on the lack of direct connection to Agricola.

[59] Ibid., pp. 43–5.

[60] Gabriel Plattes, *A Discovery of Subterraneall Treasure* (London, 1639; reprint ed., in *Scarce Treatises* [see note 52]), pp. 229–234.

testimony is retained today in the archaic English-language designation "bituminous coal" for the most abundant varieties.

## CONCLUSION

Scholarship for Agricola could not treat ancient texts in isolation from the outside world. Just as sound ancient knowledge had been fashioned from a living landscape and not merely from other books – which leads to the degradation of learning – so knowledge could come alive again and be extended only by relating the authentic ancient accounts to current observations of life, places, and things.

Every thought on coal or earthy bitumen was rooted in ancient texts of scholars such as Theophrastus, Dioscorides, Pliny, and Galen, but he did not follow any of them slavishly. His ideas developed and changed as he gathered more information, with the notable if shadowy aid and encouragement of fellow scholars such as Plateanus and Cordus. The direct role of artisans, if any, is not only more obscure, but not needed to explicate the work. Yet, as with his ancient role models, limited data was gathered, one way or another, from the field.

Information of a new type of coal helped Agricola understand that even the ancients had limited perspectives on things. His mature thoughts involved imaginative and sometimes daring syntheses of the old texts, going well beyond the fashion merely to annotate Dioscorides. Equally, his thoughts were free of widespread Neoplatonic or hermetic influences to find secret interconnections between all things of creation. Fellow scholars more devoted to such ideals, however, eagerly entered his work without necessarily recasting it in their more fashionable rhetoric. Indeed, with some differences in detail, several of the most influential writers of the day simplified and organized better what Agricola contributed.

Agricola was certainly more than a compilator or reporter. He was less than the single-handed creator of a science of all he surveyed. He was the key figure among several kindred spirits whose seminal work, alas, was not sustained by subsequent generations. That failure is surely worthy of more study. If he must be categorized, Agricola was, in keeping with his own self-image, one of the last great Greek scientists. The reality, of course, is more complex. Agricolan scholarship today must be renewed so that complexity can be more perfectly realized in all the work he accomplished. His work, moreover, was formulated and disseminated through a variegated community of scholars whose work also must be taken seriously on their own grounds, if we are to understand Agricola and his actual influence.

# 13

## Theories for the birds: an inquiry into the significance of the theory of evolution for the history of systematics

PAUL LAWRENCE FARBER

The study of the origins, the structure, the context, and the change of scientific theories has dominated the history of science for several decades. Richard S. Westfall has not only carefully considered issues in all of these categories but has also set a model of meticulous scholarship in each. In spite of the occasional expostulations of social historians, sociologists, or philosophers of science that alternative perspectives are valid and useful, the majority of historians of science have acted as Westfall has, focusing attention on scientific theories: their major concepts, methodological and metaphysical assumptions, context, impact, alteration, and transformation. The preponderance of emphasis on theory, however, has its drawbacks in understanding the nature and history of science. An example from the history of natural history will illustrate this point.

The history of taxonomy can be, and is, written from quite different perspectives. Taxonomists spend a great deal of time uncovering the historical details of the naming and the ordering of groups, and they have produced an extensive body of literature documenting the place of museum specimens as well as the revisions made in classification and nomenclature. That internal history is, indeed, part of the practice of taxonomy, and few historians have had the temerity to venture very far into it. For the most part, historians in the twentieth century have focused on the theoretical foundations of systematics, particularly on the alleged impact of evolution theory on modern classifications. One sees quite clearly, for example, this approach in historical studies on ornithology. Ornithology, of course, easily admits that perspective for it is a discipline that was central to the formation and revision of Darwin's theory of evolution. The nature and the extent of the impact of evolution theory on ornithology, however, is controversial. Alfred Newton, in his well-known historical sketch of ornithology, wrote:

there was possibly no branch of Zoology in which so many of the best and informed and consequently the most advanced of its workers sooner accepted the principles of Evolution than Ornithology, and of course the effect upon its study was very marked. New spirit was given to it. Ornithologists now felt they had something before them that was really worth investigating. Questions of Affinity, and the details of Geographical Distribution, were endowed with a real interest, in comparison with which any interest that had hitherto been taken was a trifling pastime. Classification assumed a wholly different aspect. It had up to this time been little more than the shuffling of cards, the ingenious arrangement of counters in a pretty pattern.[1]

Erwin Stresemann, in his detailed history of ornithology, however, came to a different conclusion. Although very sensitive to the importance of evolution theory for certain schools of contemporary systematics, Stresemann noted that in the nineteenth century, many ornithologists refused to revise their criteria for classification and remained firm opponents of evolutionary systematics.[2] Even a nineteenth-century naturalist like Thomas Henry Huxley (1825–95), who attempted to construct a natural system for the birds that reflected their evolutionary relationships, was equivocal when he wrote about general systematics. In his book, *An Introduction to the Classification of Animals,* he wrote:

By the classification of any series of objects, is meant the actual, or ideal, arrangement together of those which are like the separation of those which are unlike; the purpose of this arrangement being to facilitate the operations of the mind in clearly conceiving and retaining in the memory, the characters of the objects in question.

Thus, there may be as many classifications of any series of natural, or of other, bodies, as they have properties or relations to one another, or to other things; or, again, as there are modes in which they may be regarded by the mind: so that, with respect to such classifications as we are

---

[1] Alfred Newton, *A Dictionary of Birds* (London: Adam and Black, 1893–6), p. 79.
[2] See Erwin Stresemann, *Ornithology from Aristotle to the Present,* (Cambridge, Mass.: Harvard University Press, 1975).

here concerned with, it might be more proper to speak of *a* classification than of *the* classification of the animal kingdom.[3]

Faced with a spectrum of opinion concerning the historical significance of evolution theory for systematics, we might be tempted to try to resolve the apparent paradoxes by analyzing the history of taxonomy into national or institutional schools and by focusing on the reception of evolutionary ideas among ornithologists in different places and at different times. Such a study would not only be interesting and useful, but could serve as a first step in comparing the impact of Darwin's ideas in different biological disciplines. It would, and has been, the approach one would expect from historians who regard scientific theories as the principals in history. It seems to me, however, that by defining the main historical problem in the history of modern systematics in terms of the impact of evolution, we may be led to overlook issues that are of equal or greater historical importance. For in the case of ornithology, at least, there has been a steady continuity of problems and practices since the early nineteenth century. Studying the history of ornithology from the perspective of the influence of evolution theory leads to a view of discontinuous change that obscures those continuities that were quite central in the growth of ornithology. Perhaps approaching the history of systematics by focusing on the theoretical level leads to a distorted picture.

A brief look at bird classification in the early nineteenth century will reveal why. Avian systematics in this period was a part of the larger enterprise of natural history classification, for natural history had not yet fragmented into a set of separate scientific disciplines.[4] Natural history, in fact, was a relatively small endeavor, and the community of naturalists was thinly scattered over the European continent. Museums were, more often than not, *cabinets d'histoire naturelle* which contained curiosities, medals, and porcelain as well as animal, plant and mineral specimens. Systematics

---

[3] Thomas Henry Huxley, *An Introduction to the Classification of Animals,* (London: John Churchill & Sons, 1869), p. 1. In the twentieth century, there exist numerous zoologists, such as Richard Blackwelder, who question the idea that systematics should reflect historical relationships. Recently, the cladist Colin Patterson wrote "as the theory of cladistics has developed, it has been realized that more and more of the evolutionary framework is inessential, and may be dropped." See Colin Patterson, "Cladistics," *Biologist,* 27(5) (1980), 239.

[4] See Paul Farber, *The Emergence of Ornithology as a Scientific Discipline: 1760–1850* (Dordrecht: D. Reidel, 1982).

was under the influence of Carl Linnaeus (1707–78), the great ar-
biter of classification of the previous century, and the eighteenth-
century goals of natural history, description and classification, remained unaltered. Under the surface, however, a transformation
was beginning. The entire enterprise during the first half of the
nineteenth century began an acceleration in the growth of its size.
The available empirical base swelled dramatically due to a burst of
interest in local faunas, and to the indirect impetus given to natural
history by the wave of colonialization that had associated with it
extensive exploration, expansion of trading companies, and encouragement of the study of exotic flora and fauna for economic
significance. Able to utilize these riches was a new generation of
naturalists that was larger and better trained than any earlier ones.
It had come into being because natural history became fashionable, and its popularity had opened opportunities that permitted
aspiring naturalists to create careers.[5] Popular writing, illustrating,
curating, collecting specimens for sale, and obtaining private or
public patronage were avenues that made it possible for more than
a handful of nature lovers to pursue their passion for more than
personal pleasure.

What did all of this mean for avian taxonomy? Indeed, what
one might expect: a lot of work! Even in the eighteenth century,
naturalists had glimpsed the enormity of the project that a complete catalog and system of the birds was going to be. Georges-
Louis Leclerc de Buffon (1707–88), for example, whose magnifi-
cent nine-volume *Histoire naturelle des oiseaux* set a baseline for ornithological iconography and was one of the principal texts that
began the process of specialization in natural history, clearly stated
that his contribution was merely a small, first step. And he was
quite correct, for his work was soon out of date as a complete
guide to the known birds of the world. Individuals like John La-
tham (1740–1837) repeatedly attempted to provide an updated
general index, but the task grew like that of the sorcerer's apprentice. It took a Bonaparte, Charles Lucien (1803–57), to marshall
an almost complete, for the time, list at midcentury. Roughly at
the same time, the major national collections began to publish their
catalogs, which proved to be invaluable guides to ornithology.

---

[5] Ibid. For a case study of the developments in Britain, see David Allen, *The Naturalist in
Britain: A Social History* (London: Allen Lane, 1976).

These great catalogs stand as monuments to the toil of many workers and reflect a growing knowledge of the outlines of the avian world. They should also be seen as the fruit of a new and substantial international community. The work done by the generation that created the scientific discipline of ornithology was rigorous and highly specialized. Instead of general natural histories, one finds in the literature on birds after the first quarter of the century local bird lists, monographs on particular groups, and short technical articles. The literature reflects the realities of research of the time. More material and more investigators were taking classification to finer and finer resolution. New organs of communication, made available to ornithologists by the revolution in the printing industry and the innovations in the reproduction of illustrations, encouraged the process of specialization by making it possible for newly formed specialized scientific societies to publish journals and for enterprising publishers like Richard Taylor (1781–1858) to establish scholarly periodicals.[6]

What did this plethora of material and frenzy of activity mean for classification systems? The problem goes beyond the confines of the history of ornithology, for systematics is a subject in which quite different methods and assumptions are often combined in new ways to create solutions to problems. We can see this aspect in the way ornithologists handled the central issues of avian classification: the breaking up of the Linnean genera and the related but more theoretical issue of what criteria should be applied for determining the composition of groups. The problems were critical. Linnaeus described seventy-eight genera in the last edition that he published of the *Systema Naturae* (1766). By 1844, George Robert Gray (1808–72) recognized eight hundred and fifteen in his *The Genera of Birds,* which attempted to bring order to over two thousand four hundred generic names compiled from other authors.[7] In Linnaeus's day, the question of what criteria to use for grouping birds was restricted. The empirical data available were small, and the specimens on hand consisted of collections of bird skins – usually mounted and often in poor condition. Collecting was unsystematic and often dependent on chance occasions. Many

---

[6] See the interesting article by Susan Sheets-Pyenson, "From the North to Red Lion Court: The Creation and Early Years of the *Annals of Natural History,*" *Archives of Natural History,* *10*(2) (1981), 221–49.

[7] For a discussion of this issue, see Farber, *The Emergence of Ornithology,* pp. 109–14.

species were named from single specimens of unknown specific location. Under such conditions, it is hardly surprising that naturalists relied on a few single external characteristics. Mathurin-Jacques Brisson (1723–1806), from whom Linneaus borrowed extensively for his last edition of the *Systema Naturae*, gave the following account:

The bills and feet are the parts that I have chosen to establish characters. The number of toes, their position, their separation or their position, the membranes that join them together, or the absence of these membranes are the characters that I have used to create the large and primary divisions; the subdivisions are designated by the different form of bill; other particularities determine the genus, and finally, the difference in colors distinguish species.[8]

And to make each of his descriptions of approximately fifteen hundred species and varieties (comprising his one hundred and fifteen genera and twenty-six orders) parallel, Brisson described "first the size and proportions of the bird; next its color, starting with the head and finishing with the tail."[9] Brisson, of course, realized how tentative these diagnoses were. After all, he usually was working from a single specimen or relying on a written report of a single specimen. In his discussion on birds of prey, he complains, quite frankly, of the frustration of relying on such a limited amount of materials, for he was aware that not only species, but also individuals differ "at different times and during the course of their life." It is necessary, therefore, to observe many specimens, but, he lamented, "that is most often impractical."[10]

What was impractical in the eighteenth century, however, became standard in the nineteenth. Individual collectors often sent back more specimens than were in the largest ornithological collections of the eighteenth century. The field collectors of the nineteenth century, moreover, were trained, and naturalists could, with confidence, request relevant information.[11] Male and female spec-

---

[8] Mathurin-Jacques Brisson, *Orinthologie ou Méthode contenant la division des oiseaux en Ordres, Sections, Genres, Espèces & Leurs Variétés* (Paris: Bauche, 1760), I, xiv–xv. For an interesting discussion of Linnaeus's use of Brisson, see J. A. Allen, "Collation of Brisson's Genera of Birds with Those of Linnaeus," *Bulletin of the American Museum of Natural History, 28* (1910), 317–35.

[9] Brisson, *Ornithologie,* I, xv.    [10] Ibid., 308.

[11] The training of collectors has not been very well studied. For an interesting discussion that suggests the richness of the subject, see Yves Laissus, "Les voyageurs naturalistes du Jardin du roi et du Muséum d'histoire naturelle: essai de portrait-robot," *Révue d'histoire des sciences, 34*(3–4) (1981), 259–317.

imens, juveniles, and geographic variations were soon available in intimidating numbers. And consequently, for those working in the traditions of Linneaus and Brisson, there was a wealth of information to be organized.

The flood of ornithological data that arrived in Europe included not only specimens but also recorded observations on the distribution and behavior of different birds. Some ornithologists suggested that this new information could be utilized to construct a broader picture of the avian world and a more sophisticated classification system. Jules Verreaux (1807–73) and Baron Noël-Frédéric-Armand-André de LaFresnaye (1783–1861), for example, voiced strong sentiments concerning the significance of the habits of birds for an understanding of their grouping.[12]

But ornithologists did not restrict themselves to the traditional use of external features or the suggestive but small amount of new information concerning the natural history of birds. Avian systematics benefited from developments in other fields, most notably from the emergence of the discipline of comparative anatomy. Under the guidance of its professors, the Paris Muséum, already the metropole for ornithology due to the size and accessibility of its collection, became the center of a rigorous morphological approach to the living world. Comparative anatomy, which had been in the eighteenth century a part of the allied medical sciences and a mere adjunct to natural history, came to be regarded by the leading French naturalists such as Georges Cuvier (1769–1832), Étienne Geoffroy Saint-Hilaire (1772–1844), and Henri-Marie Ducrotay de Blainville (1777–1850) as the most fruitful method of discovering the fundamental principles of living beings and the key for uncovering the order in nature. Although challenged by the parallel rise of physiology for the claim of the science of life, comparative anatomy achieved a lasting influence in natural history, for it provided a broader and more rigorous method of distinguishing and characterizing groups than the use of external features had afforded.

Confidence in the morphological approach to systematics was deepened by the influence of the new discipline of comparative embryology. Embryology brought to classification not only a new dimension in empirical data but also new concepts, the most im-

---

[12] See, for example Baron Frédéric de LaFresnaye, *Essai d'une nouvelle manière de grouper les genres et les espèces de l'Ordre des Passereaux (Passeres L.) d'après leurs rapports de moeurs et d'habitation* (Falaise, France: Brée, 1838).

portant of which was the branching conception of nature that utilized the potent metaphor of a treelike arrangement of animals and was based on the claim that during embryological development characters appear in the order of their generality; the most basic first and the more specialized later. Embryology, therefore, promised to be an excellent guide for providing criteria of classification, in that one could use the study of embryos to determine the features that define class, order, etc. Naturalists like Henri Milne-Edwards (1800–85), who drew on the brilliant research of Karl Ernst von Baer (1792–1876), advocated the use of embryological studies to construct a branching arrangement in the animal kingdom, a treelike structure where large, general stems diverge and ramify into ever-specialized branches.[13] It was even suggested that the branching arrangement was reflected in the fossil record, which allegedly revealed a progression from a more general to more specific forms in major groups.[14]

The embryological approach was hailed by many as a critical breakthrough in classification. Louis Agassiz (1807–73) around midcentury stated that

in the real changes which animals undergo during their embryonic growth, in those external transformations as well as in those structural modifications within the body, we have a natural scale to measure the degree or the gradation of those full grown animals which correspond in their external form and in their structure, to those various degrees in the metamorphoses of animals, as illustrated by embryonic changes, a real foundation for zoological classification.[15]

Unfortunately, comparative embryology, perhaps due to the difficulty of obtaining data, did not prove to be as useful in nineteenth-century ornithology as it did in other zoological disciplines. Ornithologists used a variety of criteria for classification, although external morphological features, for practical reasons, continued to be of primary value. The overall increase in data, however, did make possible more complex systems. Among the

---

[13] For a good discussion, see Dov Ospovat, *The Development of Darwin's Theory: Natural History, Natural Theology, and Natural Selection, 1838–1859* (Cambridge: Cambridge University Press, 1981), pp. 115–45.

[14] For a history of the idea, see Peter Bowler, *Fossils and Progress. Paleontology and the Idea of Progressive Evolution in the Nineteenth Century* (New York: Science History Publications, 1976).

[15] Louis Agassiz, *Twelve Lectures on Comparative Embryology, Delivered before the Lowell Institute in Boston, December and January, 1848–9* (Boston: Redding, 1849), p. 29.

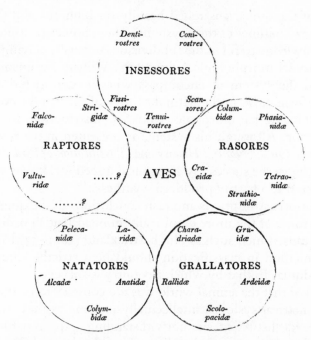

Figure 13.1. A quinarian classification of the birds by Vigors. [From Nicholas Aylward Vigors, "Observations on the Natural Affinities that Connect the Orders and Families of Birds," *Transactions of the Linnean Society, 14* (1825), 509.]

most influential of these was the famous quinarian system devised by William Sharp Macleay (1792–1865) and extended to ornithology by Nicholas Aylward Vigors (1785–1840) and William Swainson (1789–1855). Macleay believed that animals could be grouped in a series, but unlike many of his predecessors he did not hold that the series was linear. Instead he conceived of nature as ordered in a circular fashion. The affinities among animals, according to this view, could be depicted by arranging five primary animal groups in a series that returned into itself. Each group in turn was thought to be composed of five subgroups that could be depicted as a series that returned into itself, etc. Quinarians, following Macleay's original formulation, also held that each of the groups composing a taxonomic level contained "analogous" subgroups. Ornithologists, such as Vigors, were able to place the known birds into a complex series of telescoping circles and could express the relationship between two birds as either an affinity (same subgroup) or analogy (different subgroup) (see Figure 13.1).

Most of the quinarians relied heavily on Linneaus and Cuvier for the broad outlines of their systems. They, however, used more recent knowledge to fill in the subdivisions – usually relying heavily on external morphological characters. William Swainson tried to expand the system to encompass what he considered all the known information on the natural history of birds, i.e., not only internal and external morphology but also distribution, function, and behavior. Although his system, as presented in his two-volume treatise *On the Natural History and Classification of Birds* (1836), did not achieve the success Swainson had hoped for, it was a grandiose, even if too hastily produced, synthesis.[16]

The a priori assumptions and rash conclusions of the quinarians bothered some of the more empirically minded British ornithologists and caused men such as Hugh Strickland (1811–53) to argue that the method in classification should be a strictly "inductive process, similar to that by which a country is surveyed."[17] Strickland believed that the animal world was too complex to be squeezed into symmetrical systems, intellectually pleasing as they may be. There are, idealistic fantasies notwithstanding, gaps in nature that are related to environmental realities. Strickland stated that it was too early, in fact, to see the outline of the natural system of birds:

all that we can say at present is, that ramifications of affinities exist; but whether they are so simple as to admit of being correctly depicted on a plane surface, or whether, as is more probable, they assume the form of an irregular solid, it is premature to decide. They may even be of so complicated a nature that they cannot be correctly expressed in terms of space, but are like those algebraical formulae which are beyond the powers of the geometrician to depict. Without, however, going deeper into this obscure question, let us hope that the affinities of the natural system will not be of a higher order than can be expressed by a solid figure; in which case they may be shown with tolerable accuracy on a plain surface; just as the surface of the earth, though an irregular spheroid, can be protracted on a map. The natural system may, perhaps, be most truly compared to an irregularly branching tree, or rather to an assemblage of detached trees and shrubs of various sizes and modes of growth. And as we

[16] For a discussion of some of the possible reasons for Swainson's haste, see my article, "Aspiring Naturalists and Their Frustrations: The Case of William Swainson," in *Natural History in the Early Nineteenth Century* (London: Society for the History of Natural History, in press).

[17] Hugh Strickland, "On the True Method of Discovering the Natural System in Zoology and Botany," *The Annals and Magazine of Natural History*, 6 (1841), 185.

*Ann. & Mag. of Nat. Hist.* Vol.VI.Pl.VIII.

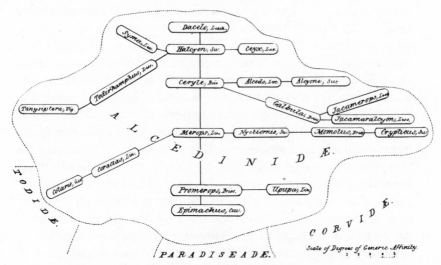

Figure 13.2.   A sketch of a possible natural system for ornithology by Strickland. [From Hugh Strickland, "On the True Method of Discovering the Natural System in Zoology and Botany," *The Annals and Magazine of Natural History*, 6 (1841), plate VIII between pp. 192 and 193.]

show the form of a tree by sketching it on paper, or by drawing its individual branches and leaves, so may the natural system be drawn on a map, and its several parts shown in greater detail in a series of maps.[18]

Strickland, like those naturalists in the German tradition stemming from Johann Friedrich Blumenbach (1752–1840), believed that ornithologists needed to examine all the essential characters of birds and that "the natural system is that arrangement in which the distance from each species to every other is in exact proportion to the degree in which the essential characters of the respective species agree"[19] (see Figure 13.2). In practice, Strickland relied heavily on external morphological characters rather than relying on a single anatomical part on which to construct a system.

[18] Ibid., 190.
[19] Ibid., 184. For an excellent account of the Blumenbach tradition in natural history, see Timothy Lenoir, "The Göttingen School and the Development of Transcendental Naturphilosophie in the Romantic Era," *Studies in the History of Biology*, 5 (1981), 111–205.

By midcentury, the literature on avian systematics was extensive and quite sophisticated. A number of different criteria were used, some effectively, and much of the research done was of lasting value. No wonder, then, that men of broad speculative faculty, like Charles Darwin, consulted classification or the adepts who could make it accessible. It was the core of natural history, and, as one might also expect, open to very different interpretations. The professors at the Paris Muséum believed that they were discovering an order that existed in nature. It was an order that they desired to define rigorously, and one that they held could be studied in a secular framework, independently of one's private religious or irreligious sentiments. British naturalists for most of the first half of the nineteenth century looked at their classifications from the perspective of natural theology. Their research in comparative anatomy and systematics, in fact, helped in the reformation of natural theology into a more sophisticated philosophy of nature than the simplistic gawking at adaptations exemplified by William Paley's *Natural Theology* (1801). Ornithologists such as Vigors, Swainson, and Strickland, as well as the leading British comparative anatomist, Richard Owen (1804–92), presented readers with a view of the order in nature that was far richer and more complex than earlier ones. As Vigors wrote in praising the ultimate value of Macleay's system:

To those persons who were induced to seek a more intimate acquaintance with the works of Nature by the noblest, and indeed only legitimate end of all such research – the desire of studying "the wisdom of God in the creation," – a new source of delight was thrown open, a new region of wonder revealed; as they were now enabled to trace that wisdom not merely in the detail – not merely in the beautifully combined mechanism of an isolated object, – but in the comprehensive system by which all are united, and the harmony that prevails throughout.[20]

In Germany, a more philosophic approach prevailed. The rigorous anatomical traditon of the Göttingen school advocated a teleological framework for systematics. Early in the century, Gottfried Treviranus (1776–1837) synthesized the numerous themes of that school and encompassed in an encyclopedic fashion a wealth of

---

[20] Nicholas Aylward Vigors, "Observations on the Natural Affinities that Connect the Orders and Families of Birds," *Transactions of the Linnean Society, 14* (1825), 398–9.

empirical information in his six-volume *Biologie* (1802–22).[21] Many of the most influential and well-known of the German ornithologists of the first half of the nineteenth century who concentrated on systematics, however, followed a considerably more speculative path and elaborated what appear today as highly fanciful systems of idealistic classification. Ludwig Reichenbach (1793–1879) and Johann Jakob Kaup (1803–73), for example, started from a priori assumptions about regularity and symmetry, which they derived from first principles, and tried, then, to arrange the data of ornithology accordingly. It is difficult for us in the twentieth century to make sense of these fantastic systems, but there was an aesthetic attraction in the endeavor that drew adherents, and since it was rooted in a more general nature philosophy, it had the appeal of integrating natural history into the broader *Wissenschaft*.

From this very brief sketch of some of the major interpretations of systematics in the first half of the nineteenth century, we can see that a wide range of opinions existed: religious and secular, boldly speculative and cautiously empirical. There was a corresponding spectrum of ideas regarding the idea of change in the animal world. Strickland and his quinarian foes agreed that species have not and could not change. The French and German transcendental anatomists shared with the rigorous cosmopolitan cataloger Bonaparte a view that animals have changed in time – an opinion shared by the English gentleman–naturalist Charles Darwin, whose interest in pigeon breeding and avian classification was central for his theoretical formulation of the mechanism of that change.[22]

A clearly discernible theoretical pluralism existed in the first half of the nineteenth century. Although evolutionary theories increased in number and importance in the second half of the century, one could trace a complex theoretical pluralism there also. What I have tried to suggest in this brief sketch of early nineteenth-century classification, however, is that historically many other important changes were taking place in ornithology. There

[21] For an excellent discussion of this tradition and a sketch of the different interpretations in natural history in nineteenth-century Germany, see Lenoir, "The Göttingen School," and his *The Strategy of Life. Teleology and Mechanics in Nineteenth-Century German Biology* (Dordrecht: Reidel, 1982).

[22] See James Secord, "Nature's Fancy: Charles Darwin and the Breeding of Pigeons," *Isis*, 72 (1981), 163–86, and Frank Sulloway, "Darwin's Conversion: The *Beagle* Voyage and Its Aftermath," *Journal of the History of Biology*, 15(3) (1982), 325–96.

was a significant continuity of problems and practices that can get overlooked in the search for historical discontinuity on the theoretical level. Ornithologists experienced an acceleration in the growth of their collections, which resulted in a scramble to get the basic job of sorting and naming done. Older classification practices and systems had to be replaced or reworked. New dimensions were opened by the increasing availability of information on behavior, comparative anatomy, embryology, and geographical distribution and variation. The institutional framework and the cultural setting of the study of birds evolved in such a way as to expand dramatically the number of individuals involved, to sharpen the level of resolution of their studies, and to increase the rigor of their writings. By midcentury, ornithology as a scientific discipline had a nascent form that continued to develop throughout the century. General theories also had an impact on the practice of avian systematics. The relationship is most obvious in the more extreme systems; it would be difficult to account for Kaup's bizarre classification of birds without understanding the theoretical position from which he argued. But the closer we approach the mainstream of practice, and when we consider the broad consensus on the general outlines of the discipline, the less obvious is the importance of the theoretical underpinnings. Most of the theoretically minded naturalists in the middle of the nineteenth century, after all, were attempting to explain the accepted facts of natural history, which included an implied order revealed in taxonomic arrangements. This point is worth stressing. For it is all too easy to describe a change in theory as the pivot around which all data, method, and ideas turn. Although theories at times have had critical importance for the selection of criteria of classification, they have at other times been secondary. Moreover, the taxonomic level on which theories have had their greatest importance has varied. Darwinian evolution had its most fertile implications for issues concerning species and subspecies, but it was highly problematic for discussions concerning the higher taxa of birds. Transcendental anatomy, in contrast, had more to say about primary divisions than geographical variation.

There is a real danger in overly stressing the impact of general theories on taxonomy, for it ignores the cumulative aspects of the history of systematics, and glosses over other changes that have occurred in natural history. It, therefore, can lead us to neglect what we might call the preconditions of those continuities and changes: the causes of the growth of the available empirical base, the expansion and transformation of natural history collections, the fragmentation of natural history itself into a set of specialized scientific disciplines, the shifting patterns of patronage and funding, and the fashion of natural history literature, to name a few of the more interesting ones. These, and other, major factors in the history of natural history can easily be overlooked if we focus too narrowly on the contrast between natural theology and the Darwinian worldview, or the impact of evolution theory on modern biology. The history of ornithology and the much broader history of biology are stories too rich to be told as the unveiling of the truths of nature, as a series of great men, great ideas, or great theories.

This brief survey of the classification of birds during the nineteenth century illustrates the complexity of the history of natural history and should make us skeptical about characterizing it in narrow terms. It also should suggest the need for reexamining the central role we usually give to theories in the history of science. For like great men, great theories have existed, not as the principal actors of an intellectual drama, but as one of the many factors we need to consider in understanding the nature and history of science.

# BIBLIOGRAPHY OF RICHARD S. WESTFALL'S WRITINGS ON THE HISTORY OF SCIENCE

## Books

Science and Religion in Seventeenth-Century England. Yale Historical Publications, Miscellany 67. New Haven, Conn.: Yale University Press, 1958.
Reprint ed. Hamden, Conn.: Archon Books, 1970.
Paperback ed. Ann Arbor, Mich.: University of Michigan Press, 1972.
Steps in the Scientific Tradition. Readings in the History of Science. New York: Wiley, 1968. Co-edited with Victor E. Thoren.
Force in Newton's Physics. The Science of Dynamics in the Seventeenth Century. London: MacDonald, 1971.
Italian translation: Newton e la dinamica del XVII secolo. Bologna: Il Mulino, 1982.
The Construction of Modern Science. Mechanisms and Mechanics. New York: Wiley, 1971.
New edition, Cambridge University Press, 1977.
Japanese translation by Masao Watanabe, Tokyo: Misuzu Shobo, 1980.
Spanish translation: La Construccion de la ciencia moderna. Barcelona: Editorial Labor, 1980.
Foundations of Scientific Methods. The Nineteenth Century. Bloomington, Ind.: Indiana University Press, 1973. Co-edited with Ronald Giere.
Never at Rest. A Biography of Isaac Newton. Cambridge: Cambridge University Press, 1980.
Paperback edition. Cambridge: Cambridge University Press, 1983.

## Articles in journals, chapters in books

"Unpublished Boyle papers relating to scientific method," Annals of Science, 12 (1956), 63–73, 103–17.
"Isaac Newton: Religious rationalist or mystic?" Review of Religion, 22 (1957–8), 155–70.
"The development of Newton's theory of color," Isis, 53 (1962), 339–58.
Reprinted in Bobbs-Merrill Reprint Series in History of Science, HS-79.

"The foundations of Newton's philosophy of nature," *British Journal for the History of Science, 1* (1962), 171–82.

"Newton and his critics on the nature of colors," *Archives internationales d'histoire des sciences, 15* (1962), 47–58.

"Newton's reply to Hooke and the theory of colors," *Isis, 54* (1963), 82–96.
  Reprinted in Russian in A. N. Bogoliubov, ed., *U istokov klassicheskoi nauki.* Moscow: Izdatel'stvo "Nauka," 1968, pp. 100–22.

"Shortwriting and the state of Newton's conscience, 1662," *Notes and Records of the Royal Society, 18* (1963), 10–16.

"Newton and absolute space," *Archives internationales d'histoire des sciences, 17* (1964), 121–32.

"Isaac Newton's Coloured Circles twixt two Contiguous Glasses," *Archive for History of Exact Sciences, 2* (1965), 181–96.

"The problem of force in Galileo's physics," in *Galileo Reappraised,* ed. Carlo L. Golino. UCLA Center for Medieval and Renaissance Studies, Contribution 2. Berkeley: University of California Press, 1966, pp. 67–95.

"Newton's Opticks: The present state of research," *Isis, 57* (1966), 102–7.

"Newton defends his first paper: The Newton–Lucas correspondence," *Isis, 57* (1966), 299–314.

"Reply" (to criticism by A. R. Hall), *Isis, 58* (1967), 403–5.

"Hooke and the law of universal gravitation: A reappraisal of a reappraisal," *The British Journal for the History of Science, 3* (1967), 245–61.

"The science of optics in the seventeenth century," *History of Science, 6* (1967), 150–6.

"Uneasily fitful reflections on fits of easy transmission," *The Texas Quarterly, 18* (1967), 86–102.
  Reprinted in *The Annus Mirabilis of Sir Isaac Newton, 1666–1966,* ed. Robert Palter. Cambridge, Mass.: M.I.T. Press, 1970, pp. 88–104.

"Newton and order," in *The Concept of Order,* ed. Paul G. Kuntz. Seattle, Wash.: University of Washington Press, 1968, pp. 77–88.

"Huygens' rings and Newton's rings: Periodicity and seventeenth century optics," *Ratio, 10* (1968), 64–77.

"A note on Newton's demonstration of motion in ellipses," *Archives internationales d'histoire des sciences, 22* (1969), 51–60.

"Science," in *Bibliography of British History. Stuart Period, 1603–1704,* ed. Godfrey Davies and Mary Frear Keeler, 2nd ed. London: Oxford University Press, 1970, pp. 406–19.

"Stages in the development of Newton's dynamics," in *Perspectives in the History of Science and Technology,* ed. Duane H. D. Roller. Norman, Okla.: University of Oklahoma Press, 1971, pp. 177–97.

"Newton and the hermetic tradition," in *Science, Medicine and Society in the Renaissance, a Festschrift in Honor of Walter Pagel,* ed. Allen G. Debus. New York: Neale Watson Academic Publications, 1972, vol. 2, pp. 183–98.

"Circular motion in seventeenth century mechanics," *Isis, 63* (1972), 184–9.

"Newton and the fudge factor," *Science, 179* (1973), 751–8.

"Circular motion as the crucial problem of 17th century mechanics," in *Actes du XIIIe Congrès Internationale d'Histoire des Sciences,* 1971 (publ. 1974), part 5, pp. 288–92.

"Comments on Professor Cohen's paper in the Colloquium on principal stages in the evolution of classical mechanics," in *Actes du XIIIe Congrès International d'Histoire des Sciences*, 1971 (publ. 1974), Résumés of the Colloquia, p. 75.

"The role of alchemy in Newton's career," in *Reason, Experiment and Mysticism in the Scientific Revolution*, ed. M. L. Righini Bonelli and William R. Shea. New York: Science History Publications, 1975, pp. 189–232.

"Isaac Newton: A sober, thinking lad," *Review* (Alumni Association of the College of Arts and Sciences – Graduate School, Indiana University), *18*, (1975–6), 17–33.

"Isaac Newton's *Index Chemicus*," *Ambix, 22* (1975), 174–85.

"The changing world of the Newtonian industry," *Journal of the History of Ideas, 37* (1976), 175–84.

"Newton's marvelous years of discovery and their aftermath: myth versus manuscript," *Isis, 71* (1980), 109–21.

"The influence of alchemy on Newton," in *Science, Pseudo-Science and Society*, eds. Marsh P. Hanen, Margaret J. Osler, and Robert G. Weyant. Waterloo, Ontario: Wilfrid Laurier University Press, 1980, pp. 145–69.

"Isaac Newton in Cambridge: the Restoration university and scientific creativity," in *Culture and Politics from Puritanism to the Enlightenment*, ed. Perez Zagor. Publication from the Clark University Professorship, UCLA, 5. Berkeley: University of California Press, 1980, pp. 135–64.

"The career of Isaac Newton. A scientific life in the seventeenth century," *The American Scholar, 50* (Summer 1981), pp. 341–53.

Reprinted in *The Norton Reader: An Anthology of Expository Prose*, ed. Arthur M. Eastman et al., 6th ed. New York: Norton, 1984, pp. 955–68.

"Newton and Christianity," in Japanese translation, in *Newton To Sono Shuhen (Newton: Lights and Shadows)*, ed. M. Watanable. Tokyo, 1981, pp. 173–89.

"Newton's theological manuscripts," in *Contemporary Newtonian Research*, ed. Z. Bechler. Dordrecht: Reidel, 1982, pp. 129–43.

"Isaac Newton's *Theologiae Gentilis Origines Philosophicae*," in *The Secular Mind. Transformations of Faith in Modern Europe. Essays presented to Franklin L. Baumer*, ed. W. Warren Wagar. New York: Holmes & Meier, 1982, pp. 15–34.

"Newton's development of the *Principia*," in *Springs of Scientific Creativity*, ed. Rutherford Aris, H. Ted Davis, and Roger H. Stuewer. Minneapolis: University of Minnesota Press, 1983, pp. 21–43.

"Robert Hooke, mechanical technology, and scientific investigation," in *The Uses of Science in the Age of Newton*, ed. John G. Burke. Berkeley: University of California Press, 1983, pp. 85–110.

"Science and patronage. Galileo and the telescope," *Isis, 76* (1985), 11–30.

## Essay reviews, encyclopedia articles, introductions, and miscellaneous writings

Introduction to the republication of Sir David Brewster, *Memoirs of the Life, Writings, and Discoveries of Sir Isaac Newton*. New York: Johnson Reprint, 1965.

"Newton, Sir Isaac," biographical article in *The New Catholic Encyclopedia*, Washington, 1967, *10*, 424–8.

"The science of optics in the seventeenth century," an essay review of A. I.

Sabra, *Theories of Light. From Descartes to Newton, History of Science 6* (1967), 150–6.

Introduction to the republication of Robert Hooke, *The Posthumous Works*. New York: Johnson Reprint, 1969.

"Brewster, Sir David," biographical article in *Encyclopedia Americana,* 1969 ed., *4,* 512.

"An essay review: The correspondence of Isaac Newton, vols. 1–4," *American Scientist, 56* (1968), 182–8.

"Hooke, Robert," biographical article in *Dictionary of Scientific Biography,* 1972, *6,* 481–8.

"History of theories of light," in *Encyclopedia Americana,* 1973 edition, *17,* 445–51.

"Measurement of the velocity of light," in *Encyclopedia Americana,* 1973 edition, *17,* 451–5.

"Pemberton, Henry," biographical article in *Dictionary of Scientific Biography,* 1974, *10,* 500–1.

"Ray, John," biographical article in *Encyclopedia Britannica,* 1974 edition, *15,* 535–6.

"Newton, Isaac," biographical article in *Encyclopedia Britannica,* 1974 edition, *13,* 16–21.

"Hooke, Robert," biographical article (translated into Italian) in *Scienziati e tecnologi* (Milan, 1975), *2,* 123–5.

"Toulmin and human understanding," *Journal of Modern History, 47* (1975), 691–8.

"Eloge: Joseph Schiller 1906–1977," by Joe D. Burchfield, Paul L. Farber, and Richard S. Westfall, *Isis, 69* (1978), 75–6.

"Newton: moving science with reason," *Toronto Globe and Mail,* 5 Jan. 1981, p. 19. (In an issue devoted to the AAAS meeting then taking place in Toronto.)

"Marxism and the history of science: Reflections on Ravetz's essay," *Isis, 72* (1981), 402–5.

"Deification and disillusionment," an essay review of Stillman Drake, *Galileo at Work, Isis, 70* (1979), 273–5.

"Newtonian absolutus," an essay review of vols. 5, 6, and 7 of the *Correspondence of Isaac Newton, British Journal for the Philosophy of Science, 30* (1979), 173–7.

"Newton enthroned: His correspondence from 1718 to 1727," an essay review of vol. 7 of the *Correspondence of Isaac Newton, British Journal for the History of Science, 12* (1979), 197–200.

"Newton's achievements: from rumour to fact," an essay review of *The Mathematical Papers of Isaac Newton,* ed. D. T. Whiteside, 8 vols., *Nature, 295* (1982), 265–6.

"Newton, Sir Isaac," in *Funk & Wagnalls New Encyclopedia.* New York, 1983, *19,* 45–7.

# INDEX